D1203822

INDUSTRIAL ECOLOGY

Environmental Chemistry and Hazardous Waste

INDUSTRIAL ECOLOGY

Environmental Chemistry and Hazardous Waste

Stanley E. Manahan
Department of Chemistry
University of Missouri
Columbia, Missouri

Library of Congress Cataloging-in-Publication Data

Manahan, Stanley E.
Industrial ecology : environmental chemistry and hazardous wastes / by Stanley E. Manahan.
p. cm.
Includes bibliographical references and index.
ISBN 1-56670-381-6 (alk. paper)
1. Industrial ecology. 2. Environmental chemistry. 3. Hazardous wastes. I. Title.
TS161.M353 1999
628.4'2—dc21 98-49415
CIP

Direct all inquiries to CRC Press LLC, 2000 Corporate Blvd., N.W., Boca Raton, Florida 33431.

Lewis Publishers is an imprint of CRC Press LLC

International Standard Book Number 1-56670-381-6
Library of Congress Card Number 98-49415
Printed in the United States of America 1 2 3 4 5 6 7 8 9 0
Printed on acid-free paper

PREFACE

Environmental chemistry is a relatively new subdiscipline of chemical science and has made significant contributions to solving a variety of problems involving pollution and hazardous wastes. In the United States and throughout the world, much legislation has been passed and many regulations promulgated to control pollution, restrict hazardous chemicals and their disposal, and clean up hazardous chemical waste disposal sites. The gains derived from these measures have been impressive with noted improvements in air and water quality, effective regulation of hazardous chemicals, and cleanup of some of the worst chemical waste sites.

It has become obvious, however, that an approach to environmental protection based primarily on regulation of pollutants and wastes has distinct limitations. This is particularly true in the area of hazardous wastes, where in some cases severe restrictions on the use and processing of hazardous materials have made problems even worse by inhibiting reasonable, cost-effective approaches to dealing with such materials. The fear of liability, overzealous regulation, or even criminal prosecution have prevented implementation of sensible measures and processes that could be employed to recycle and use potentially hazardous materials. The net result has been for generators to choose disposal of quantities of material at a high cost, whereas with less restrictive and more reasonable regulation the material could be recycled at a net profit.

During the 1990s, a relatively new concept gained acceptance, at least in principle, for dealing with pollution and limits on available material resources. This is the idea of industrial ecology, which seeks to treat industrial systems in a manner analogous to ecological systems in nature. In natural ecological systems, true waste products are very rare. Virtually all materials are recycled by some organism. There exists in nature a grand materials cycle composed of numerous subcycles. Plants fix carbon dioxide as plant material by photosynthesis, herbivores eat plants and convert the plant matter to energy and their own biomass, carnivores eat herbivores and, when they die, their biomass is degraded by bacteria. Carbon dioxide released through metabolism or through the decay of dead biomass recycles through the whole system by photosynthesis. Everything is driven by energy, which for most natural ecosystems is solar energy fixed initially as chemical energy by photosynthesis. Although the path taken by the energy in a natural ecosystem is essentially one way — high quality, readily utilized solar photochemical energy is eventually dissipated as heat at low temperature — the natural ecosystem utilizes the energy very efficiently through several levels.

Industrial ecology views an industrial system as an artificial ecosystem, with primary sources of raw materials and energy and with a number of enterprises making use of what would otherwise be waste products of other members of the system. Industrial ecology deals with total life cycles of materials and of components of manufactured devices with the objective of maximizing material utilization and minimizing waste. Although the ideal of an industrial ecosystem in which no materials are required beyond those needed to start it and no wastes are produced can never be realized in practice, industrial ecology provides an excellent framework around which to design industrial systems for maximum efficiency and minimum waste.

Industrial Ecology: Environmental Chemistry and Hazardous Wastes has as its basic theme coverage of hazardous wastes from an environmental chemistry perspective within an overall framework of industrial ecology. The first seven chapters deal with environmental science and technology and industrial ecology. Chapter 1, "The Environment and the Anthrosphere," defines what is meant by the environment. In addition to discussing the four environmental spheres that are commonly recognized — the atmosphere, hydrosphere, geosphere, and biosphere — it introduces a fifth sphere, which is all too often neglected in discussions of environmental science, the anthrosphere, which consists of the things that humans make and do. Chapter 2, "Industrial Ecology and Industrial Systems," defines and explains industrial ecology and how it relates to industrial systems in detail. Chapter 3, "Principles of Industrial Metabolism," defines the processes performed by an industrial ecosystem as it processes materials and utilizes energy. Chapter 4, "Industrial Ecosystems," explains industrial ecosystems and the potential to construct such systems in the future. Chapter 5, "Life Cycles: Products, Processes and Facilities," treats the way that life cycles of materials and components are analyzed to optimize their use and minimize wastes. Chapter 6, "Industrial Ecology and Resources," views renewable and nonrenewable resources from industrial ecology and environmental chemistry perspectives. Chapter 7, "Environmental Chemistry and Industrial Ecology," explains in detail the relationship between industrial ecology and environmental chemistry.

The remaining chapters of the book deal specifically with hazardous substances and hazardous waste as they relate to industrial ecology and environmental chemistry. Chapter 8, "Hazardous Substances and Wastes," discusses and defines hazardous materials. This coverage involves definitions and characteristics of such materials from both scientific and regulatory viewpoints. Chapter 9, "Environmental Chemistry of Hazardous Materials," treats the chemical behavior of hazardous materials in the anthrosphere, atmosphere, hydrosphere, geosphere, and biosphere. Chapter 10, "Hazardous Inorganic and Organometallic Materials," and Chapter 11, "Hazardous Organic Materials," deal with various classes of hazardous substances from the viewpoint of their chemical characteristics. Chapter 12, "Toxicological and Biological Hazards," addresses the toxicological chemistry of toxic substances and the hazards presented by various materials from biological sources. As their titles imply, Chapters 13–15 cover, respectively, waste minimization, waste treatment, and waste disposal within the context of industrial ecology and environmental chemistry.

Industrial Ecology: Environmental Chemistry and Hazardous Wastes is meant to serve as a reference work in the areas covered by the title and as a textbook in courses dealing with hazardous waste. The level is such that a reader with an understanding of general chemistry plus some exposure to organic chemistry can readily understand the material presented. It is the author's hope that this text will serve a broad clientele both as a trade book and a textbook and that it will be useful for people who are interested in the emerging area of industrial ecology and how it relates to the more established areas of environmental chemistry and hazardous wastes. Reader input is appreciated and may be addressed to the author by e-mail at manahans@missouri.edu.

The author gratefully acknowledges the excellent support of the staff of CRC Press in preparing this book. Publisher Robert Hauserman has offered continued support and encouragement in this endeavor and Editor Mimi Williams has done outstanding work in preparing the final work for publication.

AUTHOR

Stanley E. Manahan is Professor of Chemistry at the University of Missouri-Columbia, where he has been on the faculty since 1965. He received his A.B. in chemistry from Emporia State University in 1960 and his Ph.D. in analytical chemistry from the University of Kansas in 1965. Since 1968 his primary research and professional activities have been in environmental chemistry and have included the development of methods for chemical analysis of pollutant species, environmental aspects of coal conversion processes, development of coal products useful for pollution control, hazardous waste treatment, and toxicological chemistry. He teaches courses on environmental chemistry, hazardous wastes, toxicological chemistry, and analytical chemistry and has lectured on these topics throughout the U.S. as an American Chemical Society Local Section tour speaker. He serves as an advisor to the Environmental Sciences Program of the Chinese University of Hong Kong and has recently given invited lectures on his research and environmental chemistry at the National Autonomous University of Mexico in Mexico City, at Hokaido University in Sapporo, Japan, and for the 23rd Latin American Chemical Congress in Puerto Rico. He is also President of ChemChar Research, Inc., a firm working on the development of non-incinerative thermochemical and electrothermochemical treatment of mixed hazardous substances containing refractory organic compounds and heavy metals.

Professor Manahan has written books on environmental chemistry (*Environmental Chemistry*, 6th ed., Stanley E. Manahan, CRC Press/Lewis Publishers, 1994); environmental science (*Environmental Science and Technology*, Stanley E. Manahan, CRC Press/Lewis Publishers, 1997); general and environmental chemistry (*Fundamentals of Environmental Chemistry*, Stanley E. Manahan, CRC Press/Lewis Publishers, 1993); hazardous wastes (*Hazardous Waste Chemistry, Toxicology and Treatment*, 1990, Lewis Publishers, Inc.); toxicological chemistry (Toxicological Chemistry, 2nd ed., 1992, Lewis Publishers, Inc.); applied chemistry; and quantitative chemical analysis. He has been the author or co-author of approximately 90 research articles.

TABLE OF CONTENTS

1 THE ENVIRONMENT AND THE ANTHROSPHERE

1.1. INTRODUCTION

One of the most vexing problems faced by a modern industrialized society is the generation of a wide variety of potentially hazardous byproducts of industrial production. Although commonly regarded as problems of highly industrialized nations with modern manufacturing systems, the problems may be even more severe in less developed countries whose economies do not generate sufficient income to deal effectively with industrial byproducts. In recent times, the terms **hazardous waste, hazardous substances**, or **hazardous materials** have been widely used to describe a variety of potentially dangerous products and byproducts of manufacturing. Many industrial materials are potentially hazardous because of their flammability, chemical reactivity, or toxicity, and sophisticated means have been developed to handle and deal with them in a generally safe manner. This leaves, however, large quantities of hazardous waste materials that traditionally have been discharged to the atmosphere, water, or land surface or buried in landfills, often causing severe environmental problems at later times and at some distance from the disposal site.

As discussed later in this book, much legislation has been passed in the U.S. and other countries and reams of regulations have been promulgated to deal with hazardous waste. Attempts have been made to regulate materials with the potential to become hazardous waste "from cradle to grave." It is now generally recognized that this approach has been very inefficient and costly. Although impressive gains have been achieved in cleaning up waste sites, such as landfills containing hazardous chemical wastes, a disproportionate share of the money spent in dealing with hazardous waste problems has gone to support litigation, and many problems have remained unresolved.

Now it is increasingly recognized that there is a better way to deal with the hazardous waste problem through the application of an emerging science called **industrial ecology**. Industrial ecology, which is defined in detail in Chapter 2, is loosely based on natural ecosystems which have evolved to the degree that there is no such thing as a waste product; everything produced by the system serves a purpose for some organism and all materials are continually recycled. Only energy — almost entirely in the form of inexhaustible solar energy — is used dissipatively. It is unrealistic to believe that complete material recycle and production of no hazardous waste can be entirely achieved in modern industrial systems. However, the principles of industrial ecology provide a basis for a much more sustainable global industrial system than the one that now exists. And although there will always be some hazardous materials requiring some sort of disposal, they can be reduced to a very small and highly manageable fraction of their present amounts.

Recycling and practices of industrial ecology are international in scope. The Kalundborg, Denmark, community (see Section 4.4) is a classic example of the practice of industrial ecology, with a variety of products, such as gypsum to make wallboard reclaimed from a power plant flue gas cleaning operation, steam from the power plant used in a petroleum refinery, and wastes from a fish farm and from a fermentation process used to fertilize fields for growing plants. In 1991 Japan passed a law Promoting the Utilization of Recyclable Resources. A basic thrust of this law is to encourage recyclability as an integral part of product design.

This book is designed to provide an approach to hazardous waste based upon the principles of industrial ecology and environmental science and technology, particularly environmental chemistry. The first 7 chapters deal with industrial ecology and environmental science and technology. The remaining chapters discuss hazardous waste from the perspective of industrial ecology and environmental chemistry.

1.2. THE ENVIRONMENT

Traditionally, when people thought about the environment, they have considered water in its various forms and locations, air, the solid earth, and various living organisms. In other words, conventional consideration of the environment has involved the **hydrosphere**, the **atmosphere**, the **geosphere**, and the **biosphere**. This book deals with a fifth environmental sphere, the **anthrosphere**, which humans have created and which has a profound effect upon humankind and the rest of the environment.[1] In a sense humans have operated the anthrosphere "out of context" with the rest of the environment, resulting in profound environmental disruption and creating a system that can be sustained for only a brief period of time compared to humankind's time on earth. A basic theme and purpose of this book is to outline ways in which the anthrosphere can be modified and operated, particularly as it relates to the environmental chemistry of the hazardous materials and wastes that it tends to generate, so that humankind can survive indefinitely in a reasonable degree of harmony with its surroundings and the support systems upon which humans depend for their existence.

The **anthrosphere** may be defined as *that part of the environment made or modified by humans and used for their activities.* Of course, there are some ambiguities associated with that definition. Clearly, a factory building used for manufacture is part of the anthrosphere, as is an ocean-going freighter used to ship goods made in the factory. The ocean on which a ship moves belongs to the hydrosphere, but it is used by humans. A pier constructed on the ocean shore and used to load a ship is part of the anthrosphere, but closely associated with the hydrosphere.

During most of their time on Earth, humans made little impact on the planet. Humankind's small, widely scattered anthrospheric artifacts — simple huts or tents for dwellings, narrow trails worn across the land for movement, clearings in forests used to grow some food — rested lightly on the land with virtually no impact. However, with increasing effect as the industrial revolution developed, and especially during the last century, humans have built structures and modified the other environmental spheres, especially the geosphere, such that it is necessary to consider the anthrosphere as a separate area with pronounced, sometimes overwhelming influence on the environment as a whole.

The five environmental spheres are strongly interrelated. Perturbations in any particular sphere of the environment are very likely to affect other spheres. Figure 1.1 illustrates the strong interactions that occur among the five environmental spheres.

Industries at Risk from Environmental Degradation

Although everyone has an important stake in environmental protection, there are several key industries that face direct threats from environmental degradation. One such industry is tourism. Scenic places and good climate are major draws for tourists. Typically, a pristine beach washed with clean ocean water and blessed with a good climate and pure air is very attractive. Degradation of the beach with trash, polluted ocean water unfit for swimming, and smog-laden air degraded by air pollutants obviously cause tourists to stay away.

The insurance industry also has a strong interest in environmental preservation. The most obvious stake held by the insurance industry is in climate as it influences the occurrence of major weather events, such as hurricanes or floods. If, indeed, environmental factors, particularly the emission of greenhouse gases, are changing climate such that strong storms and torrential rains will become more common, the insurance industry will clearly be affected. Environmental factors, such as air pollution, that adversely affect human health can be very detrimental to the health insurance industry due to the increase in respiratory ailments and other maladies.

The finance industry is also affected by environmental conditions. In recent years financial institutions have taken substantial losses on properties on which they hold mortgages because of the discovery of hazardous waste sites that have required expensive cleanup. Environmental conditions that cause lowered property values have a direct adverse effect upon financial institutions.

1.3. COMPONENTS OF THE ANTHROSPHERE

The various spheres of the environment are each divided into several subcategories. For example, the hydrosphere consists of oceans, streams, groundwater, ice in polar icecaps, and other components. The anthrosphere, too, consists of a number of different parts. These may be categorized by considering where humans live; how they move; how they make or provide the things or services they need or want; how they produce food, fiber, and wood; how they obtain, distribute, and use

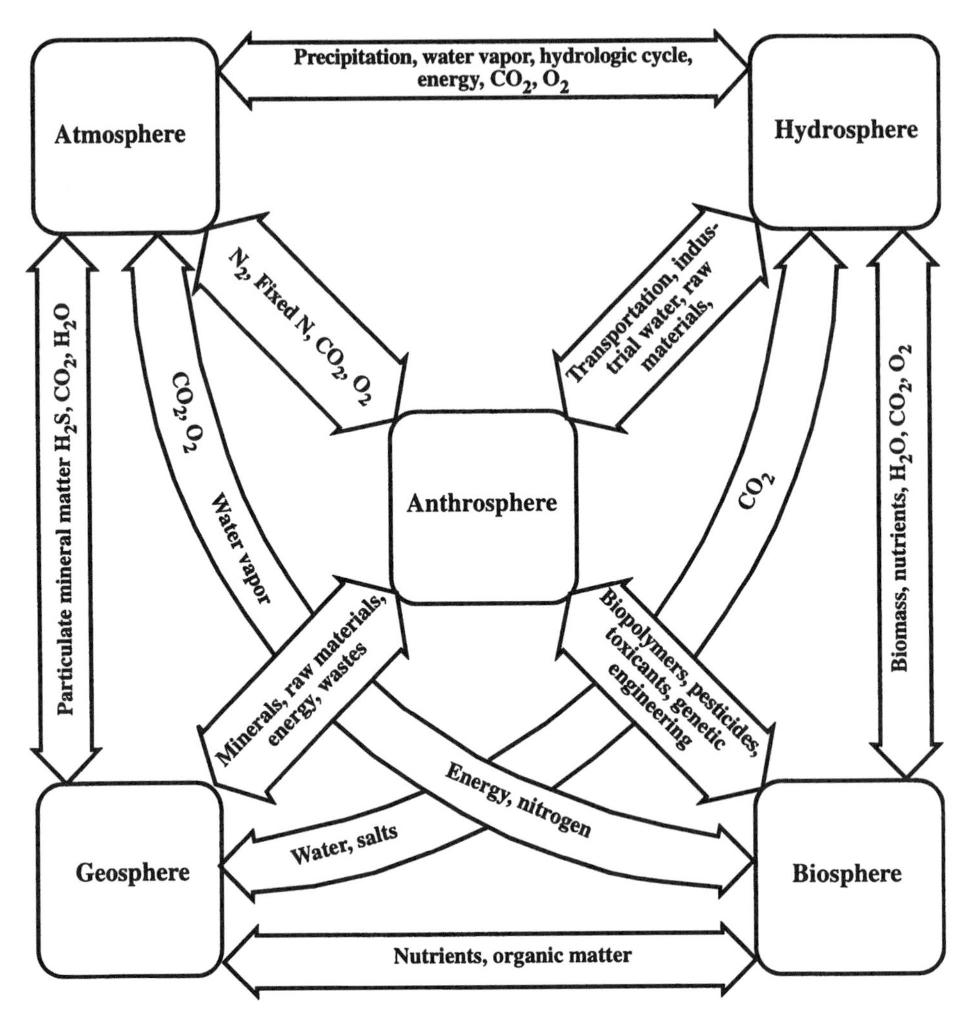

Figure 1.1. The five environmental spheres and the interchange of materials between them. Large quantities of energy may also be exchanged between spheres of the environment.

energy; how they communicate; how they extract and process nonrenewable minerals; and how they collect, treat, and dispose of wastes. With these factors in mind, it is possible to divide the anthrosphere into the following categories:

- Structures used for dwellings
- Structures used for manufacturing, commerce, education, and other activities
- Utilities, including water, fuel, and electricity distribution systems and waste distribution systems, such as sewers
- Structures and facilities used for transportation, including roads, railroads, airports, and waterways constructed or modified for water transport

- Structures and other parts of the environment modified for food production, such as fields used for growing crops and water systems used to irrigate the fields
- Machines of various kinds, including automobiles, farm machinery, and airplanes
- Structures and devices used for communications, such as telephone lines or radio transmitter towers
- Structures, such as mines or oil wells, associated with extractive industries

From the list above it is obvious that the anthrosphere is very complex with an enormous potential to affect the environment. Prior to addressing these environmental effects, several categories of the anthrosphere will be discussed in more detail.

1.4. TECHNOLOGY AND THE ANTHROSPHERE

The anthrosphere is the result of technology, so it is appropriate to discuss here **technology** as the ways in which humans do and make things with materials and energy. In modern times, technology is to a large extent the product of engineering based on scientific principles. Science deals with the discovery, explanation, and development of theories pertaining to interrelated natural phenomena of energy, matter, time, and space. Based on the fundamental knowledge of science, engineering provides the plans and means to achieve specific practical objectives. Technology uses these plans to carry out the desired objectives.

Since about 1900, advancing technology has been characterized by vastly increased uses of energy; greatly increased speed in manufacturing processes, information transfer, computation, transportation, and communication; automated control; a vast new variety of chemicals; new and improved materials for new applications; and, more recently, the widespread application of computers to manufacturing, communication, and transportation. In transportation, the development of passenger-carrying airplanes has effected an astounding change in the ways in which people get around and high-priority freight is moved. Rapid advances in biotechnology now promise to revolutionize food production and medical care.

The technological advances of the present century are largely attributable to two factors. The first of these is the application of electronics—now based upon solid state devices—to technology in areas such as communications, sensors, and computers for manufacturing control. The second area largely responsible for modern technological innovations is based upon improved materials. For example, since before World War II, airliners have been made of special strong alloys of aluminum; these are being supplanted by even more advanced composites. Synthetic materials with a significant impact on modern technology include plastics, fiber reinforced materials, composites, and ceramics.

Until very recently, technological advances were largely made without heed to environmental impacts. Now, however, the greatest technological challenge is to reconcile technology with environmental consequences. The survival of humankind and of the planet that supports it requires that the established two-way interaction between science and technology become a three-way relationship including environmental protection.

Engineering

Engineering uses fundamental knowledge acquired through science to provide the plans and means to achieve specific objectives in areas such as manufacturing, communication, and transportation. At one time engineering could be conveniently divided between military and civil engineering. With increasing sophistication, civil engineering evolved into even more specialized areas, such as mechanical engineering, chemical engineering, electrical engineering, and environmental engineering. Other engineering specialties include aerospace engineering, agricultural engineering, biomedical engineering, CAD/CAM (computer-aided design and computer-aided manufacturing) engineering, ceramic engineering, industrial engineering, materials engineering, metallurgical engineering, mining engineering, plastics engineering, and petroleum engineering.

The role of engineering in constructing and operating the various components of the anthrosphere is obvious. In the past, engineering was often applied without much, if any, consideration of environmental factors. As examples, huge machines designed by mechanical engineers were used to dig up and rearrange Earth's surface without regard for the environmental consequences, and chemical engineering was used to make a broad range of products without consideration of the wastes produced.

Fortunately, that approach is rapidly changing. Examples of environmentally friendly engineering include machinery designed to minimize noise, much improved energy efficiency in machines, and the uses of earth-moving equipment for environmentally beneficial purposes, such as restoration of strip-mined lands and construction of wetlands. Efficient generation, distribution, and utilization of electrical energy based on the principles of electrical engineering constitute one of the most promising avenues of endeavor leading to environmental improvement. Automated factories developed through applications of electronic engineering can turn out goods with lowest possible consumption of energy and materials, while minimizing air and water pollution and production of hazardous wastes. Chemical factories can be engineered to maximize the most efficient utilization of energy and materials while minimizing waste production.

1.5. INFRASTRUCTURE

The **infrastructure** is that part of the anthrosphere composed of the utilities, facilities, and systems used in common by members of a society and upon which the society depends for its normal function. The infrastructure includes both physical components — roads, bridges, and pipelines — and the instructions — laws, regulations, and operational procedures — under which the physical infrastructure operates. Parts of the infrastructure may be publicly owned, such as the U.S. Interstate Highway system and some European railroads, or privately owned, as is the case with virtually all railroads in the U.S. Some of the major components of the infrastructure of a modern society are the following:[2]

- Transportation systems, including railroads, highways, and air transport systems
- Energy generating and distribution systems
- Buildings
- Telecommunications systems
- Water supply and distribution systems
- Waste treatment and disposal systems, including those for municipal wastewater, municipal solid refuse, and industrial wastes

In general, the infrastructure refers to the facilities that large segments of a population must use in common in order for a society to function. In a sense, the infrastructure is analogous to the operating system of a computer. A computer operating system determines how individual applications operate and the manner in which they distribute and store the documents, spreadsheets, and illustrations created by the applications. Similarly, the infrastructure is used to move raw materials and power to factories and to distribute and store their output. An outdated, cumbersome computer operating system with a tendency to crash is detrimental to the efficient operation of a computer. In a similar fashion, an outdated, cumbersome, broken-down infrastructure causes society to operate in a very inefficient manner and is subject to catastrophic failure.

For a society to be successful, it is of the utmost importance to maintain a modern, viable infrastructure. Such an infrastructure is consistent with environmental protection. Properly designed utilities and other infrastructural elements, such as water supply systems and wastewater treatment systems, minimize pollution and environmental damage.

The development of new and improved materials is having a significant influence on the infrastructure. From about 1970 to 1985 the strength of steel commonly used in construction nearly doubled. During the latter 1900s significant advances were made in the properties of structural concrete. Superplasticizers enabled mixing cement with less water, resulting in a much less porous, stronger concrete product. Polymeric and metallic fibers used in concrete made it much stronger. For dams and other applications in which a material stronger than earth but not as strong as conventional concrete is required, roller-compacted concrete consisting of a mixture of cement with silt or clay has been found to be useful. The silt or clay used is obtained on site with the result that both construction costs and times are lowered.

The major challenge in designing and operating the infrastructure in the future will be to use it to work with the environment and to enhance environmental quality to the benefit of humankind. Obvious examples of environmentally friendly infrastructures are state-of-the-art sewage treatment systems, high-speed rail systems that can replace inefficient highway transport, and stack gas emission control systems in power plants. More subtle approaches with a tremendous potential for making the infrastructure more environmentally friendly include employment of workers at a computer terminal in their homes so that they do not need to commute, instantaneous electronic mail that avoids the necessity for moving letters physically, and solar electric powered installations to operate remote signals and relay stations, which avoids having to run electric power lines to them.

Whereas advances in technology and the invention of new machines and devices enabled rapid advances in the development of the infrastructure during the 1800s and early 1900s, it may be anticipated that advances in electronics and computers will have a comparable relative impact in the future. One of the areas in which the influence of modern electronics and computers is most visible is in telecommunications. Dial telephones and mechanical relays were perfectly satisfactory in their time, but have been made totally obsolete by innovations in electronics, computer control, and fiber optics. Air transport controlled by a truly modern, state-of-the-art computerized control system (which, unfortunately, is not yet fully installed in the U.S.) could enable present airports to handle many more airplanes safely and efficiently, thus reducing the need for airport construction. Sensors for monitoring strain, temperature, movement, and other parameters can be imbedded in the structural members of bridges and other structures. Information from these sensors can be processed by computer to warn of failure and to aid in proper maintenance. Many similar examples could be cited.

Although the payoff is relatively long term, intelligent investment in infrastructure yields very high returns. In addition to the traditional rewards in economics and convenience, properly designed additions and modifications to the infrastructure can pay large returns in environmental improvement as well.

Planned Communities

Planned communities represent aspects of the infrastructure with tremendous potential to enable humans to live efficiently in harmony with their surroundings in a manner consistent with good practices of industrial ecology. One of the most modern examples of such a community is that of Poundbury attached to Dorchester in southwestern England.[3] Located on 400 acres, Poundbury is designed to provide dwellings for 5,000 people, all living within easy walking distance of essential stores and services. A unique aspect of Poundbury is that the houses and cottages have been constructed to resemble those from the 1700s and 1800s. The dwelling exteriors are constructed with locally available materials, including stucco, flint, chalk, brick, and roof slate. However, the interiors are designed to take advantage of modern high-efficiency features, including thick and efficient insulation, condensing boilers for heating, and computer management of heating and electricity. Modern double-glazed windows are used to conserve energy. The water lines, sewer and drainage pipes, electrical utilities, gas lines, telephone lines, and television cables are hidden in underground channels. Rather than a clutter of individual rooftop television antennas, the community is served by a single large satellite dish, itself hidden behind masonry walls. Houses open directly onto streets, which are laid out in short, winding segments to control traffic speed.

1.6. MANUFACTURING AND ITS BYPRODUCTS

Once a device or product is designed and developed through the applications of engineering (see Section 1.4), it must be made — synthesized or manufactured. This may consist of the synthesis of a chemical from raw materials, casting of metal or plastic parts, assembly of parts into a device or product, or any of the other processes that go into producing a product that is needed in the marketplace.

Manufacturing activities have a tremendous influence on the environment. Energy, petroleum to make petrochemicals, and ores to make metals must be dug from, pumped from, or grown on the ground to provide essential raw materials. The potential for environmental pollution from mining, petroleum production, and intensive cultivation of soil is enormous. Huge land-disrupting factories and roads must be built to transport raw materials and manufactured products. The manufacture of goods carries with it the potential to cause significant air and water pollution and production of hazardous wastes. The earlier in the design and development process that environmental considerations are taken into account, the more "environmentally friendly" a manufacturing process will be.

A measure of the environmental impact of manufacturing is provided by the amounts of potentially hazardous substances released to the environment and where they are released, as provided by the U.S. Environmental Protection Agency's Toxics Release Inventory.[4] The major industrial groups and their releases are shown in Table 1.1. There are some important variables in these kinds of figures. For one thing, fortunately, the amounts of materials released are decreasing each year. Furthermore, the hazards of the materials and the susceptibility to damage of the sinks into which they are released vary appreciably. So the quantities shown in Table 1.1 do not necessarily reflect the potential environmental damage posed by each sector. However, the figures do show that there is a large potential to reduce quantities of wastes released. Assuming that the wastes contain at least some materials of value for use and recycling, the table illustrates the potential of an industrial ecology approach in conserving materials and reducing wastes.

A brief overview of the kinds of materials that may go into the toxics release inventory provides some perspectives on materials that may be recycled through industrial ecosystems. These substances include gases, vapors, particulate matter entrained in gas streams, liquids, water, solutions, sludges, and various kinds of solids as illustrated by the examples listed below.

- Air pollutant gases, particularly NO, NO_2, and SO_2. Of these, enough SO_2 is produced to make its recovery worthwhile in some cases to make sulfuric acid. Removal of NO_X is required as an air pollution control measure. Carbon dioxide, CO_2, has economic value in the pure form and can be recovered from some processes, such as fermentation, in which it is produced in large quantities in a relatively dilute form.

- Vapors. Volatile organic compounds (VOCs), such as dichloromethane or hydrohalocarbons (HCFCs), now used in place of ozone-endangering chlorofluorocarbons, have been discharged in large quantities to the atmosphere. Release of these materials may be prevented, or they may be reclaimed from air with adsorbents.

Table 1.1. Toxic Chemical Releases in the United States (1990)[1]

Industry	Quantity, kg/year	Major releases or transfers to
Chemicals	835,000	Air, surface water, underground well injection, land, off-site transfer
Metals	483,000	Air, surface water, underground well injection, land, off-site transfer
Paper	134,000	Air, surface water, land, off-site transfer
Transportation	107,000	Air, off-site transfer
Plastics	98,000	Air, off-site transfer
Furniture	63,000	Air, land, off-site transfer
Petroleum	57,000	Air, surface water, underground well injection, land, off-site transfer
Electrical	55,000	Air, land, off-site transfer
Machinery	30,000	Air, off-site transfer
Printing	26,000	Air, off-site transfer
Lumber	21,000	Air, off-site transfer
Textiles	17,000	Air, off-site transfer

1 From *1990 Toxics Release Inventory: Public Data Release*, U.S. Environmental Protection Agency, Office of Pollution Prevention and Toxics, Washington, D.C., 1992.

- Solvents. Solvents, such as those used for parts cleaning, compose some of the larger quantities of substances formerly discharged from manufacturing operations. These can be reclaimed and purified for reuse. Substitution of less volatile solvents (for example, toluene for highly volatile dichloromethane) or aqueous solutions can reduce emissions.

- Organic liquids. Various kinds of organic liquids are used in chemical synthesis and may be reclaimed as unreacted materials.

- Acids. Acids are used in a wide variety of applications, including metal ore processing and steel pickling. Processes are available to reclaim acids from waste streams.

- Process residues. Various kinds of residues consisting of liquid, sludge, or solid materials produced during manufacturing processes are generated. Measures should be taken to prevent making these materials wherever possible.

- Product residues. Product residues consist of the same kinds of materials that are incorporated into the manufactured product. A common example consists of small scraps or strips of excess material generated as "sprues" or "runners" during plastics molding. Some kinds of materials can be recycled directly to the process. Processes can be designed to minimize production of this kind of scrap.

- Packaging residues. Packaging, such as that in which components are shipped to a manufacturer, produces large quantities of materials that can create a waste problem. The best solution is to reuse the packaging to ship additional components. Paper and some kinds of plastics can be refabricated.

- Metals. Copper, lead, cadmium, zinc, and other metals used in metal fabrication, electroplating, and other applications are relatively easy to reclaim and recycle and usually have significant scrap value.

1.7. TRANSPORTATION

Few aspects of modern industrialized society have had as much influence on the environment as developments in transportation. These effects have been both direct and indirect. The direct effects are those resulting from the construction and use of transportation systems. The most obvious example of this is the tremendous effects that the widespread use of automobiles, trucks, and buses have had upon the environment. Entire landscapes have been entirely rearranged to construct highways, interchanges, and parking lots. Emissions from the internal combustion engines used in automobiles are the major source of air pollution in many urban areas.

The indirect environmental effects of widespread use of automobiles are enormous. The automobile has made possible the "urban sprawl" that is characteristic of residential and commercial patterns of development in the U.S., and in many other industrialized countries as well. The paving of vast areas of watershed and alteration of runoff patterns have contributed to flooding and water pollution. Discarded, worn-out automobiles have caused significant waste disposal problems. Vast enterprises of manufacturing, mining, and petroleum production and refining required to support the "automobile habit" have been very damaging to the environment.

On the positive side, however, applications of advanced engineering and technology to transportation can be of tremendous benefit to the environment. Modern rail and subway transportation systems, concentrated in urban areas and carefully connected to airports for longer distance travel, can enable the movement of people rapidly, conveniently, and safely, with minimal environmental damage. Although pitifully few in number in the U.S. with respect to the need for them, examples of such systems are emerging in progressive cities, showing the way to environmentally friendly transportation systems of the future.

Transportation and Hazardous Waste

Transportation plays a key role in the management of hazardous waste for the obvious reason that hazardous waste and hazardous substances must often be moved, sometimes in very large quantities. Transportation accidents, such as train derailments, are an unfortunate means by which many kinds of hazardous substances are transferred from the anthrosphere to the other spheres of the environment. Improvements in the safety and efficiency of transportation systems are very important in managing hazardous substances and wastes through the proper application of industrial ecology.

In the U.S. the transport of hazardous substances is closely regulated by both the U.S. Environmental Protection Agency (EPA) and the Department of Transportation (DOT). The regulations enforced by these agencies require that hazardous waste to be transported be described accurately, placed in appropriate containers with required labels, accompanied by appropriate documentation, and moved safely to a destination specified in advance.[5] During shipment, the waste must be properly segregated to prevent contact between incompatible wastes.

1.8. COMMUNICATIONS

It has become an overworked cliché that we live in an information age. Nevertheless, the means to acquire, store, and communicate information are expanding at an incredible pace. This phenomenon is having a tremendous impact upon society and has the potential to affect the environment in a number of ways.

The major areas to consider in respect to information are its acquisition, recording, computing, storing, displaying, and communicating. Consider, for example, the detection of a pollutant in a major river. Data pertaining to the nature and concentration of the pollutant may be obtained with a combination gas chromatograph and mass spectrometer. Computation by digital computer is employed to determine the identity and concentration of the pollutant. The data can be stored on a magnetic disk, displayed on a video screen, and instantaneously communicated all over the world by satellite and fiber optic cable.

All the aspects of information and communication listed above have been tremendously augmented by recent technological advances. Perhaps the greatest such advance has been that of silicon integrated circuits. Optical memory consisting of information recorded and read by microscopic beams of laser light has enabled the storage of astounding quantities of information on a single compact disk. The use of optical fibers to transmit information digitally by light has resulted in a comparable advance in the communication of information.

The central characteristic of communication in the modern age is the combination of telecommunications with computers called **telematics**.[6] Automatic teller machines use telematics to make cash available to users at locations far from the customer's bank. Information used for banking, for business transactions, and in the media depends upon telematics.

There exists a tremendous potential for good in the applications of the "information revolution" to environmental improvement. An important advantage is the ability to acquire, analyze, and communicate information about the environment and environmental problems. For example, such a capability enables detection of perturbations in environmental systems, analysis of the data to determine the nature and severity of the pollution problems causing such perturbations, and rapid communication of the findings to all interested parties. Modern capabilities in communication, including the Internet, World Wide Web sites, and e-mail, are very useful in the management of hazardous waste and in dealing with accidents involving hazardous substances.

1.9. FOOD AND AGRICULTURE

The most basic human need is food. Without adequate supplies of food, the most pristine and beautiful environment becomes a hostile place for human life. The industry that provides food is **agriculture**, an enterprise primarily concerned with growing crops and livestock.

The environmental impact of agriculture is enormous. One of the most rapid and profound changes in the environment that has ever taken place was the conversion of vast areas of the North American continent from forests and grasslands to cropland. Throughout most of the continental United States, this conversion took place predominantly during the 1800s. The effects of it were enormous. Huge acreages of forest lands that had been undisturbed since the last Ice Age were suddenly deprived of stabilizing tree cover and subjected to water erosion. Prairie lands put to the plow were destabilized and subjected to extremes of heat, drought, and wind that caused topsoil to blow away, culminating in the Dust Bowl of the 1930s.

In recent decades, valuable farmland has faced a new threat posed by the urbanization of rural areas. Prime agricultural land has been turned into subdivisions and paved over to create parking lots and streets. Increasing urban sprawl has led to the need for more highways. In a vicious continuing circle, the availability of new highway systems has enabled even more development. The ultimate result of this pattern of development has been the removal of once productive farmland from agricultural use.

On a positive note, agriculture has been a sector in which environmental improvement has seen some notable advances during the last 50 to 75 years. This has occurred largely under the umbrella of soil conservation. The need for soil conservation became particularly obvious during the Dust Bowl years of the 1930s, when it appeared that much of the agricultural production capacity of the U.S. would be swept away from drought-stricken soil by erosive winds. In those times and areas in which wind erosion was not a problem, water erosion took its toll. Ambitious programs of soil conservation have largely alleviated these problems. Wind erosion has been minimized by practices such as low-tillage agriculture, strip cropping in which crops are grown in strips alternating with strips of summer-fallowed crop stubble, and reconversion of marginal cultivated land to pasture. The application of low-tillage agriculture and the installation of terraces and grass waterways have greatly reduced water erosion.

Food production and consumption are closely linked with industrialization and the growth of technology. As illustrated by the examples of Japan, Taiwan, and South Korea, countries that develop high population densities prior to major industrial development experience two major changes that strongly impact food production and consumption.[7] First, they lose cropland as the result of industrialization at a rate such that increases in grain productivity cannot compensate, and total grain production decreases. Second, the increased income from industrialization raises the consumption of livestock products (meat, milk, eggs), which require much more cropland than does production of cereals, such as rice, consumed directly by humans. It may be anticipated that these effects will be particularly pronounced as China, which experienced economic growth of about

10% per year during much of the 1990s and which has a population of about 1.2 billion people, continues to grow. The growth and industrialization of this massive nation may well result in demands for grain and other food supplies that will cause disruptive food shortages and dramatic price increases.

In addition to the destruction of farmland to build factories, roads, housing, and other parts of the infrastructure associated with industrialization, there are other factors that tend to decrease grain production as economic activity increases. A major one is air pollution, which decreases crop production. Water pollution can seriously curtail fish harvests. Intensive agriculture uses large quantities of water for irrigation. If groundwater is used for irrigation, aquifers may become rapidly depleted.

Hazardous waste and agriculture are strongly interconnected in at least two main respects. One of these involves the potential effects of hazardous wastes upon the soil which provides the basis of agricultural production. For example, spreading waste sludges on land can cause heavy metal pollution that is detrimental to crop growth or that can contaminate grain or forage. The second reason is that modern industrialized agriculture carries with it the potential to generate and disperse hazardous waste. Among the listed hazardous wastes specified by U.S. Environmental Protection Agency regulations are those generated in connection with the production of agricultural products. One such material is 2,6-dichlorophenol waste from the production of herbicidal 2,4-D. Another agriculturally related waste is composed of wastewater treatment sludges generated during the production of veterinary pharmaceuticals from arsenic or organoarsenic compounds.

1.10. EFFECTS OF THE ANTHROSPHERE ON EARTH

The effects of the anthrosphere on Earth have been many and profound. Persistent and potentially harmful products of human activities have been widely dispersed and concentrated in specific locations in the anthrosphere as well as other spheres of the environment. Among the most troublesome of these are toxic heavy metals and organochlorine compounds. Such materials have accumulated in the anthrosphere in painted and coated surfaces (such as organotin-containing paints used to prevent biofouling on boats), under and adjacent to airport runways, under and along highway paving, buried in old factory sites, disposed in landfills, and incorporated into materials dredged from waterways and harbors, sometimes used as landfill on which buildings, airport runways, and other structures have been placed. In many cases productive topsoil used to grow food has been contaminated with discarded industrial wastes, phosphate fertilizers, and dried sewage sludge. Some of the portions of the anthrosphere that may be severely contaminated by human activities are shown in Figure 1.2.

Potentially harmful wastes and pollutants of anthrospheric origin have found their way into water, air, soil, and living organisms. For example, chlorofluorocarbons (Freons) have been released into the atmosphere in such quantities and are so stable that they are now constituents of "normal" atmospheric air and pose a threat to the protective ozone layer in the stratosphere. Lake sediments, stream beds, and deltas deposited by flowing rivers are contaminated with heavy metals and refractory organic compounds of anthrospheric origin. The most troubling repository of wastes in the hydrosphere is groundwater. Some organisms have accumulated high enough levels of persistent organic compounds or heavy metals to do harm to themselves or to humans that use them as a food source.

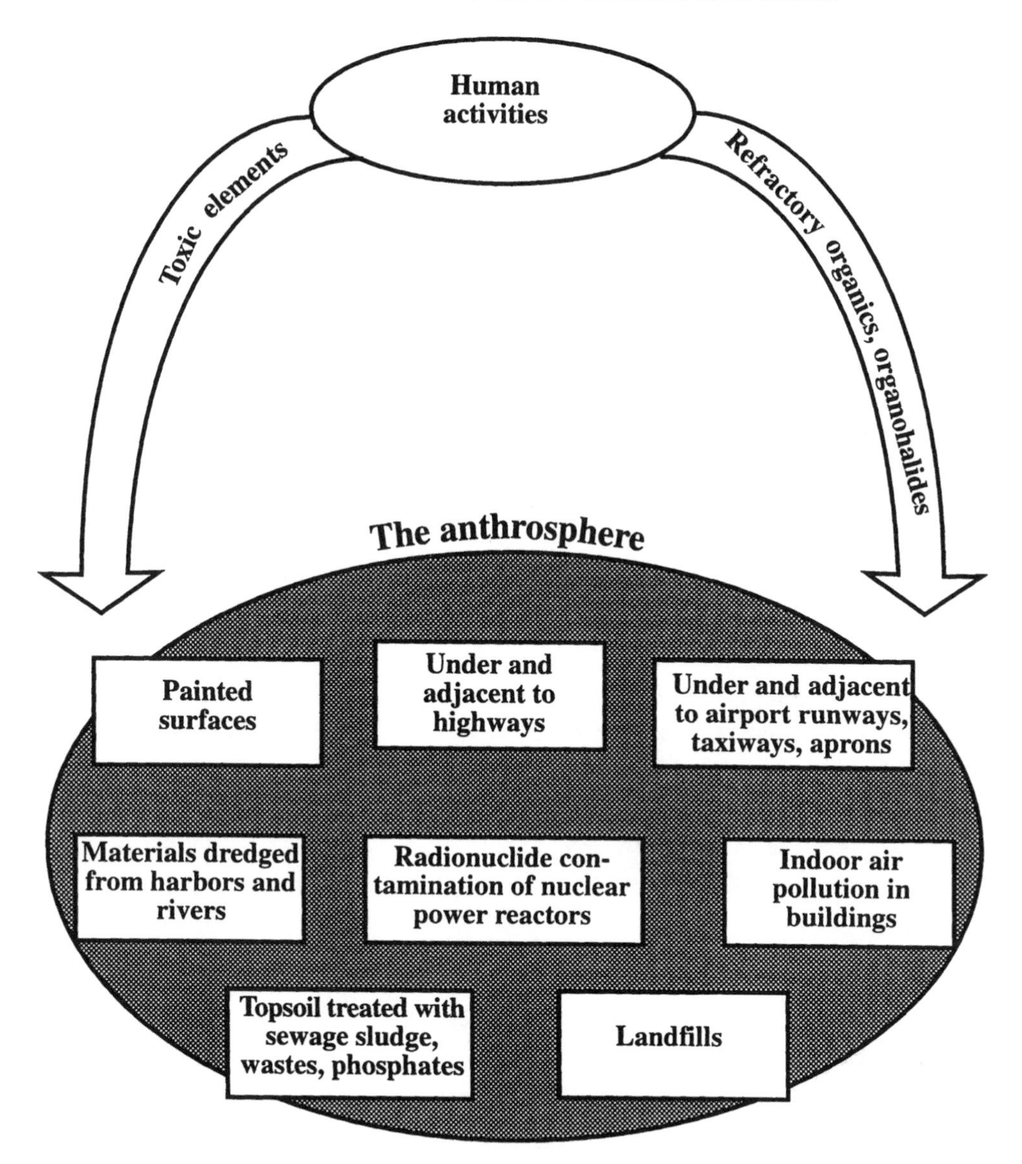

Figure 1.2. The anthrosphere is a repository of many of the pollutant byproducts of human activities.

1.11. INTEGRATION OF THE ANTHROSPHERE INTO THE TOTAL ENVIRONMENT

To a degree, the early anthrosphere created by pre-industrial humans integrated well with the other spheres of the environment and caused minimal environmental degradation. This resulted less from any noble instincts of humankind toward nature than it did from the lack of power to alter the environment. In those cases where humans had the capability of modifying or damaging their surroundings, such as by burning forests to provide cropland, the effects on the natural environment could be profound and very damaging. In general, though, pre-industrial humans integrated their anthrosphere, such as it was, with the natural environment as a whole.

The relatively harmonious relationship between the anthrosphere and the rest of the environment began to change markedly with the introduction of machines, particularly power sources, beginning with the steam engine, that greatly multiplied the capabilities of humans to alter their surroundings. As humans developed their use of machines and other attributes of industrialized civilization, they did so with little consideration of the environment and in a way that was out of synchronization with the other environmental spheres. A massive environmental imbalance has resulted, the magnitude of which has been realized only in recent decades. The most commonly cited manifestation of this imbalance has been pollution of air or water.

Because of the detrimental effects of human activities undertaken without adequate consideration of environmental consequences, significant efforts have been made to reduce the environmental impacts of these activities. Figure 1.3 shows three stages of the evolution of the anthrosphere, from an unintegrated appendage to the natural environment, to a system more attuned to its surroundings. The first approach to dealing with the pollutants and wastes produced by industrial activities — particulate matter from power plant stacks, sulfur dioxide from copper smelters, and mercury-contaminated wastes from chlor-alkali manufacture — was to ignore them. However, as smoke from uncontrolled factory fur-

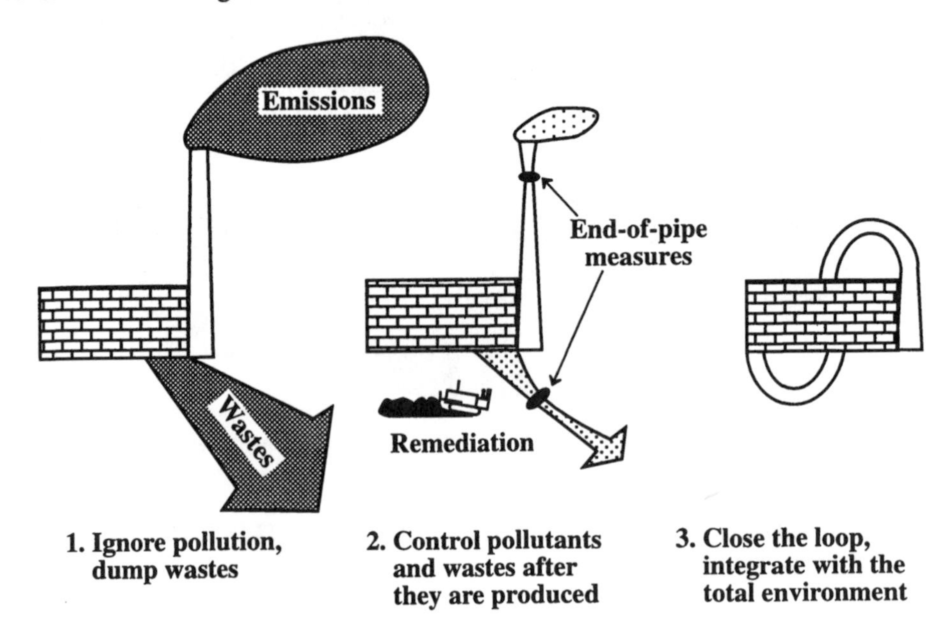

Figure 1.3. Steps in evolution of the anthrosphere to a more environmentally compatible form.

naces, raw sewage, and other byproducts of human activities became more troublesome, "end-of-pipe" measures were adopted to prevent the release of pollutants after they were generated. Such measures have included electrostatic precipitators and flue gas desulfurization to remove particulate matter and sulfur dioxide from flue gas; physical processes used in primary sewage treatment; microbial processes used for secondary sewage treatment; and physical, chemical, and biological processes for advanced (tertiary) sewage treatment. Such treatment measures are often very sophisticated and effective. Another kind of end-of-pipe treatment is the disposal of wastes in a supposedly safe place. In some cases, such as municipal solid wastes, radioactive materials, hazardous chemicals, power plant ash, and contaminated soil, disposal of sequestered wastes in a secure location is practiced as a direct treatment process. In other cases, including flue-gas desulfurization sludge, sewage sludge, and sludge from chemical treatment of industrial wastewater, disposal is practiced as an adjunct to other end-of-pipe measures. Waste disposal practices later found to be inadequate have spawned an entirely separate end-of-pipe treatment called **remediation** in which discarded wastes are dug up, sometimes subjected to additional treatment to reduce their mobilities and potential hazards, then placed in a more secure disposal site.

Although sometimes unavoidable, the production of pollutants followed by measures taken to control or remediate them to reduce the quantities and potential harmfulness of wastes is not very desirable. Such measures do not usually eliminate wastes and may, in fact, transfer a waste problem from one part of the environment to another. An example of this is the removal of air pollutants from stack gas and their disposal on land, where they have the potential to cause groundwater pollution. Clearly, it is now unacceptable to ignore pollution and to dump wastes, and the control of pollutants and wastes after they are produced is not a good permanent solution to waste problems. Therefore, it has become accepted practice to "close the loop" on industrial processes, recycling materials as much as possible and allowing only benign waste products to be released to the environment. Such an approach is the basis of industrial ecology discussed throughout this book.

1.12. ENVIRONMENTAL PROTECTION AND THE REGULATORY APPROACH TO POLLUTION

The **regulatory approach** to environmental protection is one in which laws were passed prohibiting the discharge and emission of specified pollutants. Environmental regulation has undergone three different phases in the U.S. since about 1970.

The first stage of the regulatory approach to pollution control involved the passage of basic laws regulating pollution, particularly of air and water, media in which pollutants are often most obvious. These acts, which were implemented largely in the 1970s, include the following:

- The Clean Water Act written to reduce water pollution.
- The Clean Air Act for air pollution control.

- The Toxic Substances Control Act of 1976 to regulate toxic substances, such as PCBs.
- The Resource Conservation and Recovery Act (RCRA) of 1976, amended and reauthorized by the Hazardous and Solid Wastes Amendments Act (HSWA) of 1984 to protect human health and the environment from improper management and disposal of hazardous wastes.
- Comprehensive Environmental Response, Compensation, and Liability Act (CERCLA) of 1980, extended for 5 years by the passage of the Superfund Amendments and Reauthorization Act (SARA) of 1986. This legislation pertains to actual or potential releases of hazardous materials that might endanger people or the surrounding environment at uncontrolled or abandoned hazardous waste sites.

The overall goal of these laws has been to prevent entry of certain materials into the waste stream. They have had several counterproductive effects. Compliance with the Clean Water Act resulted in the use of water treatment processes that generated solid waste sludges. Flue gas scrubbers collected toxic dusts and sludges that presented solid waste disposal problems. The laws also placed some inhibitions upon recycling because they tend to regard recycling as a form of disposal making recyclers liable for the disposal of materials.[8] Under these laws new materials were subjected to much less regulation than used ones. For example, in a regulatory sense it was much easier to buy and use newly produced acid than to recycle and use acid in steel pickling liquor. The net effect has been to inhibit reuse of materials that might otherwise have industrial applications.

Although laws such as the Clean Water Act and the Clean Air Act clearly resulted in reduced emissions and pollution, they have not been particularly cost-effective in their mission. Therefore, a second stage of the regulatory approach to pollution control developed that was oriented more toward reducing pollution within industrial sources. This was the major goal of the Pollution Prevention Act of 1990. One of the more significant provisions of this act was to require companies to report releases of pollutants through a Toxic Release Inventory. The attendant publicity resulted in significant reductions in releases and the development of less toxic substitutes for toxic materials, thus diminishing amounts and toxicities of effluents. An example is the substitution of water with appropriate cleaning agents dissolved in it for toxic organic solvents in the electronics industries and use of water-based coatings in place of solvent-based coatings.

A third stage of the regulatory approach to pollution control is the Common Sense Initiative launched by the U.S. Environmental Protection Agency in 1994. This initiative attempts to reduce pollution through cooperative efforts between government agencies and industry, relying on information and positive incentives in place of restrictive regulations. It encourages clean production strategies, such as one that employs computer-based control of manufacturing processes carefully tuned to avoid unnecessary use and release of materials. An example would be precision measurement and cutting of materials using computer control and laser measurement techniques to avoid waste of scrap materials. Clean production strategies often end up saving money because they are relatively more efficient.

Industrial Ecology and Environmental Economics

Industrial ecology can be tied very closely with **environmental economics**, the branch of economics that considers environmental resources in addition to capital, labor, and raw materials in the allocation of goods and services. Environmental resources include clean air and water, biodiversity, agreeable climate, and unpolluted, productive topsoil. It is difficult to assign monetary values to such resources, but they clearly have great value.

Natural Capital

Whereas capital has been defined traditionally as the human means of production, **natural capital** refers to Earth's ability to produce. Natural capital includes mineral resources, energy resources, topsoil, water resources, gases in the atmosphere, and Earth's biological resources. Just as capital in the form of industrial plant infrastructure may be degraded by lack of maintenance, obsolescence, overuse, and abuse, natural capital may be similarly degraded. Water may become so polluted that it is no longer useful for municipal, agricultural, or industrial applications. Topsoil can be degraded by wastes to the extent that its ability to produce crops is severely impaired. Extinction of biological species results in an irreplaceable loss of genetic diversity. Greenhouse gas pollution of the atmosphere has the potential to cause adverse changes in climate, another loss of natural capital.

The two major ways in which natural capital can be lost are resource consumption and environmental degradation. When a nonrenewable resource is consumed in such a way that it cannot be recycled, it is lost from the pool of natural capital forever. For example, mineral phosphate used as fertilizer and dispersed in soil or water runoff from the soil is a resource that is lost permanently for human use. Water in an aquifer polluted by hazardous wastes to the extent that it cannot be used also represents a loss of natural capital. Thus a highly industrialized society that emphasizes consumption puts a double burden on its environment by both consuming large quantities of resources and degrading the environment that is conducive to a high quality of life.

An interesting approach to the issue of natural capital is to be found in attempts to use conventional economic calculations, which do not place any value on "nature" as such, to assign a monetary worth to Earth, including such intangibles as the value of clean air, climate regulation, and recreational amenities. Such a study has been performed by the Institute of Ecological Economics of the University of Maryland.[9] This study estimated that Earth's "ecosystem services" had a value of $33 trillion in 1994, about twice that of the conventional global gross national product (GNP) for that year. Placing a financial value on Earth's natural capital is an exercise subject to much uncertainty, and some authorities contend that it should not be done at all. However, the assignment of monetary worth to nature may be the best way to get the attention of decision makers, who overwhelmingly are guided by costs and benefits measured in conventional monetary terms. If Earth's value can be internalized in these decision-making processes, environmental protection will almost certainly have a much higher priority.

Permit Systems for Pollution Abatement

An approach to pollution abatement that has gained favor in recent years is the use of **tradeable permits** in place of direct regulations of emissions. Under the permit system, a concern that curtails emissions of a specific pollutant by various measures may sell or trade pollution abatement to another concern that may choose to pay for the right to pollute rather than cutting back on its own emissions. A related approach is to tax pollutant emissions. Although these measures have been attacked as a "license to pollute," current evidence suggests that they have been effective in reducing pollution and increasing the financial incentives to develop pollution controls.

1.13. THE ANTHROSPHERE AND INDUSTRIAL ECOLOGY

Normally an industrial ecosystem must be based upon a large supplier of a particular kind of material and/or energy analogous to the photosynthetic "primary producer" in a natural ecosystem. One of the most common of these is a petroleum refiner, which provides both material and energy to form the basis of a large petrochemical complex.

Dealing largely with materials, one of the major goals of industrial ecology is to reduce and preferably eliminate the use of substances that cannot be converted into benign forms by biodegradation or consumption as a fuel. This is particularly true of toxic substances, such as heavy metals. Another aspect of the industrial ecology of materials is the development of optimum processes for recycling them so that less "virgin" material is required. A third way in which use of materials can be minimized is by design for reuse of components. Devices can be designed for ease of disassembly such that reusable components can be readily removed, refurbished, if necessary, then used again. One of the most common examples of this consists of auto parts, such as alternators, brake master cylinders, and water pumps, which have long been remanufactured and resold for repairs.

From the beginning, industrial ecology must consider process/product design in determining the output of materials, including the ultimate fates of materials when they are discarded. A basic part of industrial ecology is recycling, often using the waste from one process as the raw material for another. The product and the materials used to make it should be subjected to an entire life-cycle analysis. The following must be considered:

- If there is a choice, the kinds of materials to be used that will minimize waste
- Kinds of materials that can be reused or recycled
- Components that can be recycled
- Alternate pathways for the manufacturing process or for various parts of it

An important aspect of industrial ecology is separation processes employed to "un-mix" materials for recycling at the end of a product cycle. An example of this is the separation of graphite carbon fibers from epoxy resin binders in carbon fiber composites. The chemical industry provides many examples where separations are desirable. For example, the separation of toxic heavy metals from solutions or sludges can yield a valuable metal product, leaving nontoxic water that can be recycled and harmless solids that can be reused or safely disposed.

The applications of the principles of industrial ecology constitute a logical and efficient means for humans to live in harmony and in a sustainable manner with their environment, while still providing the goods and services that they need for their existence and well-being. Environmental chemistry is an integral part of any industrial ecological system and must be considered in implementing such systems. Environmental chemistry is applied to anticipate problems before they occur so that they can be avoided. Dealing with all the spheres of the environment, environmental chemistry serves a crucial function in interfacing the hydrosphere, atmosphere, geosphere, and biosphere with the anthrosphere within a framework of industrial ecology. Industrial ecology and its relationship to environmental chemistry are discussed in greater detail in Chapter 7.

LITERATURE CITED

1. Manahan, Stanley E., *Environmental Science and Technology*, CRC Press/Lewis Publishers, Boca Raton, FL, 1997.

2. Ausubel, Jesse H. and Robert Herman, Eds., *Cities and Their Vital Systems: Infrastructure Past, Present, and Future*, National Academy Press, Washington, D.C., 1988.

3. Hoge, Warren, "In Stone, a Prince's Vision of Britain," *New York Times*, June 11, 1998, p. B1.

4. "Industrial Process Residues: Composition and Minimization," Chapter 15 in *Industrial Ecology*, Thomas E. Graedel and B. R. Allenby, Prentice Hall, Englewood Cliffs, NJ, 1995, pp. 204–230.

5. McGinnis, Peter J., "Hazardous Waste Transportation," Section 14.2 in *Standard Handbook of Hazardous Waste Treatment and Disposal*, 2nd ed., Harry M. Freeman, Ed., 1998, pp. 14.13–14.41.

6. Gille, Dean, "Combining Communications and Computing: Telematics Infrastructure," Chapter 10 in *Cities and Their Vital Systems: Infrastructure Past, Present, and Future*, Jesse H. Ausubel and Robert Herman, Eds., National Academy Press, Washington, D.C., 1988, pp. 233–257.

7. Brown, Lester R., *Who Will Feed China?*, W. W. Norton and Company, New York, NY, 1995.

8. Frosch, Robert A., "Industrial Ecology: Adapting Technology for a Sustainable World," *Environment Magazine*, **37**, 16–37, 1995.

9. Roubi, Maureen, "Fallout from Pricing Earth," *Chemical and Engineering News*, June 30, 1997, p. 38.

SUPPLEMENTARY REFERENCES

Frosch, Robert A. and Nicholas E. Gallopoulos, "Strategies for Manufacturing," *Scientific American*, **261**, 94–102 (1989).

2 INDUSTRIAL ECOLOGY AND INDUSTRIAL SYSTEMS

2.1. INTRODUCTION AND HISTORY

Industrial ecology *is an approach based upon systems engineering and ecological principles that integrates the production and consumption aspects of the design, production, use, and termination (decommissioning) of products and services in a manner that minimizes environmental impact while optimizing utilization of resources, energy, and capital.* The practice of industrial ecology represents an environmentally acceptable, sustainable means of providing goods and services. The meaning of industrial ecology, a relatively new concept, has been succinctly outlined in a book dealing with industrial ecology and global change.[1] Industrial ecology mimics natural ecosystems, which, usually driven by solar energy and photosynthesis, consist of an assembly of mutually interacting organisms and their environment in which materials are interchanged in a largely cyclical manner. An ideal system of industrial ecology follows the flow of energy and materials through several levels, uses wastes from one part of the industrial ecosystem as raw material for another part, and maximizes the efficiency of energy utilization. Whereas wastes, effluents, and products used to be regarded as leaving an industrial system at the point where a product or service was sold to a consumer, industrial ecology regards such materials as part of a larger system that must be considered until a complete cycle of manufacture, use, and disposal is completed.[2]

From the discussion above and in the remainder of this book, it may be concluded that industrial ecology is all about *cyclization of materials*. This approach is summarized in a statement attributed to Kumar Patel of the University of California at Los Angles that "The goal is *cradle to reincarnation*, since if one is practicing industrial ecology correctly there is no grave." For the practice of industrial ecology to be as efficient as possible, cyclization of materials should occur at the highest possible level of material purity and stage of product development. For example, it is much more efficient in terms of materials, energy, and monetary costs to bond a new tread to a tire used on heavy earth-moving equipment than it is to try to separate the rubber from the tire and remold it into a new one. The meaning of industrial ecology, a relatively new concept, has been succinctly outlined in a book dealing with industrial ecology and global change.[3]

Industrial Ecology and Production of Goods and Services

Figure 2.1 illustrates the four major inputs required to make a product or provide a service. The relative amounts of each of the components illustrated in Figure 2.1 can vary greatly. Usually waste and environmental degradation occur as the result of making, using, or disposing of a product or in providing a service. In some cases, however, wastes are actually consumed or eliminated and environmental conditions can be improved.

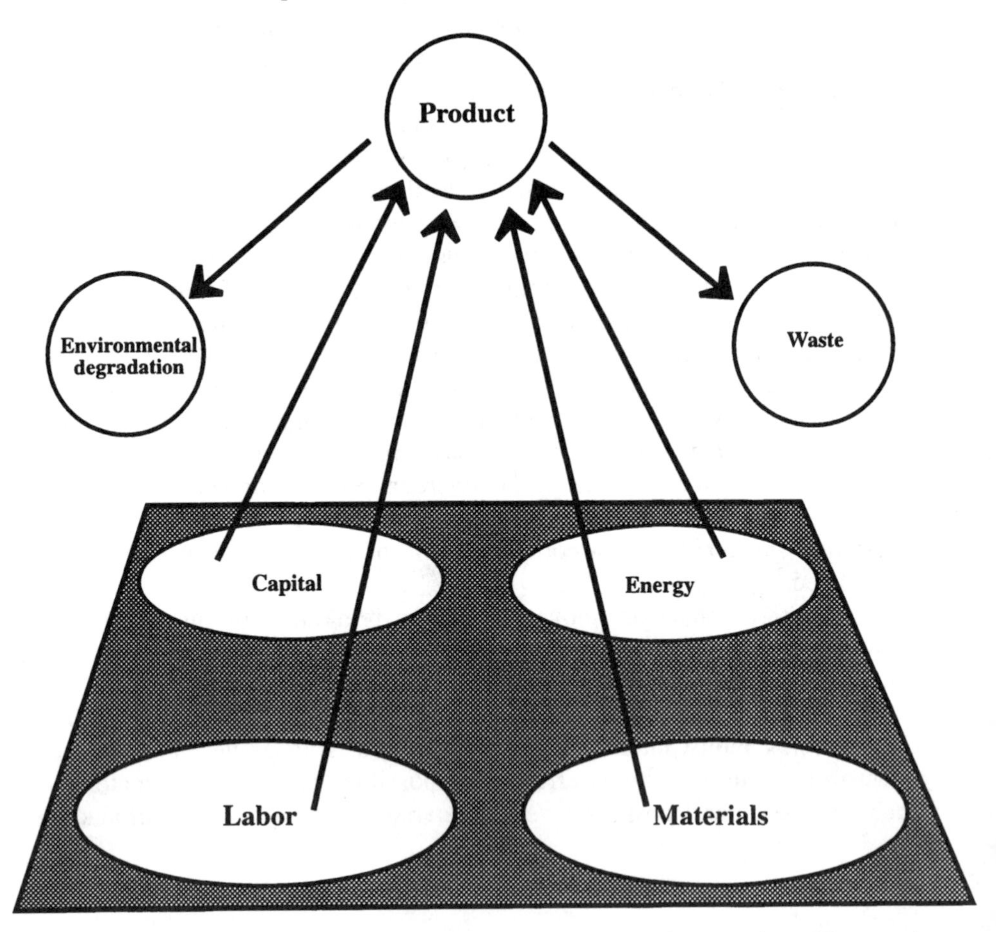

Figure 2.1. Capital, energy, materials, and labor are required to make a product. Waste and environmental degradation are usually the result.

Since the beginning of the industrial revolution, and particularly during the last two centuries, increasingly efficient machines and vastly increased use of energy, mostly from fossil sources, have resulted in spectacular increases in labor productivity. As a consequence, the consumption of natural resources has gone up enormously while the inflation-adjusted monetary costs of resources has declined steeply. In a sense, therefore, the share of labor required to provide products as shown in Figure 2.1 has decreased markedly relative to capital, energy, and materials. Such a picture can be a little misleading, however, if human ingenuity is included as part of labor. It can be argued that, whereas the share of labor measured by humans doing things directly with their hands has shrunk, the importance of human brainpower has increased, and is now more important than ever.

Two opposing schools of thought have emerged in respect to materials. One of these is that the supply of materials, such as critical metals, is finite and that humankind is quickly running out of them, which will force economic activity to decline. The opposing school of thought holds that the law of supply and demand rules, so shortages will cause higher prices, which will result in a larger supply. During the last several decades the latter model has proven to be the more accurate. However, that has led to a false sense of security in some quarters. Common sense dictates that critical resources are finite and, even though higher prices and better technology have greatly extended supplies, there are limits that can be very painful for humankind.

Minimization of labor can put a significant strain on resources and result in increased energy use. For example, a society that uses automated processes to manufacture huge quantities of consumer goods, many designed to be used by the leisure class, obviously consumes a lot of resources and energy. The manufacture of large quantities of goods consumes enormous amounts of materials. If the manufactured goods are simply discarded, large quantities of wastes are produced. Air and water may be polluted as the result of manufacturing activities, and the disposal of discarded items can also cause environmental degradation. However, labor-intensive remanufacturing of such things as automobile parts can save significant amounts of resources and energy. There is, of course a balance. For example, few would be willing to discard the modern mechanized agricultural system enjoyed in more advanced countries and return to times in which the majority of the population was employed in subsistence farming.

The traditional economically based view of an industrial system holds it apart from its surroundings. With an approach based upon industrial ecology, however, economic systems are seen to be fundamental anthrospheric constituents of a larger whole in which the anthrosphere is integrated with the geosphere, hydrosphere, atmosphere, and biosphere. A key aspect of industrial ecology is optimization of materials utilization in the pathway (1) raw material → (2) finished material → (3) component → (4) product → (5) waste. A well-designed industrial ecosystem also optimizes energy utilization. And it utilizes capital in the most efficient manner possible consistent with environmental protection.

In reality, it is probably impossible to realize a complete industrial ecosystem. However, the elements of industrial ecology can be applied to industrial systems in part to maximize the efficiency of such systems, to make them much more environmentally friendly, and to enable them to approach a condition of sustainable development. Therefore, the concepts of industrial ecology should be part of the knowledge base of any environmental scientist.

2.2. INDUSTRIAL METABOLISM

The basis of industrial ecology is provided by the phenomenon of **industrial metabolism**, which refers to the ways in which an industrial system handles materials and energy, extracting needed materials from sources, such as ores, using energy to assemble materials in desired ways, and disassembling materials and components. In this respect an industrial ecosystem operates in a manner analogous to biological organisms, which act on biomolecules to perform anabolism (synthesis) and catabolism (degradation).

Just as occurs with biological systems, industrial enterprises can be assembled into **industrial ecosystems**. Such systems consist of a number (preferably large and diverse) of industrial enterprises acting synergistically and, for the most part, with each utilizing products and potential wastes from other members of the system. Such systems are best assembled through natural selection and to a greater or lesser extent, such selection has occurred throughout the world. However, recognition of the existence and desirability of smoothly functioning industrial ecosystems can provide the basis for laws and regulations (or the repeal thereof) that give impetus to the establishment and efficient operation of such systems.

A basic understanding of industrial metabolism is essential to understanding industrial ecology. Chapter 3 is devoted to the concept of industrial metabolism.

2.3. LEVELS OF MATERIALS UTILIZATION

There are two extremes in levels of materials utilization in industrial systems. At the most inefficient level as shown in Figure 2.2, raw materials are viewed as being unlimited and no consideration is given to limiting wastes. Such an approach was typical of industrial development in the U.S. in the 1800s and early 1900s when the prevailing view was that there were no limits to ores, fossil energy resources, and other kinds of raw materials; furthermore, it was generally held that the continent had an unlimited capacity to absorb industrial wastes.

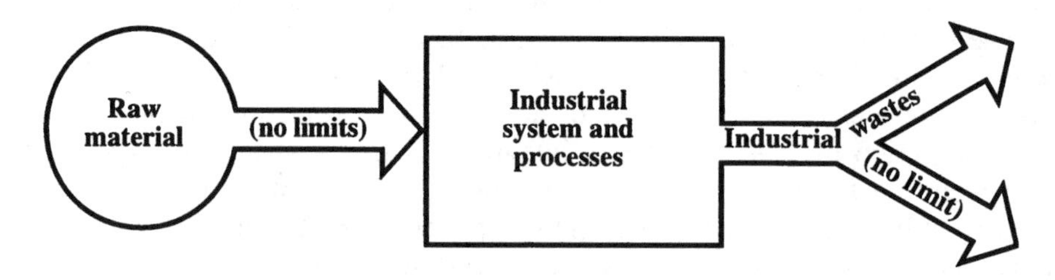

Figure 2.2. An industrial system without limits on either raw materials consumed or wastes produced.

A second kind of industrial system in which both raw materials and wastes are limited to greater or lesser extents is illustrated in Figure 2.3. Such a system has a relatively large circulation of materials within the industrial system as a whole, compared with reduced quantities of material going into the system and relatively lower production of wastes. Such systems are typical of those in industrialized nations and modern economic systems in which shortages of raw materials and limits to the places to put wastes are beginning to be felt. Even with such constraints, large quantities of materials are extracted, processed, and used, then either disposed in the environment in concentrated form (hazardous wastes) or dispersed. In recent years regulations and other constraints have markedly decreased point source pollution from industrial activity. However, the sheer volume of materials processed through industrial societies continues to increase dissipative pollution.

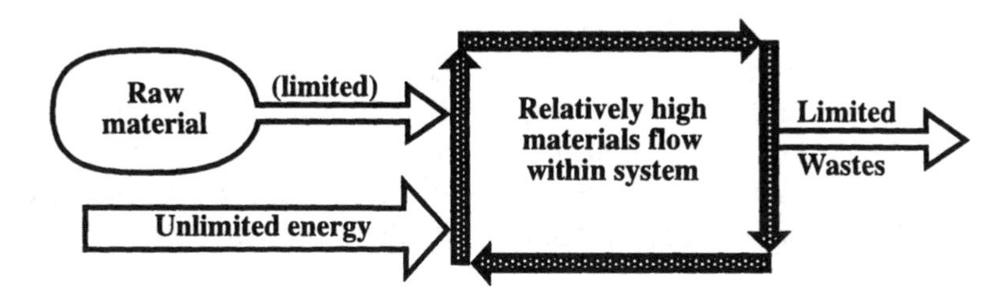

Figure 2.3. Illustration of an industrial system in which both the utilization of raw materials and the production of wastes are limited to a certain degree.

An idealized industrial ecosystem with no materials input and no wastes is illustrated in Figure 2.4. The energy requirements of such a system are rather high and the material flows within the system itself are quite high. Such a system is an idealized one that can never be realized in practice, but it serves as a useful goal around which more practical and achievable systems can be based.

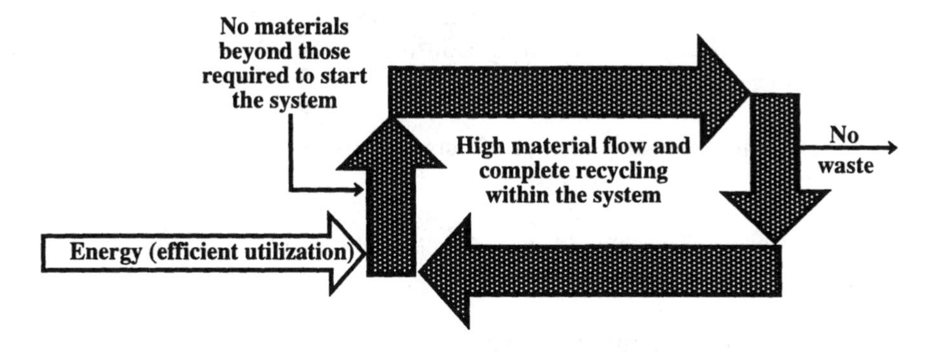

Figure 2.4. Idealized industrial ecosystem in which no materials are required for input beyond those needed to start the system. Energy requirements are relatively high, and the material flow within the system is high and continuous.

2.4. Links to Other Environmental Spheres

Having addressed industrial ecology largely from the anthrospheric viewpoint, it is now appropriate to consider how the anthrosphere and the practice of indusrial ecology influence the other four spheres of the environment—the atmosphere, the hydrosphere, the geosphere, and the biosphere. Influences of industrial activities, broadly defined to include energy and agricultural production as well as manufacture of goods and provision of essential services, can range from minor to major effects that may pose significant threats to Earth's ability to support life.

Such effects can range from highly localized ones occurring for only a brief period of time to global effects that have the potential to last for centuries. An example of the former would be an isolated incidence of water pollution by oxygen-consuming organic matter in a reservoir; only the reservoir is affected and only for the relatively short period of time required to degrade the wastes and replenish the oxygen supply. The prime example of a long term global effect is the emission of greenhouse gases, which has the potential to change Earth's entire climate for thousands of years.

The major goal of industrial ecology, therefore, must be to minimize or eliminate detrimental effects of anthrospheric activities on other spheres of the environment. Beyond environmental preservation, the practice of industrial ecology should also improve and enhance environmental conditions. Listed below are the major anthrospheric activities along with their potential effects on other environmental spheres.

Fossil fuel combustion

Atmosphere: The greatest potential effect is greenhouse warming. Emission of partially combusted hydrocarbons and nitrogen oxides can cause formation of photochemical oxidants (photochemical smog). Acid precipitation may be caused by emissions of sulfur oxides from fuel combustion. General deterioration of atmospheric quality may occur through reduced visibility.

Hydrosphere: The potential exists for water pollution from acid mine water, petroleum production byproduct brines, acid precipitation, and heating of water used to cool power plants.

Geosphere: The greatest potential effects are disturbance of land from coal mining.

Biosphere: Most effects would be indirect as the result of influences on the atmosphere, hydrosphere, and geosphere.

Industrial manufacturing and processing

Atmosphere: Greatest potential effects are due to emissions of gases, vapors, and particles. These include greenhouse gases, acid gases, particles, precursors to photochemical smog formation, and species with the potential to deplete stratospheric ozone.

Hydrosphere: Industrial activities may contaminate water with a variety of pollutants. Consumptive uses of water may put pressure on limited water supplies, especially in arid regions. Water used for cooling may be thermally polluted.

Geosphere: The greatest effect results from the extractive industries through which minerals are recovered. The geosphere may be contaminated by solid and hazardous wastes, and available landfill space may become depleted.

Biosphere: The greatest direct effect is from the distribution of toxic substances as the result of industrial activities. There may also be significant indirect effects resulting from deterioration of the atmosphere, hydrosphere, and geosphere.

Crop production

Atmosphere: A major potential effect is emission of greenhouse gases as the result of deforestation and "slash and burn" agriculture to grow more crops. Significant amounts of greenhouse gas methane are emitted to the atmosphere as the result of methane-generating bacteria growing in rice paddies.

Hydrosphere: Large quantities of water are used for irrigation. Some of the water is lost by transpiration from plants and some by infiltration to groundwater. Water returned to the hydrosphere from irrigation may have an excessively high salinity. Surface water and groundwater may become contaminated by solids, fertilizers, and herbicides from crop production.

Geosphere: Large areas of the geosphere may be disturbed by cultivation to produce crops. Topsoil can be lost from water and wind erosion. Proper agricultural practices, such as contour farming and low-tillage agriculture, minimize these effects and may even enhance soil quality.

Biosphere: Organisms are profoundly affected by agricultural practices designed to produce crops. Entire ecosystems are destroyed and replaced by other "anthrospheric" ecosystems. The greatest effect on the biosphere is loss of species diversity from the destruction of natural ecosystems and from the cultivation of only limited strains of crops.

Livestock production (domestic animals)

Atmosphere: Ruminant animals are significant producers of greenhouse gas methane as the result of methane-producing bacteria in their digestive systems.

Hydrosphere: Livestock production requires large quantities of water. Large amounts of oxygen-consuming wastes that may contaminate surface water are produced by livestock. Nitrogen wastes from the manure and urine of animals in feedlots may cause nitrate contamination of groundwater.

Geosphere: The production of a unit mass of food from livestock sources requires much more crop production than is required for grains consumed directly by humans. A major impetus behind destruction of rain forests has been to grow forage and other foods for livestock. Rangeland has deteriorated because of overgrazing.

Biosphere: A major effect is loss of species diversity. This occurs even within domestic strains of livestock where modern breeding practices have resulted in the loss of entire breeds of livestock. The ultimate loss of domestic diversity occurs when animals are cloned.

The most environmentally damaging effects of human activities are those that are cumulative. As noted previously, the most significant of these at present is likely the accumulation of greenhouse gases that have the potential to cause global warming. Some environmental problems, such as those resulting from the emission of photochemical smog-forming pollutants into the atmosphere are potentially reversible. However, by the time that global warming has been demonstrated to be a genuine problem, if such turns out to be the case, the damage will have been done, and little if anything can be done to reverse it.

2.5. Consideration of Environmental Impacts in Industrial Ecology

By its nature, industrial production has an impact upon the environment. Whenever raw materials are extracted, processed, used, and eventually discarded, some environmental impacts will occur. In designing an industrial ecological system, several major kinds of environmental impacts must be considered in order to minimize them and keep them within acceptable limits. These impacts and the measures taken to alleviate them are discussed below.

For most industrial processes, the first environmental impact is that of extraction of raw materials. This can be a straightforward case of mineral extraction, or it can be less direct, such as utilization of biomass grown on forest or crop land. A basic decision, therefore, is the choice of kind of material to be used. Wherever possible, materials should be chosen that are not likely to be in short supply in the foreseeable future. As an example, the silica used to make the lines employed for fiber optics communication is in unlimited supply and a much better choice for communication lines than copper wire made from limited supplies of copper ore.

Industrial ecology systems should be designed to reduce or even totally eliminate air pollutant emissions. Some of the most notable recent progress in that area has been the marked reduction and even total elimination of solvent vapor emissions (volatile organic carbon, VOC), particularly those from organochlorine solvents. Some progress in this area has been made with more effective trapping of solvent vapors. In other cases, the use of the solvents has been totally eliminated. This is the case for chlorofluorocarbons (CFCs), which are no longer used in plastic foam blowing and parts cleaning because of their potential to affect stratospheric ozone. Other air pollutant emissions that should be eliminated are hydrocarbon vapors, including those of methane, CH_4, and oxides of nitrogen or sulfur.

Discharges of water pollutants should be entirely eliminated, wherever possible. For many decades, efficient and effective water treatment systems have been employed that minimize water pollution. However, these are "end of pipe" measures, and it is much more desirable to design industrial systems such that potential water pollutants are not even generated.

Industrial ecology systems should be designed to prevent production of liquid wastes that may have to be sent to a waste processor. Such wastes fall into the two broad categories of water-based wastes and those contained in organic liquids. Under current conditions the largest single constituent of so-called "hazardous wastes" is water. Elimination of water from the waste stream automatically prevents pollution and reduces amounts of wastes requiring disposal. The solvents in organic wastes largely represent potentially recyclable or combustible constituents. A properly designed industrial ecosystem does not allow such wastes to be generated or to leave the factory site.

In addition to liquid wastes, many solid wastes must be considered in an industrial ecosystem. The most troublesome are toxic solids that must be placed in a secure hazardous waste landfill. The problem has become especially acute in some industrialized nations in which the availability of landfill space has been severely curtailed. In a general sense, solid wastes are simply resources that have not been properly utilized. Closer cooperation among suppliers, manufacturers, consumers, regulators, and recyclers can minimize quantities and hazards of solid wastes.

Any time that energy is expended, there is a greater or lesser degree of environmental damage. Therefore, energy efficiency must have a high priority in a properly designed industrial ecosystem. Significant progress has been made in this area in recent decades, as much because of the costs of energy as for environmental improvement. More efficient devices, such as electric motors, and approaches, such as congeneration of electricity and heat, that make the best possible use of energy resources are highly favored. An important side benefit of more efficient energy utilization is the lowered emissions of air pollutants, including greenhouse gases.

It goes without saying that design and operation for recycling must have a high priority in any modern system of industrial ecology. As noted previously in Section 2.4, there are two major pathways to recycling. The first of these, which is relatively easy to achieve, is recycling of materials within the manufacturing process. The second, much more challenging type of recycling, is that of materials and items that have been distributed in the consumer sector.

2.6. Three Key Attributes: Energy, Materials, Diversity

By analogy with biological ecosystems, a successful industrial ecosystem should have (1) renewable energy, (2) complete recyling of materials, and (3) species diversity for resistance to external shocks.[4] These three key characteristics of industrial ecosystems are addressed here.

Unlimited Energy

Energy is obviously a key ingredient of an industrial ecosystem. Unlike materials, the flow of energy in even a well-balanced closed industrial ecosystem is essentially one way, in that energy enters in a concentrated, highly usable form, such as chemical energy in natural gas, and leaves in a dilute, disperse form as waste heat. An exception is the energy that is stored in materials. This can be in the form of energy that can be obtained from materials, such as by burning rubber tire, or it can be in the form of what might be called "energy credit," which means that by using a material in its refined form energy is not consumed in making the material from its raw material precursors. A prime example of this is the "energy credit" in metals, such as that in aluminum metal, which can be refined into new aluminum objects with a fraction of the energy required to refine the metal from aluminum ore. On the other hand, recycling and reclaiming some materials can require a lot of energy, and the energy consumption of a good closed industrial ecosystem can be rather high.

Given the needed elements, any material can be made if a sufficient amount of energy is available. The key energy requirement is a source that is abundant and of high quality, that can be used efficiently, and that does not produce unacceptable byproducts.

Although energy is ultimately dissipated from an industrial ecosystem, it may go through two or more levels of use before it is wasted. An example of this would be energy from natural gas burned in a turbine linked to a generator, the exhaust gases used to raise steam in a power plant to run a steam turbine, and the relatively cool steam from the turbine used to heat buildings.

Natural ecosystems run on unlimited, renewable energy from the sun or, in some specialized cases, from geochemical sources. Successful industrial ecosystems must also have unlimited sources of energy in order to be sustained for an indefinite period of time. The obvious choice for such an energy source would seem to be solar energy. However, solar sources present formidable problems, not the least of which is that they work poorly during those times of the day and seasons of the year when the sun does not shine. Even under optimum conditions, solar energy has a low power density necessitating collection and distribution systems of an unprecedented scale if they are going to displace present fossil energy sources. Other renewable sources, such as wind, tidal, geothermal, biomass, and hydropower present similar challenges. It is likely, therefore, that fossil energy sources will provide a large share of the energy for industrial ecosystems in the foreseeable future. This assumes that a way can be found to manage greenhouse gases. At the present time it appears that injection of carbon dioxide from combustion into deep ocean regions is the only viable alternative for sequestering carbon dioxide, and this approach remains an unproven technology on a large scale.

Nuclear fusion power remains a tantalizing possibility for unlimited energy, but so far practical nuclear fusion reactors for power generation have proven an elusive target. Unattractive as it is to many, the only certain, environmentally acceptable energy source that can without question fill the energy needs of modern industrial ecology systems is nuclear fission energy. With breeder reactors that can generate additional fissionable material from essentially unlimited supplies of uranium-238, nuclear fission can meet humankind's energy needs for the foreseeable future. Of course, there are problems with nuclear fission—not technical, but political and regulatory. The solution to these problems remains a central challenge for humans in the modern era.

Recycling

For a true and complete industrial ecosystem, essentially 100% recycling of materials must be realized. In principle, given a finite supply of all the required elements and an unlimited supply of energy, essentially complete recycling can be achieved. A central goal of industrial ecology is to develop efficient technologies for recycling that reduce the need for virgin materials to the lowest possible levels. Another goal must be to implement process changes that eliminate dissipative uses of toxic substances, such as heavy metals, that are not biodegradable and that pose a threat to the environment when they are discarded. These kinds of measures are sometimes called "pollution prevention by way of clean technology."[5]

For consideration of recycling, matter can be put into four separate categories. The first of these consists of elements that occur abundantly and naturally in essentially unlimited quantities in consumable products. Food is the ultimate consumable product. Soap is consumed for cleaning purposes, discarded down the drain, precipitated as its insoluble calcium salt, then finally biodegraded. Materials in this category of recyclables are discharged into the environment and recycled through natural processes or for very low value applications, such as sewage sludge used as fertilizer on soil.

A second category of recyclable materials consists of elements that are not in short supply, but are in a form that is especially amenable to recycling. Wood is one such commodity. At least a portion of wood taken from buildings that are being razed could and should be recycled. The best example of a kind of commodity in this class is paper. Paper fibers can be recycled up to five times, and the nature of paper is such that it is readily recycled. More than 1/3 of world paper production is currently from recycled sources, and that fraction should exceed 50% within the next several decades. The major impetus for paper recycling is not a shortage of wood to make virgin paper but rather a shortage of landfill space for waste paper.

A third category of recyclables consists of those elements, mostly metals, for which world resources are low. Chromium and the platinum group of precious metals are examples of such elements. Given maximum incentives to recycle, especially through the mechanism of higher prices, it is likely that virgin sources of these metals can make up any shortfall not met by recycling in the foreseeable future.

A fourth category of materials to consider for recycling consists of parts and apparatus, such as auto parts discussed previously. In many cases such parts can be refurbished and reused. Even when this is not the case, substantial deposits at the time of purchase can provide incentives for recycling. In order for components to be recycled efficiently, they must be designed with reuse in mind in aspects such as facile disassembly. Such an approach has been called "design for environment," DFE.

Robust Character of Industrial Ecosystems

Successful natural ecosystems are very robust. Robustness means that if one part of the system is perturbed, there are others that can take its place. Consider what happens if the numbers of a top predator at the top of a food chain in a natural ecosystem are severely reduced because of disease. If the system is well balanced, another top predator is available to take its place.

The energy sector is the portion of industrial ecosystems that is most likely to suffer from a lack of robustness. Examples of energy vulnerability have become obvious with several "energy crises" during recent history. Another requirement of a healthy industrial ecology system that is vulnerable in some societies is water. In some regions of the world both the quantity and quality of water are severely limited. A lack of self sufficiency in food is a third example of vulnerability. Vulnerabililty in food and water are both strongly dependent upon climate, which is in turn tied to environmental concerns as a whole.

2.7. Industrial Ecology and Material Resources

A system of industrial ecology is successful if it reduces demand for materials from virgin sources. Strategies for reduced material use may be driven by technology, by economics, or by regulation. The four major ways in which material consumption may be reduced are (1) using less of a material for a specific application, an approach called **dematerialization**; (2) **substitution** of a relatively more abundant and safe material for one that is scarce and/or toxic; (3) **recycling**, broadly defined; and (4) extraction of useful materials from wastes, sometimes called **waste mining**. These four facets of efficient materials utilization are outlined in this section.

Dematerialization

There are numerous recent examples of reduced uses of materials for specific applications. One example of dematerialization is the transmission of greater electrical power loads with less copper wire by using higher voltages on long distance transmission lines. Copper is also used much more efficiently for communications transmission than it was in the early days of telegraphy and telephone communication. Amounts of silver used per roll of photographic film have decreased significantly in recent years. The layer of tin plated onto the surface of a "tin can" used for food preservation and storage is much lower now. In response to the need for greater fuel economy, the quantities of materials used in automobiles has decreased significantly over the last two decades, a trend reversed, unfortunately, by the more recent increased demand for large "sport utility vehicles." Automobile storage batteries now use much less lead for the same amount of capacity than they did in former years. The switch from 6-volt to 12-volt auto batteries in the 1950s enabled use of lighter wires, such as those from the battery to the electrical starter. Somewhat later, the change to steel-belted radial tires enabled use of lighter tires and resulted in much longer tire lifetimes so that much less rubber was used for tires.

One of the most commonly cited examples of dematerialization is that resulting from the change from vacuum tubes to solid state circuit devices. Actually, this conversion should be regarded as material substitution as transistors replaced vacuum tubes, followed by spectacular mass reductions as solid state circuit technology advanced.

Dematerialization can be expected to continue as technical advances, some rapid and spectacular, others slow and incremental, continue to be made. Some industries lead the way out of necessity. Aircraft weight has always played a crucial role in determining performance, so the aircraft manufacturing sector is one of the leaders in dematerialization.

Substitution of Materials

Substitution and dematerialization are complementary approaches to reducing materials use. The substitution of solid state components for electronic vacuum tubes and the accompanying reduction in material quantities has already been cited.

The substitution of polyvinylchloride (PVC) siding in place of wood on houses has resulted in dematerialization over the long term because the plastic siding does not require paint.

Technology and economics combined have been leading factors in materials substitution. For example, the technology to make PVC pipe for water and drain lines has enabled its use in place of more expensive cast iron, copper, and even lead pipe (in the last case, toxicity from lead contamination of water is also a factor to be considered).

A very significant substitution that has taken place over recent decades is that of aluminum for copper and other substances. Copper, although not a strategically short metal resource, nevertheless is not one of the more abundant metals in relation to the demand for it. Considering its abundance in the geosphere and in sources such as coal ash, aluminum is a very abundant metal. Now aluminum is used in place of copper in many high voltage electrical transmission applications. Aluminum is also used in place of brass, a copper-containing alloy, in a number of applications. Aluminum roofing substitutes for copper in building construction. Aluminum cans are used for beverages in place of tin-plated steel cans.

There have been a number of subsitutions of chemicals in recent years, many of them driven by environmental concerns and regulations resulting from those concerns. One of the greater of these has been the substitution of hydrochlorofluorocarbons (HCFCs) and hydrofluorocarbons (HFCs) for chlorofluorocarbons (Freons or CFCs) driven by concerns over stratospheric ozone depletion. Substitutions of nonhalogenated solvents, supercritical fluid carbon dioxide, and even water with appropriate additives for chlorinated hydrocarbon solvents will continue as environmental concerns over these solvents increase.

Substitutions for metal-containing chemicals promise to reduce costs and toxicities. One such substitution that has greatly reduced the possibilities for lead poisoning is the use of titanium-based pigments in place of lead for white paints. In addition to lead, cadmium, chromium, and zinc are also used in pigments, and substitution of organic pigments for these metals in paints has reduced toxicity risks. Copper, chromium, and arsenic are used in treated wood (CCA lumber). Because of the toxicity of arsenic, particularly, it would be advisable to develop substitutes for these metals in wood. It should be pointed out, however, that the production of practically indestructable CCA lumber has resulted in much less use of wood, and has saved the materials and energy required to replace wood that has rotted or been damaged by termites.

Recycling

Recycling, "broadly defined," refers to an array of strategies by which materials are reused. Recycling can mean use of the same material for the same purpose without any processing or modification. Cooling water simply returned into a cooling water system would be an example of such recycling. Recycling can involve various degress of reprocessing. Examples are lead from spent storage batteries recast into battery electrodes, scrap paper macerated to a pulp and used to make remanufactured paper, and aluminum in aluminum cans, melted and used to make aluminum stock to fabricate more cans. Finally, recycling can also mean rebuilding, repair, and refurbishment of apparatus, such as toner cartridges in laser printers.

Recycling offers several advantages. The most obvious of these is reduced need for virgin materials—ores for metals, wood for paper pulp, petroleum for plastics. There are important advantages insofar as energy is concerned. For example, rehoning and refurbishing an engine block to produce a remanufactured automobile engine obviously takes less energy than does melting iron and casting it into a new block. An advantage of recycling that is often overlooked is the lowered production of toxic byproducts. Toxic arsenic is a byproduct of the refining of some metal ores, and recycling these metals means that the arsenic that occurs with them must be dealt with only once.

It should be noted that recycling comes with its own set of environmental concerns. One of the greatest of these is contamination of recycled materials with toxic substances. In some cases, motor oil, especially that collected from the individual consumer sector, can be contaminated with organohalide solvents and other troublesome impurities. Food containers pick up an array of contaminants, and, as a consequence, recycled plastic is not generally regarded as a good material for food applications. Substances may become so mixed with use that recycling is not practical. This occurs particularly with synthetic fibers, but it may be a problem with plastics, glass, and other kinds of recyclable materials.

Combustion for energy production can be used as a form of recycling. For some kinds of materials, combustion in a power plant is the most cost-effective and environmentally safe way of dealing with materials. This is true, for example, of municipal refuse that contains a significant energy value because of combustible materials in it. Such refuse also contains a variety of items that potentially could be recycled for the materials in them. However, once such items become mixed in municipal refuse and contaminated with impurities, the best means of dealing with them is simply combustion.

Extraction of Useful Materials from Wastes

Sometimes called waste mining, the extraction of useful materials from wastes has some significant, largely unrealized potential for the reduction in use of virgin materials. Waste mining can often take advantage of the costs that must necessarily be incurred in treating wastes, such as flue gases. Sulfur is one of the best examples of a material that is now commonly recovered from wastes. Sulfur is a constituent of all coal and can be recovered from flue gas produced by coal combustion. It would not be cost-effective to use flue gas simply as a source of sulfur. However, since removal of sulfur dioxide from flue gas is now required by regulation, the incremental cost of recovering sulfur as a commodity, rather than simply discarding it, can make sulfur recovery economically feasible.

There are several advantages to recovering a useful resource from wastes. One of these is the reduced need to extract the resource from a primary source. Therefore, every kilogram of sulfur recovered from flue gas means one less kg of sulfur that must be extracted from sulfur ore sources. By using waste sources, the primary source is preserved for future use. Another advantage is that extraction of a resource from a waste stream can reduce the toxicity or potential environmental harm from the waste stream. As noted previously, arsenic is a byproduct of the refining of some other metals. The removal of arsenic from the residues of refining such metals significantly reduces the toxicities and potential environmental harm by the wastes. Coal ash, the residue remaining after the combustion of coal for power generation, has a significant potential as a source of iron (ferrosilicon), silicon, and aluminum, and perhaps several other elements as well.[6] An advantage of using coal ash in such applications is its physical form. For most power applications, the feed coal is finely ground, so that the ash is in the form of a powder. This means that coal ash is already in the physical form most amenable to processing for byproducts recovery. For a particular coal feedstock, coal ash is homogeneous, which offers some definite advantages in processing and resource recovery. A third advantage of coal ash is that it is anhydrous so that no additional energy needs to be expended in removing water from an ore.

2.8. SOCIETAL FACTORS AND THE ENVIRONMENTAL ETHIC

The "consumer society" in which people demand more and more goods, energy-consuming services, and other amenities that are in conflict with resource conservation and environmental improvement runs counter to a good workable system of industrial ecology. Much of the modern lifestyle and corporate ethic is based upon persuading usually willing consumers that they need and deserve more things and that they should adopt lifestyles that are very damaging to the environment. The conventional wisdom is that consumers are unwilling to significantly change their lifestyles and to lessen their demands on world resources for the sake of environmental preservation. However, in the few examples in which consumers have been given a chance to exercise good environmental citizenship, there are encouraging examples that they will do so willingly. A prime example of this is the success of paper, glass, and can recycling programs in connection with municipal refuse collection, implemented to extend landfill lifetimes.

Two major requirements for the kind of public ethic that must accompany any universal adoption of systems of industrial ecology are **education** and **opportunity**. Starting at an early age, people need to be educated about the environment and its crucial importance in maintaining the quality of their lives. They need to know about realistic ways, including the principles of industrial ecology, by which their environment may be maintained and improved. The electronic and print media have a very important role to play in educating the public regarding the environment and resources. Given the required knowledge, the majority of people will do the right thing for the environment.

People also need good opportunities for recycling and for general environmental improvement. It is often said that people will not commute by public transit, but of course they will not do so if public transit is not available, or if it is shabby, unreliable, and even dangerous. They will not recycle cans, paper, glass, and other consumer commodities if convenient, well-maintained collection locations are not accessible to them. There are encouraging examples, including some from the United States, that opportunity to contribute to environmental protection and resource conservation will be met with a positive response from the public.

LITERATURE CITED

1. Graedel, Thomas, "Industrial Ecology: Definition and Implementation," Chapter 3 in *Industrial Ecology and Global Change*, Robert Socolow, Clinton Andrews, Frans Berkhout, and Valerie Thomas, Eds., Cambridge University Press, New York, NY, 1994, pp. 23-41.

2. "Industrial Ecology," *Environmental Science and Technology*, **31**, 1997, p. 26A.

3. Graedel, Thomas, "Industrial Ecology: Definition and Implementation," Chapter 3 in *Industrial Ecology and Global Change*, Robert Socolow, Clinton Andrews, Frans Berkhout, and Valerie Thomas, Eds., Cambridge University Press, New York, NY, 1994, pp. 23-41.

4. Andrews, Clinton, Frans Berkhout, and Valerie Thomas, "The Industrial Ecology Agenda," Chapter 36 in *Industrial Ecology and Global Change*, Robert Socolow, Clinton Andrews, Frans Berkhout, and Valerie Thomas, Eds., Cambridge University Press, New York, NY, 1994, pp. 469-477.

5. "Introduction: Materials Perspective," Chapter 1 in *Industrial Ecology: Towards Closing the Materials Cycle*, Edward Elgar Publishers, Cheltenahm, U.K., 1996, pp. 1-17.

6. "Coal Ash: Sources and Possible Uses," Chapter 14 in Robert U. Ayres and Leslie W. Ayres, *Industrial Ecology: Towards Closing the Materials Cycle*, Edward Elgar Press, Cheltenham, U.K., 1996, pp. 259-272.

SUPPLEMENTARY REFERENCES

Ayres, Robert U. and Udo E. Simonis, Eds., *Industrial Metabolism: Restructuring for Sustainable Development*, United Nations University Press, New York, NY, 1994.

Curran, Mary Ann, Ed., *Environmental Life-Cycle Assessment*, McGraw-Hill, New York, NY, 1997.

Fiksel, Joseph, *Design for Environment: Creating Eco-Efficient Products and Processes*, McGraw-Hill, New York, NY, 1996.

Graham, John D. and Jennifer K. Hartwell, *The Greening of Industry*, Harvard University Press, Cambridge, MA, 1997.

3 PRINCIPLES OF INDUSTRIAL METABOLISM

3.1. INTRODUCTION

Recall from Chapter 2 that *industrial ecology* refers to an integrated systematic scheme primarily applied to materials that is designed to optimize an entire materials cycle from acquisition of raw materials, through processing, to consumers, and to final disposition, with the objective of minimizing environmental impact, especially in the production of wastes. A basic premise of industrial ecology is that the design and operation of industrial processes has profound effects upon, and, in turn, is affected by the environment in which these processes operate.[1] Therefore, the preservation of acceptable environmental quality requires that such anthrospheric enterprises be carefully integrated with the "natural" spheres of the environment.

At the national level, industrial firms operate within an *economic system* in which the firms share a common government, regulatory structure, and currency. A firm or collection of firms takes in materials and energy and generates goods, services, and waste products. In so doing it adds value to the materials that it processes, keeping track of materials, goods, and energy entering and leaving the industrial system, as well as within the system.

This chapter considers in some detail the topic of **industrial metabolism**, which is the process by which materials and energy flow through industrial systems from their sources, through various industrial processes, to the consumer, and finally to ultimate disposal.[2,3] Unlike the living metabolic processes that occur in natural systems where true waste products are very rare, industrial metabolism as it is now practiced has a vexing tendency to dilute, degrade, and disperse materials to an extent that they are no longer useful, but are still harmful to the environment. Indeed, waste has been defined as *dissipative use of natural resources*.[4] In addition to simple loss from dilution and dispersion in the environment, materials may be lost by being tied up in low energy forms or by being put into a chemical form from which they are very difficult to retrieve.

An example of dissipation of material resulting in environmental pollution, now, fortunately, a very much diminished problem, was the widespread use of lead in tetraethyl lead antiknock additive in gasoline. The net result of this use was to disperse lead throughout the environment with auto exhaust gas, with no hope of recovery.

As shown in Figure 3.1, an industrial system may be viewed as consisting of four main constituents. These are (1) a *primary producer* of a raw material, (2) a *manufacturer*, (3) a *consumer*, and (4) a *secondary material (waste) processor.* Raw materials or resources enter the system, shown in the figure as going into the primary producer, and wastes leave it. A successful industrial ecosystem is one that minimizes both the materials required to enter the system and the nonusable wastes leaving it. For the purposes of illustration, wastes are defined as materials that must be discarded to the environment. In this sense, therefore, wastes include benign substances such as carbon dioxide emitted to the atmosphere or fill material, such as ore tailings, that are not used for any beneficial purpose. Whereas, for the most part, raw materials enter an industrial system through the primary producer, wastes may be generated at any point in the system and are shown leaving the industrial ecosystem as a whole.

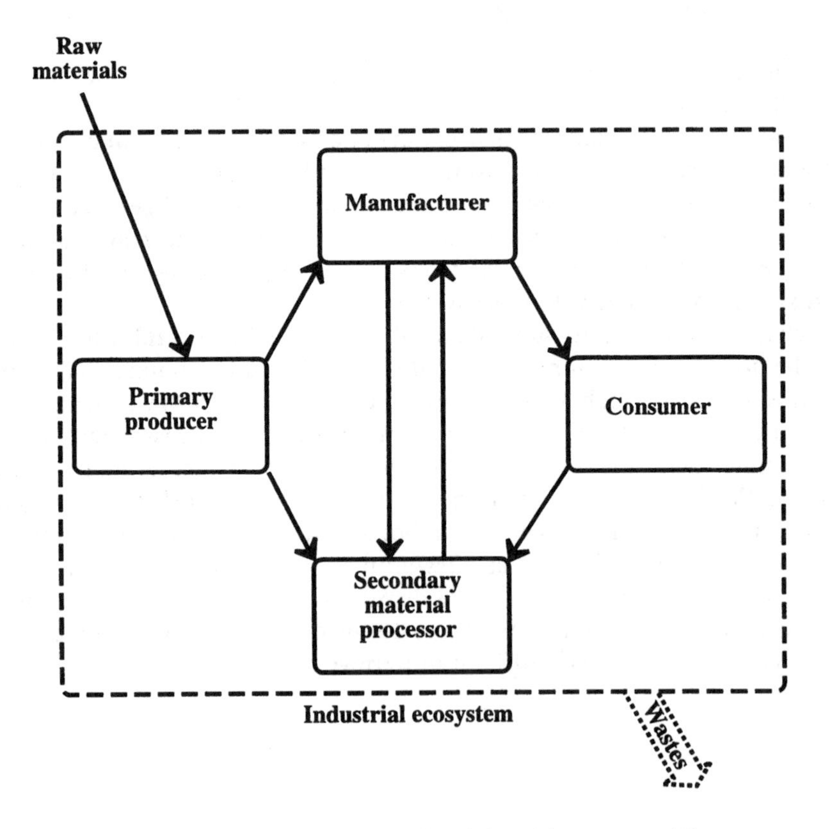

Figure 3.1. Outline of the main constituents of an industrial ecosystem. The system as a whole is outlined with a dashed line.

The major objective of a successful industrial ecosystem is to minimize the flow of raw materials in and wastes out, while maximizing the internal flow of materials within the system. In the past, and to a large extent today, industrial systems operated in essentially a one-way mode in that materials go in, often very large quantities of materials flow out, and there is little recycling of substances in the system. An ideal industrial ecosystem is one in which there is a one-time input of materials and no wastes produced. Although never realizable in practice, with that ideal as a target, it is possible to continually refine and develop processes that very significantly reduce material use.

As shown on the left in Figure 3.2, a conventional, mechanical, open industrial system is largely linear, meaning that it imports large quantities of materials, exports most of its production, and produces large quantities of wastes and potential pollutants, that either enter the environment or are controlled by end-of-pipe measures. In contrast, as shown on the right in Figure 3.2, a well-balanced, closed industrial ecosystem takes in less material, produces very little waste or pollutants, maintains a strong material cycle within the ecosystem, and uses most of its product within the industrial ecosystem. A large percentage of the materials that would be wastes discharged from an open industrial system are process wastes that stay within a closed industrial ecosystem and are eventually used in it. Even the most well-balanced, closed industrial ecosystem generates some wastes, including waste byproducts from the extraction of new materials, some wastes lost from the recycling pathways, and some wastes lost with products.

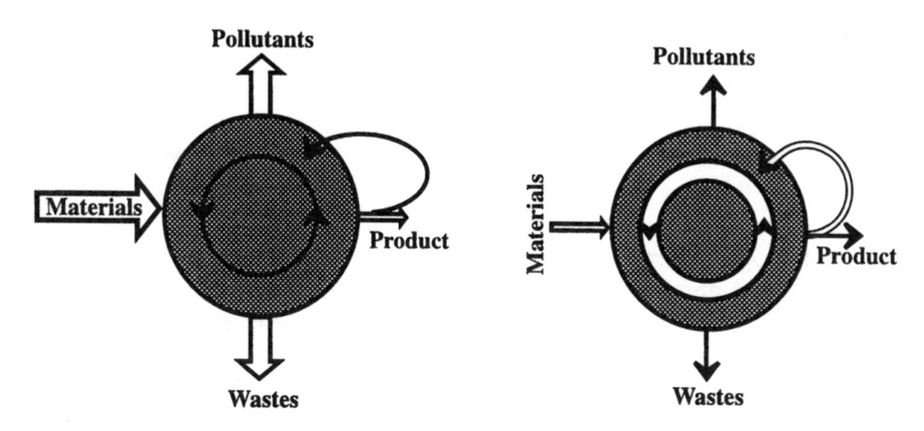

Figure 3.2. Models of a conventional mechanical linear (open) industrial system (left) and a well-balanced (closed) industrial ecosystem (right). The well-balanced industrial ecosystem is characterized by low materials inflow, low production of wastes and pollutants, maximum use of product within the system, and a high degree of materials recycle within the system.

Basis of Industrial Metabolism

Before considering industrial metabolism in detail, it is useful to look at the basis and rationale for practicing it. A good foundation is provided from perspectives on industrial ecology as enunciated by Socolow.[5] This approach emphasizes interrelationships among various producers and consumers.

The long-term goal of studying and understanding industrial metabolism is to ensure **habitability** of the planet for an indefinite period. Whereas conventional responses to environmental concerns have focused on short-term problems, such as pollution of streams by oxygen-consuming manure runoff from livestock feedlots, the practice of industrial ecology looks at longer term concerns. One of the major threats to long-term habitability comes from **environmental toxication** by persistent toxic substances. For example, while the feedlot pollution mentioned above is readily remedied by the application of suitable measures within a year or two at most, soil contamination by heavy metals has the potential to be permanent without any practical remedies. In that respect the **persistence** of environmental pollutants is of prime concern. This is the case, for example, with the large quantities of polychlorinated biphenyls (PCBs) dumped into the Hudson River and accumulated in its sediments from past years of electrical equipment manufacture. The properties of extreme chemical, thermal, and biochemical stability that made the PCBs so useful as transformer fluids and in capacitors are the same properties that have made them so undesirable in the environment. The same may be said of chlorofluorocarbons (CFCs) that threaten stratospheric ozone destruction.

Another problem threatening long-term habitability is **depletion of nonrenewable resources**. Relatively inexpensive methods of mining and processing essential metals, such as copper or chromium, have resulted in the very rapid depletion of the most available sources. Petroleum, of most value as the raw material for the manufacture of a broad range of products, including plastics, polymers, and rubber, has been largely depleted for use as a fuel for motor vehicles, a very inefficient means of moving people and goods.

The **physical and chemical degradation** of the environmental support systems upon which humans depend for their welfare is another threat to long-term habitability. A prime example of this is the degradation of topsoil, which has been lost physically in many locales as the result of human-induced erosion, or degraded chemically by the loss of nutrients and ion-exchanging, water-retaining organic humus or by contamination with heavy metals.

In the biological arena, the species loss, lowering of species diversity, and drastically decreased numbers of important species have threatened long-term habitability. Extinction of species is a common, well-known environmental problem. Whenever a species becomes extinct, an irreplaceable segment of the total gene pool of living organisms is lost forever. The tendency to raise single breeds of livestock has lowered species diversity and increased vulnerability to disease. "Improvements" in fishing methods have resulted in lowered numbers of some kinds of seafood species to the extent that it is no longer practical to even harvest them.

In considering long-term habitability, it is essential to take a **global view of the environment**. Whereas the common approach is to regard problems on a localized basis, as would be the case with the feedlot pollution cited earlier, it is more important to take a broader, global view. This means considering problems, such as greenhouse-gas atmospheric carbon dioxide, that affect the globe as whole, or considering problems that are individually small in scope, but globally widespread, such as safe and effective means of disposing of sewage.

Detrimental **perturbation of natural processes** is an important problem that threatens long-term habitability. This occurs, for example, when human activities upset major biogeochemical cycles, such as may occur by the introduction of large quantities of fixed nitrogen into the biological nitrogen cycle by manufacturing processes.

Vulnerability is another threat to long-term habitability. By concentrating on the cultivation of only a few strains of food crops, humans make the food supply vulnerable to pests that are particularly successful in attacking that strain. Another example is provided by ambitious programs of dam and levee building for flood control. Initially, and often for many decades, such measures may appear to be successful until the inevitable record or near-record flood occurs and the flood control measures introduced by humans are overwhelmed, actually making the damage worse than it otherwise would have been. Ill-conceived attempts to control the flow of China's Yangtze River, accompanied by other "ecological sins," such as extensive clear-cutting of timber in the river's watershed and drainage of wetlands and lakes in its flood plain have been blamed in part for catastrophic floods in China in 1998.[6]

3.2. INDUSTRIAL METABOLISM AND BIOLOGICAL ANALOGIES

The strong analogy between natural ecosystems and efficient industrial systems was first clearly stated in 1989 by Frosch and Gallopoulos.[7] A natural ecosystem, which is usually driven by solar energy and photosynthesis, consists of an assembly of mutually interacting organisms and their environment in which materials are interchanged in a largely cyclical manner.[8] It is possible to visualize an analogous industrial ecosystem in which materials are cycled, driven by an energy source.

In order to apply the principles of industrial ecology, it is necessary to have a basic understanding of industrial metabolism. Biological metabolism is defined as biochemical processes that involve the alteration of biomolecules. Metabolic processes can be divided into the two major categories of anabolism (synthesis) and catabolism (degradation). It is useful to view the metabolic processes of an ecosystem as a whole, rather than just observing each individual organism. An industrial ecosystem likewise synthesizes substances, thus performing anabolism, and it degrades substances, thereby performing in a manner analogous to biological catabolism. Typically, a large amount of a material, such as an ore or petroleum source, is metabolized to yield a relatively small quantity of a finished product. The objective of a properly designed and operated industrial ecosystem is to perform industrial metabolism in the most efficient manner possible so that the least possible raw material is used, the maximum amounts of materials are recycled through the system, and the most efficient possible use is made of the energy that sustains the industrial ecosystem.

An ideal biological ecosystem involves many organisms living in harmony with their environment without any net consumption of resources or production of waste products. The only input required for such an ecosystem is solar energy. This energy is captured and used by photosynthetic primary producers to convert carbon dioxide, water, and other inorganic materials into biomass. Herbivores ingest this biomass and use it for their energy and to synthesize their own biomass. Carnivores, of which there may be several levels, consume herbivores, and a food chain exists that may consist of several levels of organisms. Parasites exist on or in other organisms. Saprophytes and bacteria and fungi responsible for decay utilize and degrade biomass, eventually converting it back to simple inorganic constituents through the process of mineralization. Symbiotic and synergistic relationships abound in a natural ecosystem. Thus, an ideal ecosystem exists indefinitely in a steady-state condition without causing any net degradation of its environment.

A natural ecosystem can be visualized as having compartments in which various *stocks* of materials are kept, connected by *flows* of materials. Examples of such compartments include soil, which is a repository of plant nutrients, a body of water, such as a lake, and the atmosphere, which is a repository of carbon dioxide required for photosynthesis. In an undisturbed natural ecosystem, the quantities of the materials in each of these compartments remains relatively stable because such systems are inherently recycling. In contrast, the quantities of materials in the compartments of an industrial system are not constant. Reservoirs of raw materials, such as essential minerals, are constantly diminishing, although with new discoveries of mineral resources, they may *appear* to increase. Furthermore, reservoirs of wastes in an industrial system continually increase as materials traverse an essentially one-way path through the system. Sustainable industrial systems maximize recycling so that quantities of materials in the reservoirs remain constant insofar as possible.

In a manner analogous to natural ecosystems, industrial systems can be designed in principle to operate in a similar steady-state manner, ideally neither consuming nonrenewable resources nor producing useless waste products. It is natural and logical, therefore, to look to natural ecosystems as models for the design of anthropogenic systems that cause minimal environmental harm. Such a model is that of an **industrial ecosystem** consisting of a group of industries powered by an energy source existing synergistically and consuming only those materials produced within the ecosystem. Isolated cases exist in which such a system is approached — Amish farming communities that grow their own food, including that consumed by horses used for power, come to mind.

Systems of biological metabolism are self-regulating. At the level of the individual organism, regulation is accomplished internally by biological regulatory mechanisms, such as those that employ hormones. At the ecosystem level, regulation occurs through competition among organisms for available resources. Industrial ecosystems are also self-regulating. In this case the economic system operating under the laws of supply and demand is the regulatory mechanism.

A comparison of the metabolisms of natural ecosystems with that of industrial systems as they are commonly encountered shows a marked contrast. These contrasts are highlighted in Table 3.1.

Attractive as the idea may sound, in a modern society and especially in a "global economy," a complete industrial ecosystem is not practical or even desirable. Essentially all modern communities produce at least one product that is exported to the outside, and must bring in materials and energy sources from elsewhere. A more realistic model of an industrial ecosystem is one in which raw materials are imported from the outside and at least one major product is exported from the system, but in which a number of enterprises coexist synergistically, utilizing each other's products and services to mutual advantage. In such a system, typically, raw materials flow into a primary raw material processor, which converts them to a processed material. The processed material then goes to one or more fabricators that make a product for distribution and sale. Associated with the enterprise as a whole are suppliers of a variety of materials, items, or services, and processors of secondary materials or wastes. To meet the criteria of an industrial ecosystem, as many of the byproducts and wastes as possible must be utilized and processed within the system.

Table 3.1. Metabolic Characteristics of Natural Ecosystems and Industrial Systems

Characteristic	Natural ecosystems	Current industrial systems
Basic unit	Organism	Firm
Material pathways	Closed loops	Largely one way
Recycling	Essentially complete	Often very low
Material fate	Tend to concentrate, such as atmospheric CO_2 converted to biomass by photosynthesis	Dissipative to produce materials too dilute to use, but concentrated enough to pollute
Reproduction	A major function of organisms is reproduction	Production of goods and services is the prime objective, not reproduction *per se*

Although industrial systems are self-regulating, they are not necessarily so in a manner conducive to sustainability; indeed, the opposite is frequently the case. Left to their own devices and operating under the principles of traditional economics, industrial systems tend toward a state of equilibrium or maximum entropy in which essential materials have been exploited, run through the system, and dissipated to the environment in dilute, useless forms. A central question is, therefore, the time scale on which this irreversible dissipation will occur. If it is a few decades, modern civilization is in real trouble; if it is on a scale of thousands of years, there is ample time to take corrective action to maintain sustainability. A challenge to modern industrialized societies is to modify industrial systems to maximize the time spans under which sustainability may be achieved.

3.3. MATERIAL AND ENERGY FLOW IN INDUSTRIAL METABOLISM

Material flow and **energy flow** are two key aspects of industrial metabolism. Of these two, energy flow is much easier to analyze and optimize. Basically, energy enters an industrial ecosystem as high grade energy, in current industrial systems, usually as chemical energy in fossil fuels, and leaves the system as low grade heat, which has no further uses. The objective of a properly designed industrial ecosystem insofar as energy is concerned is to extract as much usable energy as possible during the one-way path of energy through the system. This may be done, for example, by burning petroleum in a gas turbine coupled to an electrical generator, using the hot exhaust from the gas turbine to raise steam for a steam turbine, and using the steam exhaust from the steam turbine for district heating. Sometimes material flow and energy flow are intermixed, as is the case with a wood processing operation in which the waste wood products are burned to produce energy.

Material flows in industrial metabolic processes are of crucial importance in describing or managing the system and may be quite complex. Unlike energy, which in a sense is "destructable" in being reduced to low-grade heat, matter is indestructable. Whereas it is acceptable (and unavoidable) that energy will ultimately be dissipated in an industrial ecosystem, the same approach used with materials can cause severe problems. Indeed, most environmental problems — sulfur dissipated to the atmosphere as stack gas emissions of sulfur dioxide, oxygen demand dissipated to the hydrosphere as biodegradable organic matter, and lead dissipated to the anthrosphere as lead-based paint primers, have resulted from the assumption that matter can simply be thrown away and allowed to dissipate in a dilute form to the environment.

A material that is universally used, often wastefully so, in industrial systems is water. In the U.S, there is comparatively little reuse of water, and the reuse that is practiced is generally for irrigation, a relatively low level application. In many cases significant fractions of water are lost in distribution systems. This occurs, for example, in Mexico City, where a huge fraction of the water entering the distribution system is lost from broken pipes (breakage caused in large part by settling resulting from pumping water out of the aquifers beneath the city).

Because of their particular importance and persistence in the environment, the materials that receive the most attention in an analysis of industrial metabolism are those that do not degrade (heavy metals, such as lead, cadmium, and mercury) or those that are extremely persistent, such as PCBs or chlorofluorocarbons. These kinds of materials should not be released to the environment at all or in only very limited quantities under very carefully controlled conditions where the potential benefits may outweigh the potential harm.

Material flows from the anthrosphere into the environment at large are relatively easy to control from **point sources**, such as smokestacks or sewage treatment plant outflows. Such sources are regulated rather well in industrialized countries. However, **nonpoint sources**, such as those arising from herbicide applications to crops, have proven much harder to analyze and regulate. Such sources are often **dissipative sources** in which material becomes dissipated to the environment in a dilute form, from which it is impossible to retrieve. Whereas pollution from point sources is likely to be the result of the manufacture of a product, that from nonpoint or dissipative sources is much more likely to be from the product itself and from uses of the product.

Once materials have been removed from their sources and enter an industrial system, their two major fates are recycling to the system and dissipation to the environment. As recycling increases, dissipation decreases. The desirability of recycling vs. dissipation depends upon the material. In the case of plastics dissipative fates are of relatively little importance because the amount of material consumed to manufacture plastics is not huge, the amount of room taken by disposed plastics is relatively small, and the hazard from the disposed form is very low. By way of contrast, lead is a scarce resource that needs to be conserved. Furthermore, lead dispersed to the environment poses a severe toxicity hazard to organisms. In general, heavy metals constitute a class of substances for which recycling has a top priority.

Some uses of materials are inherently dissipative, in that there is no practical way in which the material may be recycled. An example of such a use is primer and paint put on surfaces, such as bridge structural members. Other materials in this category include pesticides, disinfecting agents, chemical reagents, fertilizers, and lubricants. In the case of many, perhaps most, materials currently used, recycling is technically feasible, but is not practiced because of economics or simple inertia. This occurs, for example, with a significant fraction of steel, which is certainly recyclable, but often ends up as waste. Other examples include packaging materials, solvents, and a large fraction of water used in industrial applications. This category of materials presents the greatest opportunity for recycling through changes in economic and regulatory incentives. A third, unfortunately rather small, category of materials consists of those that are recyclable within the current framework of economics and regulation. Lead in storage batteries, structural steel and aluminum, and industrial catalyst materials generally fall in this category. An example of a type of material that has been moved largely from the category of potentially recyclable to recycled materials by the implementation of stringent regulations consists of chlorofluorocarbons (CFCs) used as refrigerants.

The degree to which materials are recycled or recyclable varies significantly with the element. Precious metals, such as gold and platinum, are recycled to a very large extent. This is because of their high economic value, common uses in the pure elemental state, and high stability as elements. Phosphorus in contrast has a very low rate of recycling because of its uses in fertilizer, food, and insecticides, all inherently dissipative uses. One of the elements used dissipatively in largest quantities is sulfur. Most sulfur goes to manufacture sulfuric acid, which in turn is used dissipatively for chemical manufacture, pulp and paper production, nonferrous metals extraction and refining, petroleum refining, and phosphate fertilizers. Most nitrogen that enters industrial systems is also used dissipatively as fertilizers and in chemicals.

3.4. INDUSTRIAL METABOLISM

Industrial metabolism in its entirety follows the flows of materials and energy from their initial sources, through an industrial system, to the consumer, and to their ultimate disposal. In biological systems, metabolism may be studied at any level ranging from the molecular processes that occur in individual cells, through the multiple processes and metabolic cycles that occur in individual organs, and to the overall process of metabolism that takes place in the organism as a whole. Similarly, industrial metabolism can be examined as individual unit operations within an industrial operation, at the factory level, at the industry level, and globally. In terms of an industrial ecology approach it is often most useful to view industrial metabolic processes at the regional level, large enough to have a number of industries with a variety of potential waste products that might be used by other industries, but small enough to permit transport and exchange of materials between various industries. From the standpoint of minimization of pollution it can be useful to consider units consisting of environmental domains, such as atmospheric basins or watersheds.

In order for a system of industrial metabolism to function properly, it is necessary that it have several key components of the industrial ecosystem in which it operates.[9] These are the following:

- At least one primary producer operating on a large scale
- At least one secondary material processor that utilizes large-volume wastes from the primary producer, as well as wastes from other manufacturers and consumers
- A firm mechanism to ensure cooperation and information exchange among participants in the system

The primary producer and the secondary material processor may be individual firms or industrial sectors.

Scale and Industrial Metabolism

A concept that was largely responsible for the growth of economic systems during the industrial revolution and that made possible the widespread availability of relatively inexpensive and high quality goods is that achieved by **standardization** and **economies of scale**. What these mean are that items for manufacture are standardized so that components are interchangeable and are manufactured in large quantities. One of the classic examples of production utilizing these concepts is the mass production of the Model T Ford automobile, which became progressively less expensive in inflation-adjusted dollars during the course of its production run extending over many years. Economies of scale have also been responsible for the relatively low prices of commodities such as steel, cement, bulk chemicals, and, notably in the late 1990s, petroleum.

Standardization and economies of scale have been so successful in mass producing consumer goods that many markets in wealthier nations have become relatively saturated in the goods and devices that humans need for a comfortable existence. Therefore, the emphasis has begun to turn away from standardization and more toward manufacturing that satisfies individual needs. One area in which this has occurred is in the manufacture of computers, where units custom made to specifications of individual consumers have become more popular.

Scale is a huge factor in determining whether or not wastes can be converted to economically valuable products. In general, the larger the scale of the conversion process, the more likely it is to be economical. Another important factor is the scale of demand. Normally a high local demand is required because it is frequently not economic to ship waste materials long distances for conversion to useful products.

3.5. OPTIMIZING INDUSTRIAL METABOLISM

The optimization of industrial metabolism is the key to establishing a successful, efficient system of industrial ecology. There are several key measures that may be taken to optimize industrial metabolic processes. They may be divided into the two main categories of (1) assessment and (2) implementation.

Assessment of Industrial Metabolism

In order to assess the efficiency of an industrial metabolic process, it is necessary to measure its productivity. This can be done on the basis of materials productivity,

$$\text{Materials productivity} = \frac{\text{Economic output}}{\text{Material input}} \tag{3.5.1}$$

and of energy productivity,

$$\text{Energy productivity} = \frac{\text{Economic output}}{\text{Energy input}} \tag{3.5.2}$$

The degree of recycling is a useful indication of the total efficiency of materials utilization. It can be expressed by the following:

$$\text{Recycling} = \frac{\text{Amount of recycled materials}}{\text{Amount of recycled plus virgin materials}} \tag{3.5.3}$$

In addition to measuring the fraction of materials actually recycled, it is useful to assess the potential quantity of recycled material. Rather than viewing economic output as simply the monetary value of goods sold, a broader measure is that of services provided to the ultimate consumer. Therefore, the overall effectiveness of an industrial metabolic process can be viewed as a ratio of these services to the total quantity of resources consumed in providing them. This ratio is increased by lower material input (more efficient utilization), lower energy input, and a higher fraction of recycled materials.

In order to properly assess the metabolism of an industrial ecosystem, it is first necessary to define the boundaries of the system. These can be in terms of both space — *where* a material or energy enters or leaves the system — and time — *when* a material or energy enters or leaves the system. All material and energy sources must be identified and quantified. The identities and fluxes of individual material and energy flows within the system must be established. The identities and quantities of material and energy wastes going to the atmosphere, the hydrosphere, the geosphere, and waste treatment facilities must be established.

Implementation of an Optimum System of Industrial Metabolism

On the basis of as complete knowledge as possible of a system of industrial metabolism, it is possible to optimize the system for maximum efficient production, minimum waste, and minimum environmental pollution. To a large extent this is done by internalization of the materials cycle, which is discussed in detail in the following section. Overall, dissipation of materials, particularly toxic and poorly degradable ones, is to be minimized and recycle and reuse are to be maximized. In many cases significant gains are to be achieved by simplifying the system, reducing the number of steps required to turn out an end product or service. When possible, particularly toxic materials required for an industrial process should be produced on site and on demand. This is the common practice with explosively reactive chlorine dioxide, ClO_2, a water disinfectant, which can be generated where it is needed by chemical processes such as the reaction between sodium hypochlorite and chlorine.

$$2NaClO_2(s) + Cl_2(g) \rightarrow 2ClO_2(g) + 2NaCl(s) \quad (3.5.4)$$

Extremes of temperature, pressure, and mechanical agitation (mixing) tend to put a system under strain and should be avoided whenever possible. In this respect, biological systems provide good examples in that they accomplish, under benign conditions, chemical conversions that often require very severe conditions when carried out by humans. Hence, there is a considerable interest in enzyme processes for carrying out various chemical operations.

3.6. INTERNALIZATION (CLOSING) THE MATERIALS CYCLE

A key aspect of a successful system of industrial metabolism is known as **internalization of the materials cycle**. What this means is that the materials cycle is closed insofar as possible so that materials need not be shipped long distances to be used. That means that local markets have to be developed for potential waste materials, or such materials need to be locally upgraded to higher value products. Many examples can be given of materials that can be used economically to produce energy, particularly by generation of electricity on site, but only if they do not need to be shipped very far. Carbon monoxide, CO, is such a substance, as is elemental hydrogen gas, H_2. These gases can be used to generate electricity near the site where they are produced, and the electricity, in turn, can be moved economically over long distances by high voltage transmission systems. Both carbon monoxide and elemental hydrogen have applications in chemical synthesis, if they can be used locally. Sulfur dioxide, SO_2, can be used locally to produce a low grade of sulfuric acid. Carbon dioxide, CO_2, is generated in large quantities at relatively high purities from fermentation processes. This product can be used locally or shipped some distance. Carbon dioxide produced by processes such as combustion and diluted with nitrogen and other gases has relatively fewer uses and must be used locally (see the example of alumina leachate recarbonation in Figure 3.5).

Although the most likely candidates for internalization of materials flows are process wastes from a particular industry, there are numerous other possibilities as well. One example of a material that can be used to close the materials cycle is municipal refuse (garbage). By introducing recycling measures at the point of collection (segregation of recyclable paper, glass, plastic), quantities of materials designated as waste can be significantly reduced. Combustion of the remaining material for electricity generation can further reduce the waste and return something of economic value. In principle at least, the waste solids from combustion of municipal refuse can be used as fill material, such as in road construction, thus reducing to zero the amount of material that must be put in a waste landfill.

3.7. SYSTEMS INTEGRATION AND INDUSTRIAL METABOLISM

Systems integration of industrial systems refers to the systematic correlation of all aspects of an industrial production system to attain maximum efficiency and profit. Systems integration is required for any industrial ecological system. Systems integration largely involves closing or internalization of the materials cycle. This means that to the maximum extent possible all materials that would otherwise be wastes are used within an integrated complex. Many potential wastes are high-volume, low-value materials, so that shipping them long distances outside the plant is too expensive. An example of such a substance is low quality sulfuric acid made from sulfur recovered from flue gases in copper smelting. This acid can be used to leach copper from copper oxides in copper ores, followed by electrolytic recovery of the leached copper, a process that is expected to yield more than 25% of world copper production soon after the year 2000. In addition to materials, energy can be retained and used in an integrated system. One reason why integrated steel production starting with iron ore and going to finished steel is economically favored is that the heat from molten pig iron is retained and the iron does not have to be remelted as it would if it were shipped long distances. Elemental hydrogen, which can be produced in large quantities in petroleum refining, can serve as both an energy source and raw material in an integrated system, a use that is favored on the site where it is produced.

There are many examples of systems integration (although that is probably not what it was called) in the history of industrial development. Some of the major ones are listed below.

- **Coke byproducts:** Coke, the solid carbon product of the pyrolysis of coal, was widely produced in the 1800s for the manufacture of iron. Koppers in Germany developed a coking system in which combustible gases, liquid hydrocarbons, coal tar products, and ammonium sulfate were reclaimed from coke ovens. These byproducts found numerous uses, such as ammonia from ammonium sulfate, and greatly increased the economic value of coke production. A whole organic chemicals industry (especially in the area of dyes) developed from utilization of coal tars.

- **Meat packing:** Huge meat packing plants in Chicago exemplified systems integration in the early 1900s. These operations were so large that it was possible to utilize every part of the animal to produce some marketable product. Lower grade portions and scraps of meat went into sausage, lard was saponified to make soap, cattle hides were used to make leather, hog bristles were salvaged to make brushes, bone meal was marketed for fertilizer, and gelatin was extracted from hooves. The meat packing industry remains an excellent example of efficient systems integration.

- **Wood products:** In addition to wood itself, the wood products industry produces a number of products with significant economic value. One of the major wood products consists of cellulose, a high molecular mass carbohydrate used to make paper or chemically processed to make products such as nitrocellulose and rayon. The lignin byproduct of the extraction of cellulose from wood is normally burned for its fuel value.

- **Corn products:** The modern corn products industry is one of the most diverse of all industries, producing starch, oil, ethanol, and a variety of other products.

3.8. ECO-EFFICIENCY

A measure of the success of a system of industrial metabolism is the achievement of a maximum state of **eco-efficiency**. The concept of eco-efficiency has been discussed at length on a book dealing with the topic.[10] This concept has been developed by the World Business Council for Sustainable Development, and thus is oriented toward business, that area of human endeavor devoted to satisfying human needs and quality of life through the delivery of goods and services at (generally) affordable prices. Eco-efficiency differs from the conventional mode of operation of business in that it constantly strives to reduce environmental impacts and material consumption. In so doing, an eco-efficient system of business emphasizes provision of services that satisfy human needs and enhance quality of life, rather than simply increasing the output of goods. Such a system is operated from a process viewpoint in consideration of the entire life cycle of products. Furthermore, it takes into account Earth's carrying capacity to support business and industrial enterprises.

An enterprise that delivers goods and services within a framework of eco-efficiency is operated with several key goals in mind. First of all, such an enterprise is designed to minimize material consumption and to maximize the efficiency of energy consumption. In order to reduce consumption of nonrenewable materials, an eco-efficient system strives to increase the proportion of materials that come from renewable sources utilized in a sustainable manner. An example of such a source is wood obtained from forests that are grown in a sustainable manner. An eco-efficient system is designed to produce materials that are recyclable to a maximum extent and to manufacture components that are durable and amenable to refurbishing and reuse. In addition, an eco-efficient system maximizes the intensity with which goods are used and services provided. An example of this would be advanced air traffic control systems which enable aircraft to land and take off at more frequent intervals on a runway system, thus avoiding the need to devote more land and materials to additional runways.

Of greater importance than the achievement of a state of complete eco-efficiency — an impossibility in the real world — are the incremental achievements attained in targeting a goal of eco-efficiency. In this respect, there is a very encouraging recent history of achievement in implementation of the manufacturing concept of "zero defects." The American automobile industry has seen particularly spectacular progress toward the goal of zero defects. During the 1970s, all too many of the products of the U.S. auto industry suffered from a variety of defects, and the domestic industry was losing market share at a frightening rate to higher quality Japanese imports. Programs of rigorous quality control involving worker participation at all levels have resulted in much higher quality U.S. products with the achievement of low rates of defects that would have been thought impossible in previous times. Major concerns are implementing much the same approach to reduce air and water pollution, hazardous waste, and energy consumption. As with zero defects, a goal of total eco-efficiency will never be achieved, but efforts to reach it can be expected to produce excellent results.

Reduction of Toxic Substance Dispersion

A goal of eco-efficiency that is particularly pertinent to the theme of this book is the reduction of the dispersion of toxic substances. Human activities inevitably involve the use of toxic substance, which results in a strong tendency to disperse such materials throughout the anthrosphere and other spheres of the environment, where they may come into contact with living organisms and cause significant harm. Therefore, a primary goal of eco-efficiency is to reduce the dispersion of such substances.

Much of the progress made to date in this area has been driven by regulations against the pollution of air and water and improper disposal of hazardous wastes. In the U.S., major reductions in the release of toxic substances have resulted from compliance with the Environmental Protection Agency's requirement of Toxic Chemical Release Inventory Reporting implemented in 1988 as mandated by Title III ("Community Planning and Community Right-to-Know Act of 1986") of the "Superfund Amendments and Reauthorization Act of 1986."[11] This regulation requires annual reports of the kinds and quantities of toxic substances released, intentionally or accidentally, to air, water, or land, or discharged to water treatment facilities (publicly owned treatment works, POTW).

A common example of the reduction of toxic substance dispersion has been the decrease in volatile organic compound (VOC) releases from industrial operations, such as parts degreasing. More efficient use of pesticides by carefully targeted applications has reduced their dispersion to the environment. One clever approach is to apply limited quantities of insecticides along with sex attractants (pheromones) specific to the targeted insects so that they are attracted directly to the pesticides designed to kill them. Substitution of water-based paints for those based on organic solvents has reduced organic compound dispersal and emissions. The replacement of lead in corrosion resistant primers has resulted in reductions in the dispersal of this heavy metal.

Emphasis on the Industrial Enterprise

The achievement of a high degree of eco-efficiency must focus on individual industrial enterprises — manufacturing concerns, energy generation and distribution systems, large farming operations — as well as on individuals or the public sector. In part this is because such enterprises have the capabilities and the need to achieve a state of maximum eco-efficiency. In a system firmly based on the principles of industrial ecology, industries are involved in the decision-making process and do not simply respond to regulations.

Increased involvement of private firms through the application of industrial ecology can mean a greater emphasis on providing services as opposed to simply marketing goods. For example, in the area of water treatment a concern may contract to provide clean water rather than simply selling the coagulants, activated carbon, filters, and other chemicals and apparatus required to produce clean water. Rather than selling pesticides, an agribusiness service provider may contract to control pests and increase agricultural productivity through a variety of measures including, in addition to pesticides, pest-resistant crops, and biological control. Dry cells and rechargeable batteries are major vectors for the distribution of heavy metals, such as lead, cadmium, and mercury, into the environment. A service that provides replacement batteries and takes back worn out units could be very helpful in minimizing the harm caused by improper battery disposal.

Measures of Efficiency

In evaluating eco-efficiency, it is useful to have a quantitative measurement of the efficiency of materials use. To do that, it is useful to consider three distinct material masses: those of product, the total mass of byproduct, and the total mass of wastes. Using these values the percentage of waste is given by the equation

$$\text{Percentage waste} = \frac{\text{mass waste}}{\text{Mass product} + \text{mass byproduct} + \text{mass waste}} \quad (3.8.1)$$

A smaller percentage waste indicates a more efficient process.

LITERATURE CITED

1. Graedel, Thomas, "Industrial Ecology: Definition and Implementation," Chapter 3 in *Industrial Ecology and Global Change*, Robert Socolow, Clinton Andrews, Frans Berkhout, and Valerie Thomas, Eds., Cambridge University Press, New York, NY, 1994, pp. 23–42.

2. Ayres, Robert U., "Industrial Metabolism," in *Technology and Environment*, J. H. Ausubel and H. E. Sladovich, Eds., National Academy Press, Washington, D.C., 1989, pp. 23–49.

3. Ayres, Robert U., "Industrial Metabolism: Theory and Policy," in *The Greening of Industrial Ecosystems*, National Academy Press, Washington, D.C., 1994, pp. 23–37.

4. "Industrial Ecology Methods and Tools," Chapter 3 in *Discovering Industrial Ecology*, John L. Warren and Stephen R. Moran, Battelle Press, Columbus, OH, 1997, pp. 37–73.

5. Socolow, Robert, "Six Perspectives from Industrial Ecology," Chapter 1 in *Industrial Ecology and Global Change*, Robert Socolow, Clinton Andrews, Frans Berkhout, and Valerie Thomas, Eds., Cambridge University Press, New York, NY, 1994, pp. 3–16.

6. Eckholm, Erik, "China Admits Ecological Sins Played Role in Flood Disaster," *New York Times*, August 26, 1998, p. 1.

7. Frosch, Robert A. and Nicholas E. Gallopoulos, "Strategies for Manufacturing," *Scientific American*, **261**, 94–102, 1989.

8. Manahan, Stanley E., *Environmental Science and Technology*, CRC Press, Boca Raton, FL, 1997.

9. "On Industrial Ecosystems," Chapter 15 in *Industrial Ecology: Towards Closing the Materials Cycle*, Robert U. Ayres and Leslie W. Ayres, Edward Elgar Publishing Co., Cheltenham, U.K., 1996, pp. 273–293.

10. DeSimone, Livio D., and Frank Popoff, *Eco-efficiency: The Business Link to Sustainable Development*, The MIT Press, Cambridge, MA, 1997.

11. Halblieb, Wayne T., "Emergency Planning and Community Right-To-Know Act," Chapter 9 in *Environmental Law Handbook*, 13th ed., Thomas F. P. Sullivan, Ed., Government Institutes, Inc., Rockville, MD, 1995, pp. 278–307.

SUPPLEMENTARY REFERENCES

Curran, Mary Ann, Ed., *Environmental Life-Cycle Assessessment*, McGraw-Hill, New York, NY, 1997.

Fiksel, Joseph, Ed., *Design for Environment: Creating Eco-Efficient Products and Processes*, McGraw-Hill, New York, NY, 1996.

Leff, Enrique, *Green Production: Toward an Environmental Rationality*, Guilford Press, New York, NY, 1995.

Nemerow, Nelson, *Zero Pollution for Industry: Waste Minimization Through Industrial Complexes*, John Wiley & Sons, New York, NY, 1995.

Peck, Steven, and Elaine Hardy, *The Eco-Efficiency Resource Manual*, Economic Developers Council of Ontario, Fergus, Ontario, Canada, 1997.

4 INDUSTRIAL ECOSYSTEMS

4.1. INTRODUCTION

In Chapter 2 the concept of industrial ecology, "the science of sustainability,"[1] and how it relates to industrial systems was discussed in some detail. Chapter 3 defined and covered industrial metabolism. In these chapters it was pointed out that industrial systems may be organized in principle around the concept of industrial ecology so that potential wastes from one firm serve as raw materials for another firm. Thus a group of firms may practice industrial ecology through a system of industrial metabolism that is efficient in the use of both materials and resources. Such a group of firms constitutes a functional **industrial ecosystem**,[2] which may be defined as a regional cluster of industrial firms and other entities linked together in a manner that enables them to utilize byproducts, materials, and energy between various concerns in a mutually advantageous manner. Chapter 4 discusses and explains industrial ecosystems.

Figure 4.1 shows the main attributes of a functional industrial ecosystem. In the simplest sense, such a system processes materials, powered by a relatively abundant source of energy. Materials enter the system from a raw materials source and are put in a usable form by a primary materials producer. From there the materials go into manufacturing goods for consumers. Associated with various sectors of the operation are waste processors that can take byproduct materials, upgrade them, and feed them back into the system. An efficient, functional transportation system is required for the system to work efficiently, and good communications links must exist among the various sectors. A key material in the system is water.

A successfully operating industrial ecosystem provides several benefits. Such a system *reduces pollution*. It results in *high energy efficiency* compared to systems of firms that are not linked and it *reduces consumption of virgin materials* because it *maximizes materials recycle*. *Reduction of amounts of wastes* is another advantage of a functional system of industrial ecology. Finally, a key measure of the success of a system of industrial ecology is *increased market value of products* relative to material and energy consumption.

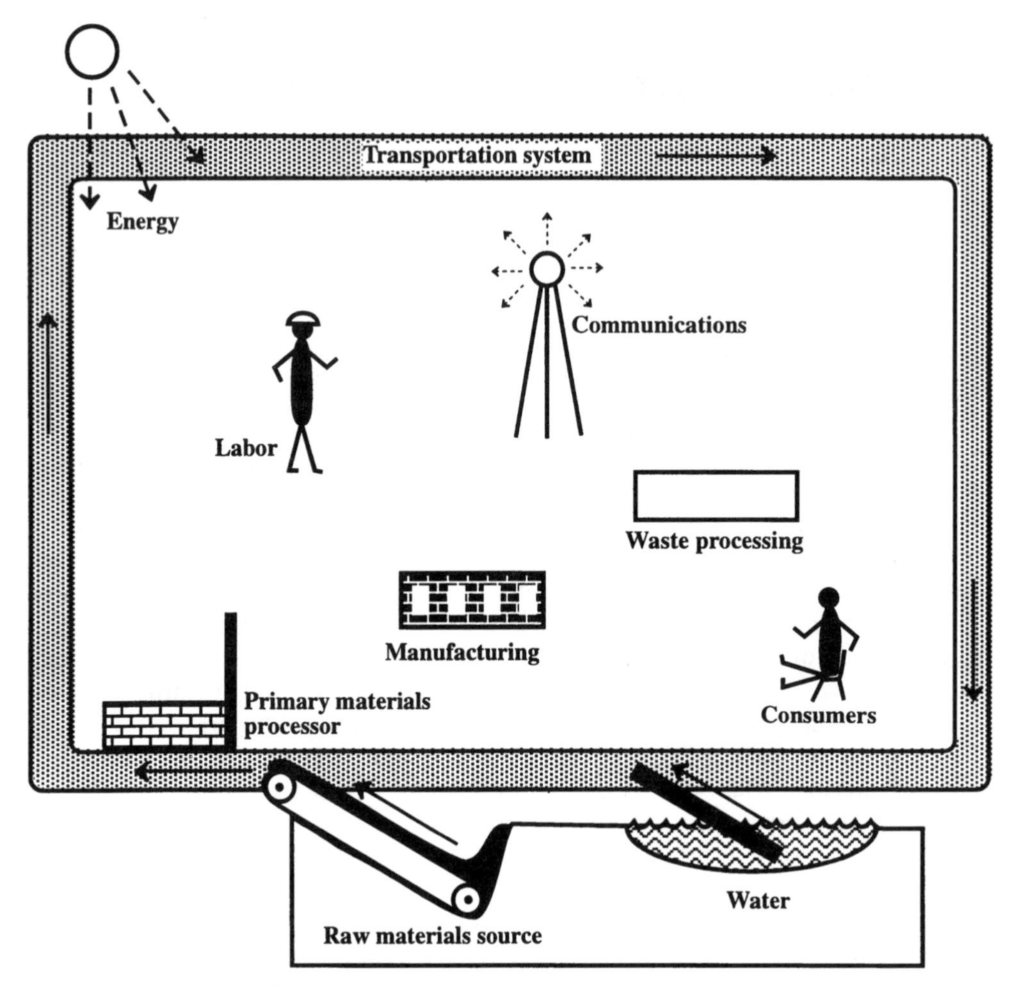

Figure 4.1. Major components required for an industrial system. When these components exist symbiotically utilizing waste materials from one concern as feedstock for another, they comprise a functioning industrial ecosystem.

In setting up an industrial ecosystem, there are two basic, complementary approaches that may be pursued. Within an industry, emphasis may be placed upon product and service characteristics that are compatible with the practice of industrial ecology. Products may be designed for increased durability and amenability to repair and recycle. Instead of selling products, a concern may emphasize leasing, so that it can facilitate recycling. The second approach emphasizes interactions between concerns so that they operate in keeping with good practice of industrial ecology. This approach facilitates materials and energy flow, exchange, and recycle between various firms in the industrial ecosystem.

An important aspect of an industrial ecosystem is the practice of a high degree of **industrial symbiosis**. Symbiotic relationships in natural biological systems occur when two very dissimilar organisms live together in a mutually advantageous manner, each contributing to the welfare of the other. Examples are nitrogen-fixing *Rhizobium* bacteria living in root nodules on leguminous plants or lichens consisting of fungi and bacteria living on a rock surface. Analogous symbiotic relationships in which firms utilize each others' residual materials form the basis of relationships between firms in a functional industrial ecosystem. Examples of industrial symbiosis are cited below in the discussion of the Kalundborg, Denmark, industrial ecosystem.

An important consideration in the establishment and function of an industrial ecosystem is the geographical scope of the system. Often a useful way to view such a system is on the basis of a transportation system, such as a length of a navigable river or an interconnected highway system. The Houston Ship Channel, which stretches for many kilometers, is bordered by a large number of petrochemical concerns that exist to mutual advantage through the exchange of materials and energy. The purification of natural gas by concerns located along the channel yields lower molecular mass hydrocarbons, such as ethane and propane, that can be used by other concerns, for example, in polymers manufacture. Sulfur removed from natural gas and petroleum can be used to manufacture sulfuric acid.

4.2. THE FOUR MAJOR COMPONENTS OF AN INDUSTRIAL ECOSYSTEM

Industrial ecosystems can be broadly defined to include all types of production, processing, and consumption. These include, for example, agricultural production as well as purely industrial operations. It is useful to define five major components of an industrial ecosystem as shown in Figure 4.2. These are (1) a primary materials producer, (2) a source or sources of energy, (3) a materials processing and manufacturing sector, (4) a waste processing sector, and (5) a consumer sector.

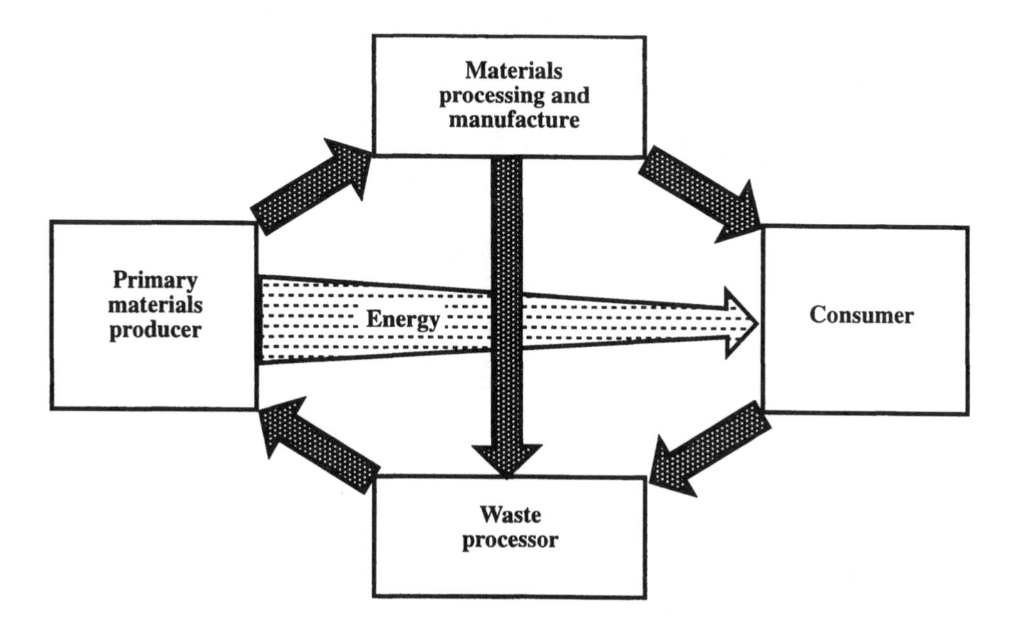

Figure 4.2. The major constituents or "hubs" of an industrial ecosystem.

In such an idealized system, the flow of materials between the four major hubs is very high. Each constituent of the system evolves in a manner that maximizes the efficiency with which the system utilizes materials and energy.

Primary Materials and Energy Producers

It is convenient to consider the primary materials producers and the energy generators together because both materials and energy are required in order for the industrial ecosystem to operate. The primary materials producer or producers may consist of one or several enterprises devoted to providing the basic materials that sustain the industrial ecosystem. Most generally, in any realistic industrial ecosystem a significant fraction of the material processed by the system consists of virgin materials. In a number of cases, and increasingly so as pressures build to recycle materials, significant amounts of the materials come from recycling sources.

The processes that virgin materials entering the system are subjected to vary with the kind of material, but can generally be divided into several major steps. Typically the first step is extraction, designed to remove the desired substance as completely as possible from the other substances with which it occurs. This stage of materials processing can produce large quantities of waste material requiring disposal, as is the case with some metal ores in which the metal makes up only a few percent or even less of the ore that is mined. In other cases, such as corn grain providing the basis of a corn products industry, the "waste," in this specific example the cornstalks associated with the grain, can be left in place (cornstalks returned to soil serve to add humus and improve soil quality). A concentration step may follow extraction to put the desired material into a more pure form. After concentration, the material may be put through additional refining steps that may involve separations. Following these steps, the material may be subjected to additional processing and preparation leading to the finished materials. Throughout the various steps of extraction, concentration, separation, refining, processing, preparation, and finishing, various physical and chemical operations are used, and wastes requiring disposal may be produced. Recycled materials may be introduced at various parts of the process, although they are usually introduced into the system following the concentration step.

The extraction and preparation of energy sources can follow many of the steps outlined above for the extraction and preparation of materials. For example, the processes involved in extracting uranium from ore, enriching it in the fissionable uranium-235 isotope, and casting it into fuel rods for nuclear fission power production include all of those outlined above for materials. On the other hand, some rich sources of coal are essentially scooped from a coal seam and sent to a power plant for power generation with only minimal processing, such as sorting and grinding.

Recycled materials added to the system at the primary materials and energy production phase may be from both pre- and postconsumer sources. As examples, recycled paper may be macerated and added at the pulping stage of paper manufacture. Recycled aluminum may be added at the molten metal stage of aluminum metal production.

Materials Processing and Manufacture Sector

Finished materials from primary materials producers are fabricated to make products in the materials processing and manufacture sector. The latter is often very complex with a large number of materials involved. For example, the manufacture of an automobile requires steel for the frame, plastic for various components, rubber in tires, lead in the battery, and copper in the wiring, along with a large number of other materials. Typically, the first step in materials manufacturing and processing is a forming operation. For example, sheet steel suitable for making automobile frames may be cut, pressed, and welded into the configuration needed to make a frame. At this step some wastes may be produced that require disposal. An example of such wastes consists of carbon fiber/epoxy composites left over from forming parts such as jet aircraft housings. Finished components from the forming step are fabricated into finished products that are ready for the consumer market.

The materials processing and manufacturing sector presents several opportunities for recycling. At this point it may be useful to define two different streams of recycled materials.

- **Process recycle streams** consisting of materials recycled in the manufacturing operation itself

- **External recycle streams** consisting of materials recycled from other manufacturers or from post-consumer products

Materials suitable for recycle can vary significantly. Generally, materials from the process recycle streams are quite suitable for recycling because they are the same materials used in the manufacturing operation. Recycled materials from the outside, especially those from postconsumer sources, may be quite variable in their characteristics because of the lack of effective controls over recycled postconsumer materials. Therefore, manufacturers may be reluctant to use such substances.

The Consumer Sector

In the consumer sector, products are sold or leased to the consumers who use them. The duration and intensity of use vary widely with the product; paper towels are used only once, whereas an automobile may be used thousands of times over many years. In all cases, however, the end of the useful lifetime of the product is reached and it is either (1) discarded or (2) recycled. The success of a total industrial ecology system may be measured largely by the degree to which recycling predominates over disposal.

Waste Processing Sector

Recycling has become so widely practiced that an entirely separate waste processing sector of an economic system may now be defined. This sector consists of enterprises that deal specifically with the collection, separation, and processing of recyclable materials and their distribution to end users. Such operations may be entirely private or they may involve cooperative efforts with governmental sectors. They are often driven by laws and regulations that provide penalties against simply discarding used items and materials, as well as positive economic and regulatory incentives for their recycle.

4.3. OVERVIEW OF AN INTEGRATED INDUSTRIAL ECOSYSTEM

Figure 4.3 provides an overview of an integrated industrial ecosystem including all the components defined and discussed in the preceding section. Such a system may be divided into three separate, somewhat overlapping sectors controlled by the following: (1) The raw materials supply and processing sector, (2) the manufacturing sector, and (3) the consumer sector.

There are several important aspects of a complete industrial ecosystem. One of these is that, as discussed in the preceding section, there are several points at which materials may be recycled in the system. A second aspect is that there are several points at which wastes are produced. The potential for the greatest production of waste lies in the earlier stages of the cycle in which large quantities of materials with essentially no use associated with the raw material, such as ore tailings, may require disposal. In many cases, little if anything of value can be obtained from such wastes and the best thing to do with them is to return them to their source (usually a mine), if possible. Another big source of potential wastes, and often the one that causes the most problems, consists of postconsumer wastes generated when a product's life cycle is finished. With a properly designed industrial ecology cycle, such wastes can be be minimized and, ideally, totally eliminated.

In general, the amount of waste per unit output decreases in going through the industrial ecology cycle from virgin raw material to final consumer product. Also, the amount of energy expended in dealing with waste or in recycling decreases farther into the cycle. For example, waste iron from the milling and forming of automobile parts may be recycled from a manufacturer to the primary producer of iron as scrap steel. In order to be used, such steel must be remelted and run through the steel manufacturing process again, with a considerable consumption of energy. However, a postconsumer item, such as an engine block, may be refurbished and recycled to the market with relatively less expenditure of energy.

At the present time, the three major enterprises in an industrial ecology cycle, the materials producer, the manufacturer, and the consumer, act largely independently from each other. As raw materials become more scarce, there will be more economic incentives for recycling and integration of the total cycle. Furthermore, there is a need for better, more scientifically based regulatory incentives leading to the practice of industrial ecology.

4.4. THE KALUNDBORG EXAMPLE

The most often cited example of a functional industrial ecosystem is that of Kalundborg, Denmark. The various components of the Kalundborg industrial ecosystem are shown in Figure 4.4. To a degree, the Kalundborg system developed spontaneously without being specifically planned as an industrial ecosystem. It is

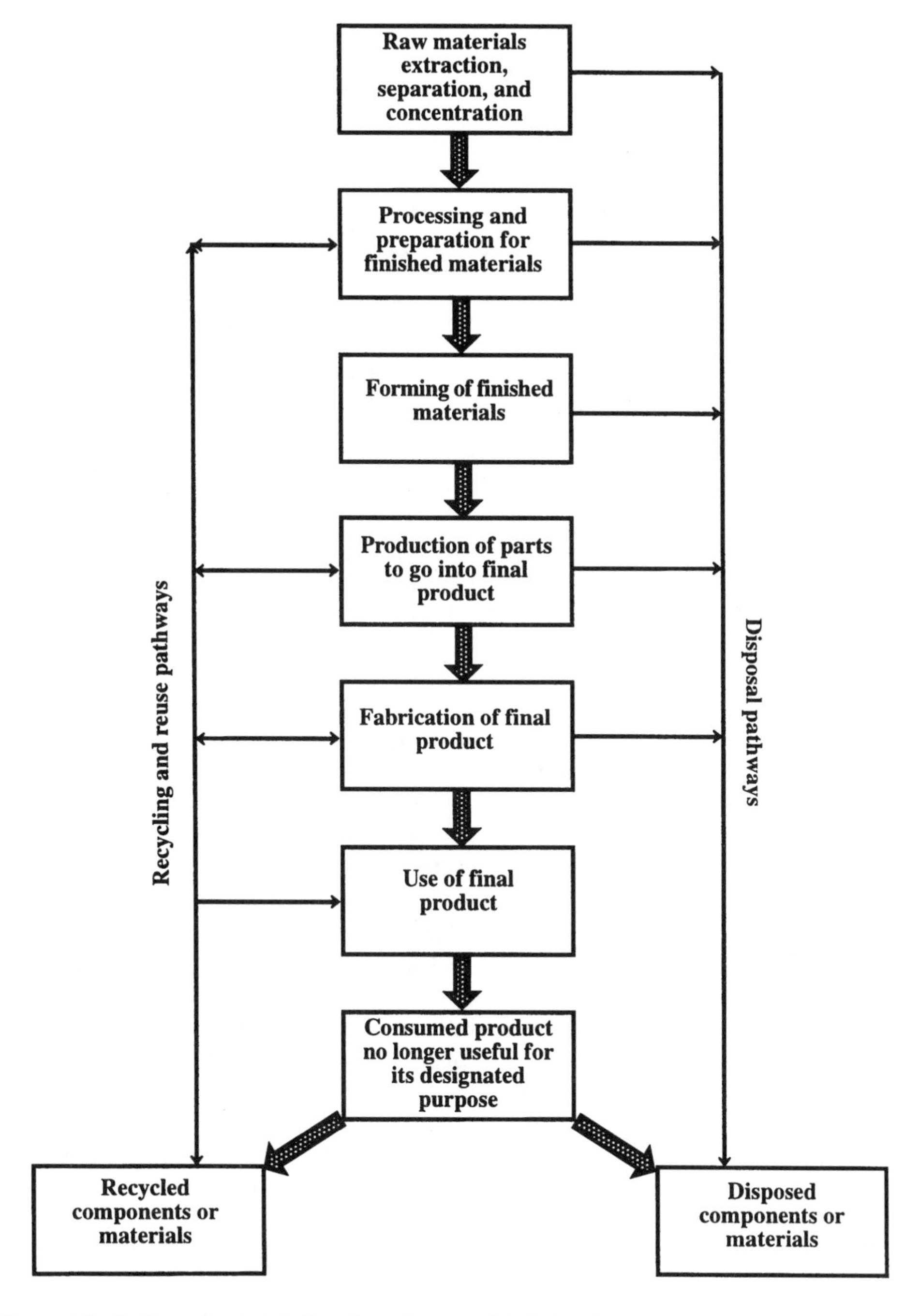

Figure 4.3. Outline of materials flow through a complete industrial ecosystem.

based upon two major energy suppliers, the 1,500-megawatt ASNAES coal-fired electrical power plant and the 4–5 million tons/year Statoil petroleum refining complex, each the largest of its kind in Denmark. The electric power plant sells pro-

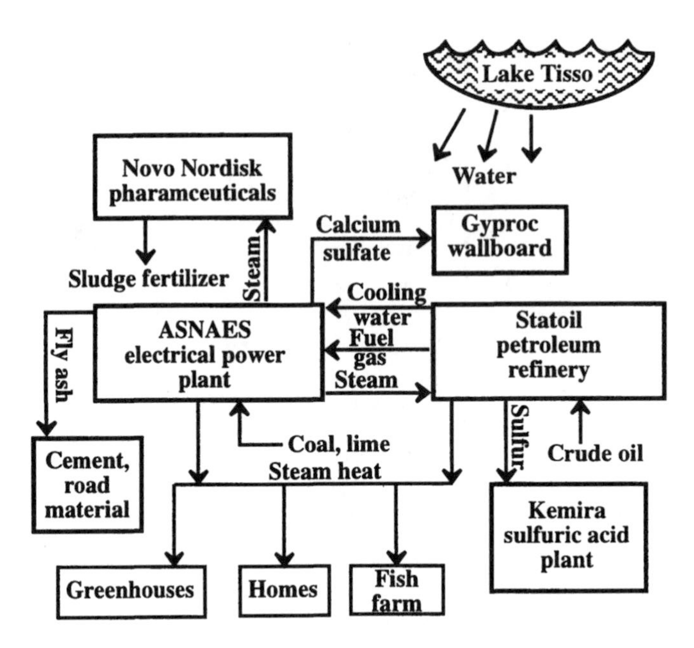

Figure 4.4. Schematic of the industrial ecosystem in Kalundborg, Denmark.

cess steam to the oil refinery, from which it receives fuel gas and cooling water. Sulfur removed from the petroleum goes to the Kemira sulfuric acid plant. Byproduct heat from the two energy generators is used for district heating of homes and commercial establishments, as well as to heat greenhouses and a fish farming operation. Steam from the electrical power plant is used by the $2 billion/year Novo Nordisk pharmaceutical plant, a firm that produces industrial enzymes and 40% of the world's supply of insulin. This plant generates a biological sludge that is used by area farms for fertilizer. Calcium sulfate produced as a byproduct of sulfur removal by lime scrubbing from the electrical plant is used by the Gyproc company to make wallboard. The wallboard manufacturer also uses clean-burning gas from the petroleum refinery as fuel. Fly ash generated from coal combustion goes into cement and roadbed fill. Lake Tisso serves as a fresh water source. Other examples of efficient materials utilization associated with Kalundborg include use of sludge from the plant that treats water and wastes from the fish farm's processing plant for fertilizer and blending of excess yeast from Novo Nordisk's insulin production as a supplement to swine feed.

The development of the Kalundborg complex occurred over a long period of time, beginning in the 1960s, and provides some guidelines for the way in which an industrial ecosystem can grow naturally. The first of many synergistic (mutually advantageous) arrangements was cogeneration of usable steam along with electricity by the ASNAES electrical power plant. The steam was first sold to the Statoil petroleum refinery, then, as the advantages of large-scale, centralized production of steam became apparent, steam was also provided to homes, greenhouses, the pharmaceutical plant, and the fish farm. The need to produce electricity more cleanly than was possible simply by burning high sulfur coal resulted in two more synergies. Installation of a lime scrubbing unit for sulfur removal on the power plant stack resulted in the production of large quantities of calcium sulfate, which found a market in the manufacture of gypsum wallboard. It was also found that a clean-burning gas byproduct of the petroleum refining operation could be substituted in part for the coal burned in the power plant, further reducing pollution.

The implementation of the Kalundborg ecosystem occurred largely because of the close personal contact between the managers of the various facilities in a relatively close social and professional network over a long period of time. All the contracts have been based upon sound business fundamentals and have been bilateral. Each company has acted upon its perceived self-interest and there has been no master plan for the system as a whole. The regulatory agencies have been cooperative, but not coercive in promoting the system. The industries involved in the agreements have fit well, with the needs of one matching the capabilities of the other in each of the bilateral agreements. The physical distances involved have been small and manageable; it is not economically feasible to ship commodities such as steam or fertilizer sludges for long distances.

4.5. OTHER INDUSTRIAL ECOSYSTEMS

Kalundborg happens to be the most well-documented industrial ecosystem and perhaps the most well-organized and integrated one on a community basis. However, other systems exist. Indeed, any time that two or more concerns have a symbiotic relationship in which wastes from one are utilized by another, they can be regarded as constituting at least a partial industrial ecosystem, hundreds of which exist throughout the world. One such large and diverse system has been described in the Austrian province of Styria.[3] The investigation of this system showed involvement of more than 50 concerns in an elaborate materials and energy web that had developed in the province. In addition to those dealing specifically with wastes and waste recycling, the concerns are from a variety of industries including plastics, textiles, paper, building materials, wood products, agricultural products, and food processing. Energy is another commodity utilized within the system.

In the Styrian case, a large variety of materials were utilized and recycled. These include the commonly recycled, generally low-value products such as used lubricating oil, solvents, scrap iron and other metals, paper, and gypsum produced from power plant desulfurization with limestone. In some cases users found that the quality of recycled materals was at least equal to, and sometimes exceeded that of materials obtainable from primary sources.

An interesting aspect of the Styrian industrial ecosystem is that its existence was not even known until it was investigated. It did not develop as part of a master plan, but was motivated by the profits to be had by producers selling byproducts and by the lower prices and, in some cases, higher quality of materials available to consumers, as well as restrictions on landfill that made disposal of wastes (potential byproducts) increasingly costly.

Both the Kalundborg and Styrian industrial ecosystems, as well as perhaps hundreds of others not documented, developed spontaneously, primarily through bilateral agreements and symbiotic relationships between firms. This approach is similar to the way in which natural ecosystems have developed and operate and is the way that such systems have developed. Centralized planning does not seem to have been involved to any significant extent, and one need only look at the failed industrial systems of the former Communist Bloc nations of Eastern Europe to see the pitfalls and failures of centralized planning. However, the intriguing possibility remains that intelligent centralized planning, regulation, and incentives can be successfully used to promote the development of viable systems of industrial ecology. Indeed, there are examples of the tremendous success of centralized planning in the development of industrial systems. Perhaps the most prominent of these is the federally financed Tennessee Valley Authority (TVA) power projects that have had a tremendous impact in developing a formerly backward and poverty stricken region into one of the most prosperous and dynamic economic systems in the U.S. Although its overall environmental impact has been adverse on many respects, the U.S. interstate highway system has undoubtedly had an enormous influence on economic development. Clearly, governmental investment and regulation has the potential to greatly influence and facilitate the development of systems of industrial ecology.

4.6. COAL-BASED INDUSTRIAL ECOSYSTEMS

Arguably, the first industrial ecosystems to be implemented were those based around coal coking for iron production. Since coal is an abundant, versatile material, it is not suprising that it has served as the basis for some elaborate proposals for industrial ecosystems. In this respect, coal has an advantage of serving as a source of both material and energy. In the material area, coal can provide organic materials, whereas its ash can act as a source of several minerals, including alumina for aluminum production and calcium silicate for cement manufacture. Experience with coal conversion processes for the generation of hydrocarbon liquids, methane (CH_4), and synthesis gas (a mixture of CO and H_2) goes back almost two centuries, so there is a large body of technology to draw upon.

The most straightforward approach to converting coal into useful byproducts consists of **pyrolysis**, or **coking**, in which coal is heated in the absence of air. This process takes advantage of the fact that coal is actually a hydrocarbon with an empirical formula of approximately $CH_{0.7}$, so when it is heated, it disproportionates into hydrocarbon vapors, liquids, and tars and leaves a carbon solid (coke). A significant fraction of the sulfur in coal that is pyrolyzed is evolved as hydrogen sulfide gas, H_2S, which can be removed and processed to produce sulfuric acid, H_2SO_4. The coke residue consisting of carbon and mineral matter can then be used as fuel in boilers to generate electricity. Hydrocarbon liquids, such as benzene and toluene, released by pyrolysis can be used as solvents, for chemical synthesis, or in gasoline.

More possibilities for coal utilization occur when the coal is reacted with oxygen and steam to produce combustible gases. The heat required for coal gasification is generated by the exothermic reaction of carbon with oxygen to produce noncombustible carbon dioxide,

$$C + O_2 \rightarrow CO_2 + \text{heat} \tag{4.6.1}$$

Steam reacting with carbon heated by the preceding reaction generates a combustible mixture of carbon monoxide and elemental hydrogen,

$$C + H_2O + \text{heat} \rightarrow CO + H_2 \tag{4.6.2}$$

Of these two gases, elemental hydrogen is much more useful for chemical synthesis, and the H_2/CO ratio may be increased by the shift reaction,

$$H_2O + CO \rightarrow H_2 + CO_2 \tag{4.6.3}$$

over a shift catalyst. Methane, a premium gaseous fuel, may be produced by the methanation of carbon monoxide over a suitable catalyst,

$$CO + 3H_2 \rightarrow CH_4 + H_2O \tag{4.6.4}$$

or different catalysts and lower H_2/CO ratios may be used to produce other hydrocarbons, such as those composing gasoline or diesel fuel.

The production of elemental hydrogen in synthesis gas opens up another avenue of synthesis beginning with the fixation of nitrogen to produce ammonia,

$$N_2 + 3H_2 \rightarrow 2NH_3 \tag{4.6.5}$$

leading to the synthesis of a wide variety of inorganic and organic nitrogen compounds.

Coke from coal is the reducing agent used to produce iron metal from iron oxide ores.

$$3C + 2Fe_2O_3 \rightarrow 4Fe + 3CO_2 \tag{4.6.6}$$

Some coal ashes have iron contents high enough to be of interest as sources of iron. So, an industrial ecosystem in which coal is pyrolyzed to produce coke, the coke used in part as fuel to generate electricity and in part to generate synthesis gas, the coke (coal) ash used as a source of iron, and part of the coke used to make pig iron from the iron oxides separated from the coke ash forms an interesting basis for a coal-based industrial ecosystem. Furthermore, electricity generated from coke or coal can be used in electric arc furnaces to produce steel or in electrolytic processes to generate aluminum metal (see below).

Figure 4.5 shows part of an industrial ecosystem based on coal in which the coal ash is utilized. This particular system runs on high-ash coal that can serve as both a fuel and as a source of alumina from which aluminum metal may be electrolytically refined using electricity from the power plant. Several systems have been proposed using this basic approach.

As shown in Figure 4.5, the coal is not used simply as an energy source, but is first pyrolyzed (heated in the absence of air) to drive off volatile gases, liquids, tars, and sulfur, leaving a coke residue. Significant economic returns may be recovered from the coal byproducts. Some of the liquid products, such as benzene and toluene, can be blended with gasoline or used in chemical synthesis. The coal gases, including CO, H_2, and CH_4, can be used to fire a gas turbine or for chemical synthesis, such as to make ammonia, NH_3. The carbonaceous solid coke product of the gasifier can be used as a fuel in the power plant, where the fact that it has less sulfur than the parent coal is an advantage.

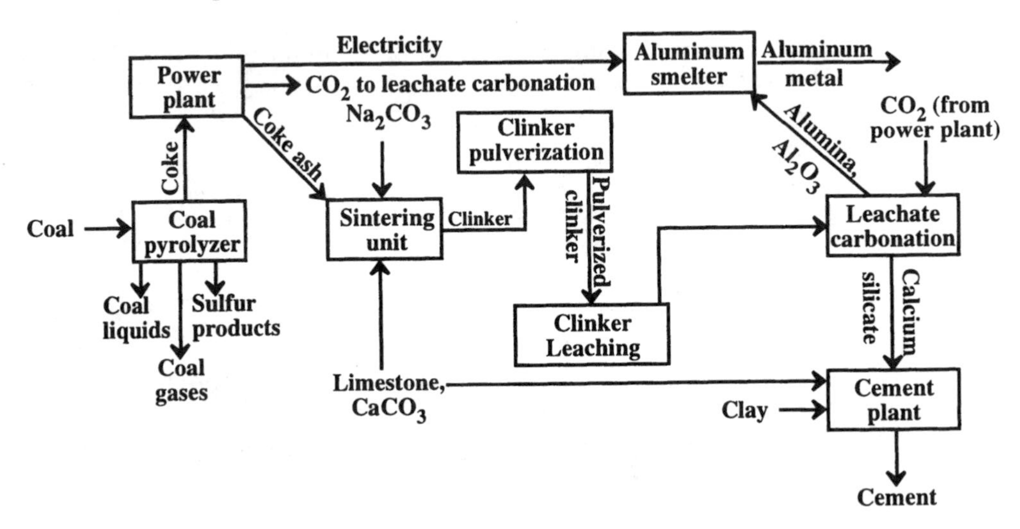

Figure 4.5. Model of a system that carries out industrial metabolism. The coal liquids, coal gases, sulfur products of coal pyrolysis, aluminum, and cement can go into other parts of the total industrial ecosystem.

The coal ash from power generation is mixed with soda ash (Na_2CO_3) and limestone ($CaCO_3$), to serve as feedstock to a sintering unit, where it is heated and fused to yield a clinker product. The clinker is pulverized and leached with hot Na_2CO_3 solution to produce soluble sodium aluminate and a precipitate of calcium silicate. Carbonation of the sodium aluminate solution precipitates aluminum hydroxide and regenerates the sodium carbonate solution. The aluminum hydroxide is heated to convert it to alumina, Al_2O_3, from which metallic aluminum is refined in the electrolytic aluminum smelter. The calcium silicate then goes to make cement.

The scheme outlined above gives some idea of the numerous cycles and complex interactions involved in a system of industrial metabolism. It is easy to visualize many additional subsystems, such as those based on sulfuric acid manufacture (from the sulfur recovered from the coal), the coal liquid byproducts, the coal gas byproducts, the aluminum, and the cement.

4.7 THE PETROLEUM INDUSTRY AS AN EXAMPLE OF SYSTEMS INTEGRATION

One of the best modern examples of industrial ecosystems and systems integration is the modern petrochemical industry as illustrated in Figure 4.6. A petrochemical complex is fed by crude oil, a complex mixture consisting of an enormous

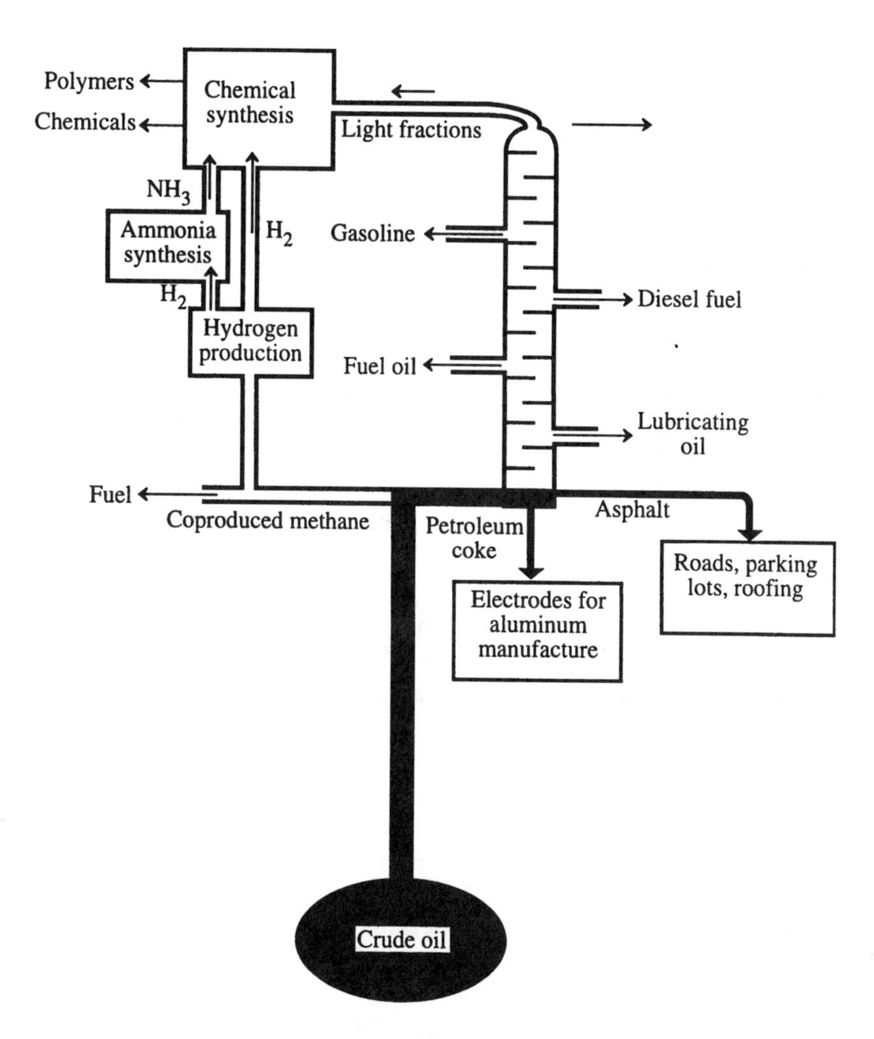

Figure 4.6. Some aspects of petroleum refining and petrochemical complex.

number of chemical species, most of which are hydrocarbons. Natural gas composed predominantly of methane, CH_4, is usually co-produced with crude oil from which it is separated. Heavier gaseous hydrocarbons including ethane, propane, and butane (C_2H_6, C_3H_8, and C_4H_{10}, respectively) can be separated from natural gas and used as "liquid petroleum fuels" or for chemical synthesis. Methane is an excellent fuel, but it has more value as a source of elemental hydrogen, H_2, used in chemical synthesis and carbon black used in tire manufacture. Hydrogen gas made from methane can be combined with nitrogen to synthesize ammonia, NH_3, the basic chemical required to make a large range of nitrogen compounds. Ethane, propane, and butane can be converted into alkenyl hydrocarbons, ethylene (ethene), propylene (propene), and butadiene, which have chemical formulas of C_2H_4, C_3H_6, and C_4H_6, respectively. These, in turn, can be polymerized to make polyethlene and polypropylene plastics, as well as butadiene rubber.

Liquid crude oil is separated into its components by distillation, which can be taken all the way to solid petroleum coke, a material commonly used in electrodes. The heavy liquid residue from petroleum distillation, called asphalt, is used as a paving material and to make a variety of items, such as roofing shingles. Lubricating oils are derived from distillation of petroleum, as are somewhat lighter fuel oils, diesel fuel, and kerosene. Gasoline is obviously an important fraction, and the hydrocarbons from petroleum are subjected to a variety of chemical processes to increase the yield and quality of gasoline. Even lighter hydrocarbons are recovered for uses as solvents, and gaseous hydrocarbons, such as butane, are recovered for fuel or chemical synthesis.

Crude oil and the gas associated with it provide the energy and raw materials to make an enormous variety of petrochemicals, many more than can be described here. A major inorganic byproduct of petroleum refining is sulfur, which also provides the basis of a large synthetic chemical industry, usually based upon sulfuric acid, H_2SO_4. Through the application of industrial ecological principles, the materials loop can be largely closed in a large petrochemical complex, thus maximizing production of higher value products and minimizing air and water pollution.

Vertical Integration

It is possible for all of the constituents of an industrial ecosystem to be part of the same industrial concern. Such a concept has been called **vertical integration**, and has been widely practiced in some large industries, such as the auto industry, in which a single concern has the means of producing most of the materials and parts needed to make automobiles. In practice, however, such systems have not been particularly effective because they tend to be inflexible. Given a captive market, internal suppliers often become very costly and inflexible, losing their innovative edge. Therefore, independent concerns working cooperatively with each other represent the best approach for an industrial ecosystem.

4.8. IMPLEMENTATION AND ORGANIZATION OF INDUSTRIAL ECOSYSTEMS

An industrial ecosystem consists of a group of industries linked together in such a way that they produce, process, and consume goods, energy, and services provided by each other in a manner analogous to natural ecosystems.[4] Most commonly, although not necessarily, a single large enterprise serves to underpin an industrial ecosystem. Normally an industrial ecosystem must be based upon a large supplier of a particular kind of material and/or energy analogous to the photosynthetic "primary producer" in a natural ecosystem. This enterprise may process large quantities of a single raw material, as is the case with a huge concern that uses corn as a commodity to make a number of products. Its primary function may be to harness energy and convert it to usable forms, such as is done by large hydroelectric power plants. Another possibility is that of waste recycling. In a well-balanced industrial ecosystem, the basic industry is tied to suppliers of goods and services, its wastes are processed by one or more waste treatment/recycling enterprises, and it provides the materials for secondary industries. A well-balanced industrial ecosystem "internalizes" materials so that the outflow of materials from the system, particularly in the form of waste products, is minimized.

Industrial ecology recognizes the strong temporal, spatial, and economic linkages that exist among the various components of an industrial ecosystem. It considers the total environmental effects of the industrial system as a whole rather than looking at individual constituents, such as the emission of a specific air pollutant from a particular source, as is the case with a purely regulatory approach.

Organization of Industrial Ecosystems

Inherent to the organization of a successful industrial ecosystem is the conflict between the need for a well-organized, integrated system and the need to encourage the self-development of the individual components of such a system. Natural ecosystems are self-organizing, and self-organization must occur in a successful industrial ecosystem as well. It is a challenge, therefore, to set up a framework of regulations and incentives for an industrial ecosystem within which individual enterprises may develop in a way that the detailed organization of the system develops naturally and spontaneously.

Requirements for the Implementation of Industrial Ecology

Figure 4.7 illustrates the steps in establishing a system of industrial ecology. The first requirement is, of course, that it be technologically feasible to establish the system. If this cannot be done, there is no point in proceeding. If the system appears to be technologically feasible, a determination must next be made if it can be implemented economically. If it passes that test, the next step is to determine if

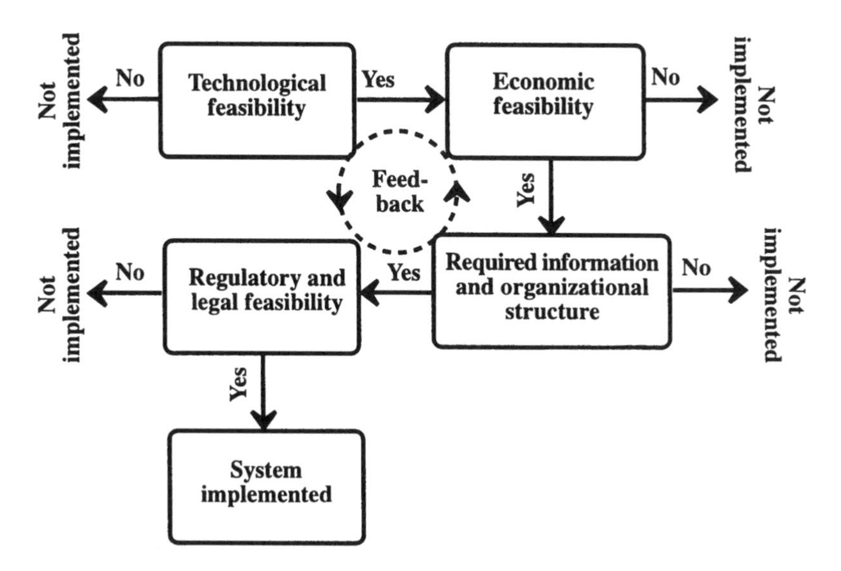

Figure 4.7. Steps in implementing an industrial ecosystem. Failure of any one of these steps can mean that the system cannot be established.

there is sufficient information and a suitable organizational structure. Examples of required information are costs, including the environmental costs of a product and its disposal, information regarding who has a particular waste and who can use it, and the identity and content of low level toxic impurities that may affect a material's use. Finally, the regulatory and legal feasibility must be established. If a system cannot be allowed, there is no point in pursuing it. The greatest legal barrier to the implementation of an industrial ecology system is the potential legal liability for a material after it has left the producer. Under current U.S. law, a producer may be liable for effects and misuse of a material for an indefinite period of time.

As shown in Figure 4.7, there is a lot of potential feedback among the various steps in the feasibility study. Advances in technology can enable a system to become economically feasible. If restrictive, even nonsensical regulations are a barrier, there is the potential to change them based upon technological, economic, and informational input. Or, regulations may be imposed that put a financial penalty on wasteful practices so that the economics of recycling become much more favorable. A dearth of information may trigger additional work in the technological or economic areas so that the required information may be obtained.

An interesting concept that can be taken into account when planning a system of industrial ecology is that of **social capital** consisting of structured interactions, communication, and cooperation among companies.[5] Although social capital has long been a staple of industrial supplier/customer interaction, its practice on a larger scale is very beneficial to the establishment of industrial ecosystems.

The use of sophisticated computer programs to model systems of industrial ecology before they are actually built can greatly enhance the opportunities to successfully establish such systems. Such modeling is already performed routinely for large, complex, integrated industrial complexes, such as petrochemical refining operations. A very significant opportunity exists to prepare models of entire industrial ecosystems incorporating a number of enterprises.[6] In addition, sophisticated computer programs can also reveal existing interrelationships that can provide the basis for industrial ecosystems.

LITERATURE CITED

1. Lifset, Reid, "Relating Industry to Ecology," *Journal of Industrial Ecology*, **1**(2), 1–2 (1997).

2. Cote, Ray, "Industrial Ecosystems: Evolving and Maturing," *Journal of Industrial Ecology*, **1**(3), 9–11 (1998).

3. Schwarz, Erich and Karl W. Steininger, *The Industrial Recycling-Network: Enhancing Regional Development*, Research Memorandum No. 9501, Insititute of Innovation Management, Karl-Franzens University of Graz, Graz, Austria, April, 1995.

4. Lowe, Ernest A. John L. Warren, and Stephen R. Moran, *Discovering Industrial Ecology: An Executive Briefing and Sourcebook*, Battelle Press, Columbus, OH, 1997.

5. Frosch, Robert A., "Industrial Ecology: Adapting Technology for a Sustainable World," *Environment Magazine*, **37**, 16–37, 1995.

6. Ausubel, Jesse, "The Virtual Ecology of Industry," *Journal of Industrial Ecology*, **1**(1), 10–11 (1997).

SUPPLEMENTARY REFERENCES

Ayres, Robert U., *The Greening of Industrial Ecosystems*, National Academy Press, Washington, D.C., 1994.

Ayres, Robert U. and Leslie W. Ayres, *Industrial Ecology: Towards Closing the Materials Cycle*, Edward Elgar Press, Cheltenham, U.K., 1996.

Graedel, Thomas E. and B. R. Allenby, *Industrial Ecology*, Prentice Hall, Englewood Cliffs, NJ, 1995, pp. 204–230.

Socolow, Robert, Clinton Andrews, Frans Berkhout, and Valerie Thomas, Eds., *Industrial Ecology and Global Change*, Cambridge University Press, New York, NY, 1994.

5 LIFE CYCLES: PRODUCTS, PROCESSES, AND FACILITIES

5.1. LIFE CYCLES: EXPANDING AND CLOSING THE MATERIALS LOOP

In a general sense, the traditional view of product utilization is the one way process of extraction → production → consumption → disposal shown in the upper portion of Figure 5.1. Materials that are extracted and refined are incorporated into the production of useful items, usually by processes that produce large quantities of waste byproducts. After the products are worn out, they are discarded. This essentially one-way path results in a relatively large exploitation of resources, such as metal ores, and a constant accumulation of wastes. As shown at the bottom of Figure 5.1, however, the one-way path outlined above can become a cycle in which manufactured goods are used, then recycled at the end of their life spans. As one aspect of such a cyclic system, it is often useful for manufacturers to assume responsibility for their products, to maintain "stewardship." Ideally, in such a system a product and/or the material in it would have a never-ending life cycle; when its useful lifetime is exhausted, it is either refurbished or converted into another product.

In considering life cycles, it is important to note that commerce can be divided into the two broad categories of products and services. Whereas most commercial activity used to be concentrated on providing large quantities of goods and products, demand has been largely satisfied for some segments of the population, and the wealthier economies are moving more to a service-based system. Much of the commerce required for a modern society consists of a mixture of services and goods. The trend toward a service economy offers two major advantages with respect to waste minimization. Obviously, a pure service involves little material. Secondly, a service provider is in a much better position to control materials to ensure that they are recycled and to control wastes, ensuring their proper disposal. A commonly cited example is that of photocopy machines. They provide a service, and a heavily used copy machine requires frequent maintenance and cleaning. The parts of such a machine and the consumables, such as toner cartridges, consist of materials that eventually will have to be discarded or recycled. In this case it is often reasonable for the provider to lease the machine to users, taking responsibility for its maintenance and ultimate fate. The idea could even be expanded to include recycling of the paper processed by the copier, with the provider taking responsibility for recyclable paper processed by the machine.

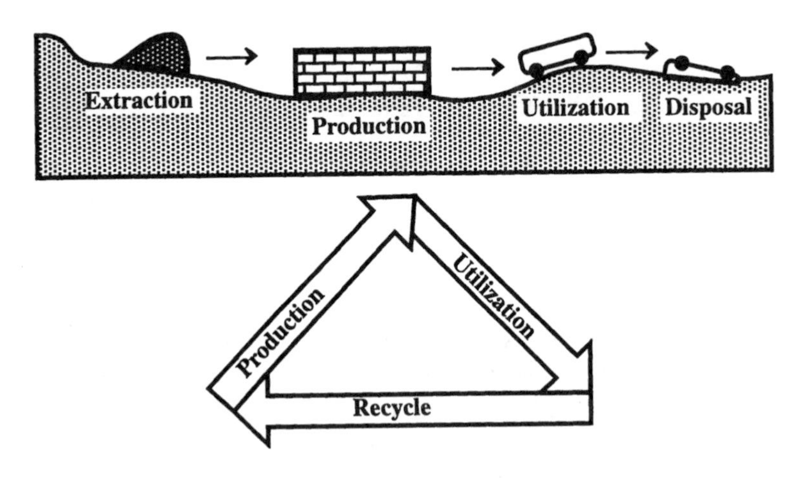

Figure 5.1. The one-way path of conventional utilization of resources to make manufactured goods followed by disposal of the materials and goods at the end consumes large quantities of materials and makes large quantities of wastes (top). In an ideal industrial ecosystem (bottom), the loop is closed and spent products are recycled to the production phase.

It is usually difficult to recycle products or materials within a single, relatively narrow industry. In most cases, to be practical, recycling must be practiced on a larger scale than simply that of a single industry or product. For example, recycling plastics used in soft drink bottles to make new soft drink bottles is not allowed because of the possibilities for contamination. However, the plastics can be used as raw material for auto parts. Usually different companies are involved in making auto parts and soft drink bottles.

Product Stewardship

The degree to which products are recycled is strongly affected by the custody of the products. For example, batteries containing cadmium or mercury pose significant pollution problems when they are purchased by the public, used in a variety of devices, such as calculators and cameras, then discarded through a number of channels, including municipal refuse. However, when such batteries are used within a single organization, it is possible to ensure their virtually complete return for recycling. In cases such as this, systems of stewardship can be devised in which marketers and manufacturers exercise a high degree of control of the product. This can be done through several means. One is for the manufacturer to retain ownership of the product, as is commonly practiced with photocopy machines.

Another mechanism is one in which a significant part of the purchase price is refunded for tradein of a spent item. This approach could work very well with batteries containing cadmium or mercury. The normal purchase price could be doubled, then discounted to half with the tradein of a spent battery.

Embedded Utility

Figure 5.2 can be regarded as an "energy/materials pyramid" showing that the amounts of energy and materials involved decrease going from the raw material to the finished product. The implication of this diagram is that significantly less energy, and certainly no more materials, are involved when recycling is performed near the top of the materials flow chain rather than near the bottom.

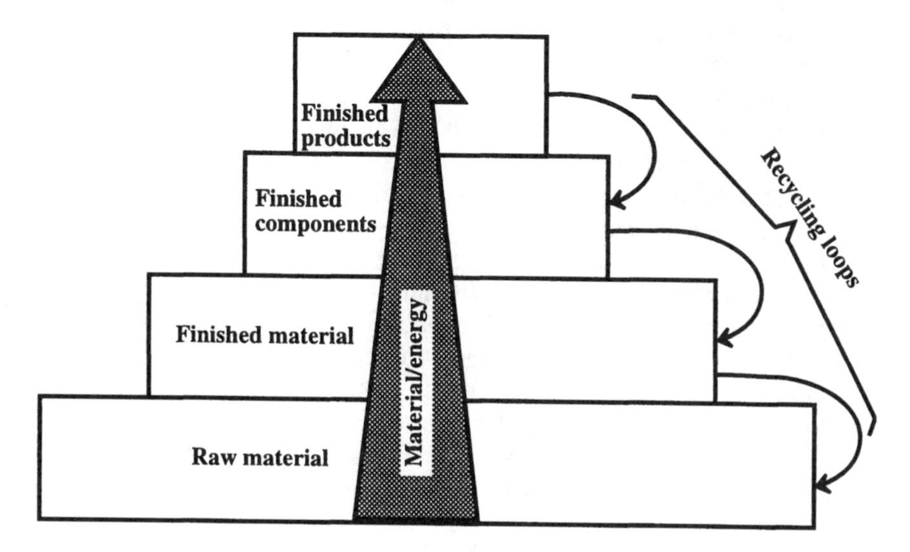

Figure 5.2. A material flow chain or energy/materials pyramid. Less energy and materials are involved when recycling is done near the end of the flow chain, thus retaining embedded utility.

To give a simple example, relatively little energy is required to return a glass beverage bottle from the consumer to the bottler, whereas returning the bottle to the glass manufacturer where it must be melted down and refabricated as a glass container obviously takes a greater amount of energy.

From a thermodynamic standpoint, a final product is relatively more ordered and it is certainly more usable for its intended purpose. The greater usability and lower energy requirements for recycle of products higher in the order of material flow are called **embedded utility**.[1] One of the major objectives of a system of industrial ecology, and, therefore, one of the main reasons for performing life-cycle assessments, is to retain the embedded utility in products by measures such as recycling as near to the end of the material flow as possible and replacing only those components of systems that are worn out or obsolete. An example of the latter occurred during the 1960s when efficient and safe turboprop engines were retrofitted to still serviceable commercial aircraft airframes to replace relatively complex piston engines, thus extending the lifetime of the aircraft by a decade or more.

5.2. LIFE-CYCLE ASSESSMENT

From the beginning, industrial ecology must consider process/product design in the management of materials, including the ultimate fates of materials when they are discarded. The product and materials in it should be subjected to an entire **life-cycle assessment** or analysis. A life-cycle assessment applies to products, processes, and services through their entire life cycles from extraction of raw materials, through manufacturing, distribution, and use to their final fates from the viewpoint of determining, quantifying, and ultimately minimizing their environmental impacts.[2] It takes account of manufacturing, distribution, use, recycling, and disposal. Life-cycle assessment is particularly useful in determining the relative environmental merits of alternative products and services. At the consumer level, this could consist of an evaluation of paper vs. styrofoam drinking cups. On an industrial scale, life-cycle assessment could involve evaluation of nuclear vs. fossil energy-based electrical power plants.

A basic step in life-cycle analysis is **inventory analysis** which provides qualitative and quantitative information regarding consumption of material and energy resources (at the beginning of the cycle) and releases to the anthrosphere, hydrosphere, geosphere, and atmosphere (during or at the end of the cycle). It is based upon various materials cycles and budgets and it quantifies materials and energy required as input and the benefits and liabilities posed by products. The related area of **impact analysis** provides information about the kind and degree of environmental impacts resulting from a complete life cycle of a product or activity. Once the environmental and resource impacts have been evaluated, it is possible to do an **improvement analysis** to determine measures that can be taken to reduce impacts on the environment or resources.

In making a life-cycle analysis the following must be considered:

- If there is a choice, selection of the kinds of materials that will minimize waste
- Kinds of materials that can be reused or recycled
- Components that can be recycled
- Alternate pathways for the manufacturing process or for various parts of it

Although a complete life-cycle analysis is expensive and time consuming, it can yield significant returns in lowering environmental impacts, conserving resources, and reducing costs. This is especially true if the analysis is performed at an early stage in the development of a product or service. Improved computerized techniques are making significant advances in the ease and efficacy of life-cycle analyses. Until now, life-cycle assessments have been largely confined to simple materials and products, such as reusable cloth vs. disposable paper diapers. A major challenge now is to expand these efforts to more complex products and systems, such as aircraft or electronics products.

Scoping in Life-Cycle Assessment

A crucial early step in life-cycle assessment is **scoping** the process by determining the boundaries of time, space, materials, processes, and products to be considered. Consider as an example the manufacture of parts that are rinsed with an organochloride solvent in which some solvent is lost by evaporation to the atmosphere, by staying on the parts, during the distillation and purification process by which the solvent is made suitable for recycling, and by disposal of waste solvent that cannot be repurified. The scope of the life-cycle assessment could be made very narrow by confining it to the process as it exists. An assessment could be made of the solvent losses, the impacts of these losses, and means for reducing the losses, such as reducing solvent emissions to the atmosphere by installation of activated carbon air filters or reducing losses during purification by employing more efficient distillation processes. A more broadly scoped life-cycle assessment would be to consider alternatives to the organochloride solvent. An even broader scope would consider whether the parts even need to be manufactured; are there alternatives to their use?

5.3. MATERIALS AND PRODUCT BUDGETS

Although some of the aspects of life-cycle analysis are uncertain and poorly developed, this is not necessarily true of materials and energy balances. Generally, industrial concerns have means of getting very accurate information on materials and energy flows, if for no other reason than the importance of these flows in saving costs and making the manufacturing process profitable. A typical materials balance might be conducted on a hydrocarbon solvent used as part of an organic chemicals synthesis as shown in Figure 5.3. In this case there is only one input of solvent to the system as a whole, the solvent from the supplier. There are several sources of solvent loss. “Fugitive” loss of solvent may occur in the mixing, synthesis reactor, solvent separator, and solvent purifier stages, such as by venting to relieve pressure. Some of the solvent may react to form other materials in the synthesis reactor. Some solvent may also be incorporated with the product and lost when the product leaves the system. Finally, a certain amount of solvent may be lost to disposal, such as solvent incorporated into distillation bottoms. If the fresh solvent from the supplier is denoted as makeup solvent, the material balance becomes

$$\begin{aligned}\text{Makeup solvent} = {} & \text{Solvent loss in processing} + \text{Solvent loss by reaction} + \\ & \text{Solvent loss with product} + \text{Solvent loss to disposal}\end{aligned} \tag{5.3.1}$$

In doing a materials balance on the solvent using the flow diagram shown in Figure 5.3 as a model, the rate of inflow of makeup solvent can be measured exactly, as can the flux of recycled solvent. These flows combined with a measure of the flow of chemically analyzed spent solvent will enable calculation of the sum of the solvent loss in processing from mixing, the synthesis reactor, and the solvent separator plus the solvent loss with product, but it will not give the individual flow of each. In principle, the solvent loss to disposal should be easy to quantify. It may be easy to establish that some of the solvent flows are negligible. In assessing the environmental impact of the process, the most important losses are likely to be those vented to the atmosphere. There may be some environmental impact from the solvent loss to disposal. In any case, it should be minimized for economic reasons because it represents the loss of a material with economic value, and its disposal is costly. Solvent loss with the product, if it occurs, may cause product quality problems.

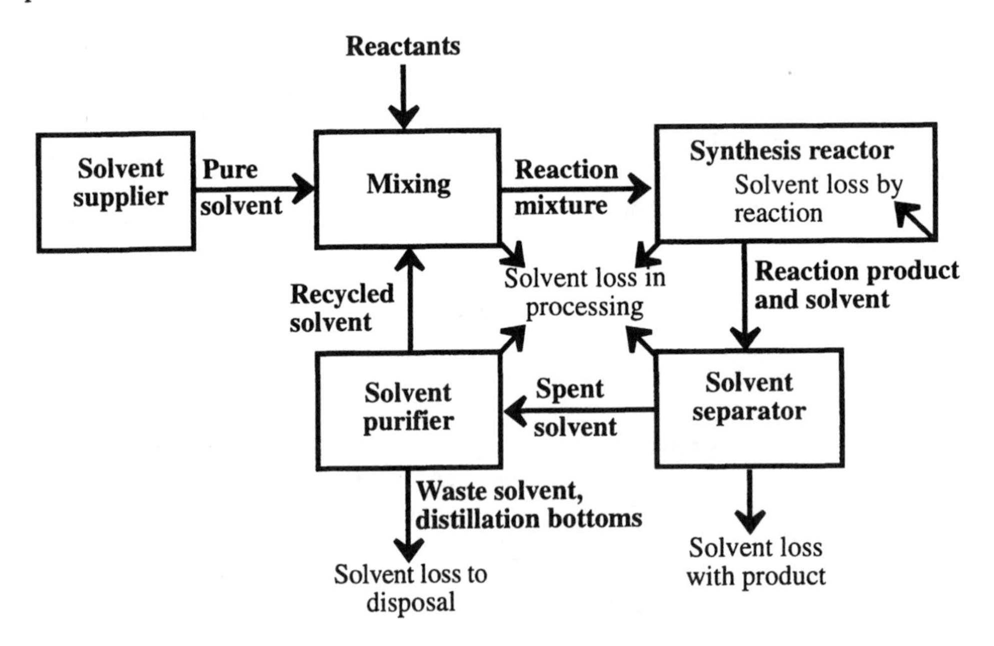

Figure 5.3. System for which a materials balance may be made for a solvent used as a reactant medium for a synthesis reaction.

Product budgets are based upon whole products rather than individual materials in the overall process in which products are manufactured. In reality, there are often too many materials to permit a complete product budget. When that is the case, it becomes desirable to choose one or several materials for doing the budget. Logical choices are the main material in the product or the material or materials most likely to cause environmental harm. Sometimes the main constituent material is also the most hazardous, as is the case of lead in lead storage battery manufacture.

In assessing the materials flow in a manufacturing process, it is useful to recognize four distinct kinds of flow. These are illustrated for two steps of a manufacturing process in Figure 5.4. Materials input (In_X) may occur at the beginning or at later stages of the process. Product flow (Pr_X) occurs between stages and as the final product leaving the system. Discard stream flow (Drd_X) may occur anywhere in the system, as may recycle flow (Rcl_X). In general, it is desirable to maximize product and recycle flow and minimize discard and input flow.

5.4. CONSUMABLE, RECYCLABLE, AND SERVICE (DURABLE) PRODUCTS

In industrial ecology, most treatments of life-cycle analysis make the distinction between **consumable products**, which are essentially used up and dispersed to the environment during their life cycle and **service or durable products**, which remain in essentially their original form after use. Gasoline is clearly a consumable product, whereas the automobile in which it is burned is a service product. It is useful, however, to define a third category of products that clearly become "worn out" when employed for their intended purpose, but which remain largely undispersed to the environment. The motor oil used in an automobile is such a substance in that most of the original material remains after use. Such a category of material may be called a **recyclable commodity**.

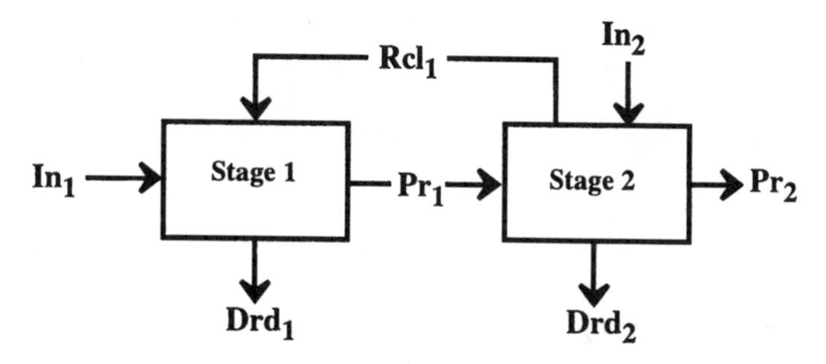

Figure 5.4. Two stages of a manufacturing process showing materials input (In_X), product flow (Pr_X), discard stream flow (Drd_X), and recycle flow (Rcl_X).

Desirable Characteristics of Consumables

Consumable products include laundry detergents, hand soaps, cosmetics, windshield washer fluids, fertilizers, pesticides, laser printer toners, and all other materials that are impossible to reclaim after they are used. The environmental implications of the use of consumables are many and profound. In the late 1960s and early 1970s, for example, nondegradable surfactants in detergents caused severe foaming and esthetic problems at water treatment plants and sewage outflows, and the phosphate builders in the detergents promoted excessive algal growth in receiving waters resulting in a condition known as eutrophication. Lead in consumable leaded gasoline was widely dispersed to the environment when the gasoline was burned. These problems have now been remedied with the adoption of phosphate-free detergents employing biodegradable surfactants and the mandatory use of unleaded gasoline.

Since they are destined to be dispersed to the environment, consumables should meet several "environmentally friendly" criteria, including the following:

- **Degradability**. This usually means biodegradability, such as that of household detergent constituents which occurs in waste treatment plants and in the environment. Chemical degradation may also occur.

- **Nonbioaccumulative.** Lipid-soluble, poorly biodegradable substances, such as DDT and PCBs tend to accumulate in organisms and to be magnified through the food chain. This characteristic should be avoided in consumable substances.
- **Nontoxic.** To the extent possible, consumables should not be toxic in the concentrations that organisms are likely to be exposed to them. In addition to their not being acutely toxic, consumables should not be mutagenic, carcinogenic, or teratogenic (cause birth defects).

Desirable Characteristics of Recyclables

The term recyclables is used here to describe a class of materials that are not used up in the sense that laundry detergents or photocopier toners are consumed, but are not durable items. In this context, recyclables can be understood to be chemical substances and formulations. The hydrochlorofluorocarbons (HCFCs) used as refrigerant fluids fall into this category, as does ethylene glycol mixed with water in automobile engine antifreeze/antiboil formulations (although rarely recycled in practice).

Insofar as possible, recyclables should be minimally hazardous with respect to toxicity, flammability, and other hazards. For example, both volatile hydrocarbon solvents and organochloride (chlorinated hydrocarbon) solvents are recyclable after use for parts degreasing and other applications requiring a good solvent for organic materials. The hydrocarbon solvents have relatively low toxicities, but may present flammability hazards during use and reclamation for recycling. The organochloride solvents are less flammable, but may present a greater toxicity hazard. An example of such a solvent is carbon tetrachloride, which is so nonflammable that it was once used in fire extinguishers, but whose current applications are highly constrained because of its high toxicity.

An obviously important characteristic of recyclables is that they should be designed and formulated to be amenable to recycling. In some cases there is little leeway in formulating potentially recyclable materials; motor oil, for example, must meet certain criteria regardless of its ultimate fate. In other cases formulations can be modified to enhance recyclability. For example, the use of bleachable or removable ink in newspapers could enhance the recyclability of the newsprint.

For some commodities, the potential for recycling is enormous. This can be exemplified by lubricating oils. The volume of motor oil sold in the U.S. each year for gasoline engines is about 2.5 billion liters, a figure that is doubled if all lubricating oils are considered.[3] A particularly important aspect of utilizing recyclables is their collection. In the case of motor oil, collection rates are low from consumers who change their own oil, and they are responsible for the dispersion of large amounts of waste oil to the environment.

Desirable Characteristics of Service Products

Since, in principle at least, service products are destined for recycling, they have comparatively lower constraints on materials and higher constraints on their ultimate disposal. A major impediment to the recycling of service products is the lack of convenient channels through which they can be put into the recycling loop. Television sets and major appliances, such as washing machines or ovens, have many recyclable components, but often end up in landfills and waste dumps simply because there is no handy means for getting them from the user and into the recycling loop. In such cases government intervention may be necessary to provide appropriate channels. One partial remedy to the disposal/recycling problem consists of leasing arrangements or payment of deposits on items such as batteries to ensure their return to a recycler. The term "de-shopping" has been used to describe a process by which service commodities would be returned to a location such as a parking lot where they could be collected for recycling.[4] According to this scenario, the analogy to a supermarket would be a facility in which service products are disassembled for recycling.

Much can be done in the design of service products to facilitate their recycle. One of the main characteristics of recyclable service products must be ease of disassembly so that remanufacturable components and recyclable materials, such as copper wire, can be readily removed and separated for recycling.

5.5. DESIGN FOR ENVIRONMENT

Design for environment is the term given to the approach of designing and engineering products, processes, and facilities in a manner that minimizes their adverse environmental impacts and, where possible, maximizes their beneficial environmental effects. In modern industrial operations, design for environment is part of a larger scheme termed "design for X," where "X" can be any one of a number of characteristics, such as assembly, manufacturability, reliability, and serviceability. In making such a design, numerous desired characteristics of the product must be considered, including ultimate use, properties, costs, and appearance. Design for environment requires that the designs of the product, the process by which it is made, and the facilities involved in making it conform to appropriate environmental goals and limitations imposed by the need to maintain environmental quality. It must also consider the ultimate fate of the product, particularly whether or not it can be recycled at the end of its normal life span.

Products, Processes, and Facilities

In discussing design for environment, the distinctions among products, processes, and facilities must be kept in clear perspective. **Products** — automobile tires, laundry detergents, and refrigerators — are items sold to consumers. **Processes** are the means of producing products and services. For example, tires are made by a process in which hydrocarbon monomers are polymerized to produce rubber molded in the shape of a tire with a carcass reinforced by synthetic fibers and steel wires. A **facility** is where processes are carried out to produce or deliver products or services. In cases where services are regarded as products, the distinction between products and processes becomes blurred. For example, a lawn care service delivers products in the forms of fertilizers, pesticides, and grass seeds, but also delivers pure services including mowing, edging, and sod aeration.

Although *products* tend to get the most public attention in consideration of environmental matters, *processes* often have more environmental impact. Successful process designs tend to stay in service for many years and to be used to make a wide range of products. While the product of a process may have minimal environmental impact, the process by which the product is made may have marked environmental effects. An example is the manufacture of paper. The environmental impact of paper as a product, even when improperly discarded, is not terribly great, whereas the process by which it is made involves harvesting wood from forests, high use of water, potential emission of a wide range of air pollutants, and other factors with profound environmental implications.

Processes develop symbiotic relationships when one provides a product of service utilized in another. An example of such a relationship is that between the process for the production of oxygen required in the basic oxygen process by which carbon and silicon impurities are oxidized from molten iron to produce steel. The long lifetimes and widespread applicability of popular processes make their design for environment of utmost importance.

The nature of a properly functioning system of industrial ecology is such that processes are even more interwoven than would otherwise be the case, because byproducts from some processes are used by other processes. Therefore, the processes employed in such a system and the interrelationships and interpendencies among them are particularly important. A major change in one process may have a "domino effect" on the others.

Key Factors in Design for Environment

Two key choices that must be made in design for environment are those involving materials and energy. The choices of materials in an automobile illustrate some of the possible tradeoffs. Steel as a component of automobile bodies requires relatively large amounts of energy and involves significant environmental disruption in the mining and processing of iron ore. Steel is a relatively heavy material, so more energy is involved in moving automobiles made of steel. However, steel is durable, has a high rate of recycling, and is produced initially from abundant sources of iron ore. Aluminum is much lighter than steel and quite durable. It has an excellent percentage of recycling. Good primary sources of aluminum, bauxite ore, are not as abundant as iron ores, and large amounts of energy are required in the primary production of aluminum. Plastics are another source of automotive components. The light weight of plastic reduces automotive fuel consumption, plastics with desired properties are readily made, and molding and shaping plastic parts is a straightforward process. However, plastic automobile components have a low rate of recycling.

Three related characteristics of a product that should be considered in design for environment are durability, repairability, and recyclability. **Durability** simply refers to how well the product lasts and resists breakdown in normal use. Some products are notable for their durability; ancient two-cylinder John Deere farm tractors from the 1930s and 1940s are legendary in farming circles for their durability, enhanced by the affection engendered in their owners who tend to preserve them. **Repairability** is a measure of how easy and inexpensive it is to repair a product. A product that can be repaired is less likely to be discarded when it ceases to function for some reason. **Recyclability** refers to the degree and ease with which a product or components of it may be recycled. An important aspect of recyclability is the ease with which a product can be disassembled into constituents consisting of a single material that can be recycled. It also considers whether the components are made of materials that can be recycled. Design for recycling is considered in more detail in Section 5.6.

Hazardous Materials in Design for Environment

A key consideration in the practice of design for environment is the reduction of the dispersal of hazardous materials and pollutants. This can entail the reduction or elimination of hazardous materials in manufacture, an example of which was the replacement of stratospheric ozone-depleting chlorofluorocarbons (CFCs) in foam blowing of plastics. If appropriate substitutes can be found, somewhat toxic and persistent chlorinated solvents should not be used in manufacturing applications, such as parts washing. The use of hazardous materials in the product, such as batteries containing toxic cadmium, mercury, and lead, should be eliminated or minimized. Pigments containing heavy metal cadmium or lead should not be used if there are any possible substitutes. The substitution of hydrochlorofluorocarbons and hydrofluorocarbons for ozone-depleting CFCs in products (refrigerators and air conditioners) is an example of a major reduction in environmentally damaging materials in products. The elimination of extremely persistent polychlorinated biphenyls (PCBs) from electrical transformers removed a major hazardous waste problem due to the use of a common product (although PCB spills and contamination from the misuse and disassembly of old transformers has remained a persistent problem even up to the present).

5.6. DESIGN FOR RECYCLING

The concept of **design for recycling**, meaning that products and components are planned and made with the objective of ultimately reusing them, has been summarized in a chapter of a book dealing with industrial ecology.[5] As an illustration of the importance of recycling, and one with which most readers can identify, the authors cite a 1991 Carnegie Mellon University research project that estimated that by the year 2005 as many as 150 million obsolete personal computers would be destined for landfills, consuming as much as 8 million cubic meters of landfill space with disposal costs of many $ millions. One can argue that the estimates of the numbers of computers, the amount of landfill space that they will consume, and the costs were all too high. However, it is certainly true that the astounding rate of progress in computer technology has made them one of the fastest products to become obsolete, making recycle of components, such as circuit boards, impractical. It is also true that the materials in computers are mixed with other materials making their reclamation impossible. Nevertheless, the quantities and costs of disposal are high *for just this one commodity*. If one considers other electronic devices, appliances, automobiles, and virtually any other item found on the shelves of discount mass marketers, the amounts of materials disposed and the costs of disposal point to the need for design of products for recycling if viable economic systems are to be maintained.

In devising a design for recycling, it is important to keep in mind the concept discussed in Section 5.1 and illustrated in Figure 5.2 that recycling should be done as far along the product flow as possible in order to facilitate recycling and minimize energy consumption. The greater usability and lower energy requirements for recycle of such products is a measure of their embedded utility.

Hierarchy of Recycling

There exists a hierarchy of desirability in recycling that should be considered in design for recycling. From the materials standpoint, the best alternative is **product lifetime extension** by repairing and maintaining the product so that it can be used for a very long time. An obvious limitation to this approach occurs when products become obsolete. Next in order of desirability is **subassembly recycle** in which groups of components comprising a subassembly are replaced with a renovated or new subassembly as is done with toner cartridges in laser printers. To a degree, subassembly replacement can be used to update products and thus extend their lifetimes. This has occurred, for example, with the installation of computerized navigational and control equipment in U.S. Air Force B-52 bombers, which were designed and built at a time when computer technology was still in its infancy. **Component recycle** can be practiced in which rebuilt components are installed in products. Rebuilt automobile parts are the prime example of this approach. At the lowest level is **material recycle** in which materials are removed from equipment, separated, and recycled to make completely new items. At all these levels proper design for recycle can facilitate recycling.

Important Considerations in Recycling

There are several key considerations in designing products and components for recycling. The main ones are listed below:

- **Simplicity**. The simpler a product, the easier it is to recycle components and materials. Therefore, the number of components in a product should be reduced to a minimum, making it easy to repair and to disassemble for component recycle. It is also desirable to minimize the number of kinds of materials.
- **Modularity**. The construction of products as modules greatly enhances their recyclability. This has been done very successfully in the modern electronics industry for the manufacture of items such as video equipment, computers, and stereo equipment which are constructed with circuit boards. Using computerized test equipment, a faulty board may be detected and simply replaced. Components that are most likely to need replacement or become obsolete should be designed as modules.
- **Repairability**. Products should be readily repaired, avoiding, if possible, the need for special skills, complex procedures, or special tools.

- **Minimizing kinds of materials**. Recycling is greatly facilitated if relatively fewer kinds of materials are used. Therefore, the kinds of materials in a product should be kept to a minimum.
- **Avoid bonding dissimilar materials**. Recycling becomes much more difficult when dissimilar materials are intimately associated. This is particularly true of composites or coatings, such as metal coatings on plastics.
- **Avoid toxics**. The presence of toxic substances in components can be a serious deterrent to recycling. Heavy metals, such as cadmium coatings on metal parts are a prime example.
- **Identification of toxics**. Where toxics must be used, they should be identified and easy to separate from the rest of the product.
- **Avoid plated metals**. Metals plated onto base metals are very difficult to separate and may introduce harmful impurities when the base metal is recycled.
- **Avoid coatings, fillers, and threaded metal inserts in plastics**. These kinds of materials can be a serious deterrent to recycling.

Closed Loop and Open Loop Recycling of Materials

A distinction made in recycling is whether a recycled material is used to make the same thing repeatedly, **closed loop recycling**, or is used to make a different product, a process called **open loop recycling**. Mixtures of these two kinds of approaches are common. Lead provides an example of closed loop recycling because the lead may be used repeatedly for the same purpose, as occurs with automobile storage batteries. The most commonly cited example of open loop recycling, or *cascade* recycling, is paper. Paper may start out as high quality bond paper, which is recycled to newsprint, then to lower grade applications such as grocery bags.

Impurities and Recycling

Impurities, especially toxic ones, may pose substantial barriers to recycling. An example is the presence of zinc in sewage sludge, which prevents the use of the sludge as fertilizer and soil conditioner. Lead used to facilitate the machining of brass can make this metal difficult to recycle. Toxic cadmium used to enable polymerization of plastics can hinder recycling of plastics and restrict the use of the recycled products. Impurity elements such as these that are detrimental to recycling are called "tramp" elements.

Remanufacturing

Remanufacturing, a process by which a product is run through an entire manufacturing process to make a newly renovated, fully functional product is a common means of recycling. Remanufacturing has two major functions. One of these is to replace or rehabilitate worn components. A second function is to replace obsolete components with ones of newer design that make the product work more effectively. Remanufacturing is feasible because product components do not entirely wear out at the same time. For example, a hydraulic elevator may start leaking oil from the cylinder that lifts it. The cylinder and the piston that works in it have to be replaced, whereas the rest of the unit remains fully functional. The feasibility of remanufacturing strongly depends upon the degree to which design for recyclability has been successful.

5.7. KINDS OF MATERIALS RECYCLED

Materials vary in their amenability to recycling. Arguably the most recyclable materials are metals in a relatively pure form. Such metals are readily melted and recast into other useful components. Among the least recyclable materials are mixed polymers or composites, the individual constituents of which cannot be readily separated. The chemistry of some polymers is such that, once they are prepared from monomers, they are not readily broken down again and reformed to a useful form. This section briefly addresses the kinds of materials that are recycled or that are candidates for recycling in a functional system of industrial ecology.

An important aspect of industrial ecology applied to recycling materials consists of the separation processes that are employed to "un-mix" materials for recycling at the end of a product cycle. An example of this is the separation of graphite carbon fibers from the epoxy resins used to bind them together in carbon fiber composites. The chemical industry provides many examples where separations are required. For example, the separation of toxic heavy metals from solutions or sludges can yield a valuable metal product, leaving nontoxic water and other materials for safe disposal or reuse.

Metals

Pure metals are easily recycled; usually the greatest challenge is to get the metals separated into a pure state. The recycling process commonly involves reduction of metal oxides to the metal. One of the more difficult problems with metals recycle is the mixing of metals, such as occurs with metal alloys, when a metal is plated onto another metal, or with components made of two or more metals in which it is hard to separate the metals. A common example of the latter is contamination of iron with copper from copper wiring or other components made from copper. As an impurity, copper produces steel with inferior mechanical characteristics. Another problem is the presence of toxic cadmium used as plating on steel parts.

Recycling metals can take advantage of the technology developed for the separation of metals that occur together in ores. Examples of byproduct metals recovered during the refining of other metals are gallium from aluminum, arsenic from lead or copper, precious metal iridium, osmium, palladium, rhodium, and ruthenium from platinum, and cadmium, germanium, indium, and thorium from zinc.

Plastics

Much attention has been given to recycling of plastics in recent years. Compared to metals, plastics are much less recyclable because recycling is technically difficult and plastics are less valuable. There are two general classes of plastics, a fact that has a strong influence upon their recyclability. Thermoplastics are those that become fluid when heated and solid when cooled. Since they can be heated and reformed multiple times, thermoplastics are generally amenable to recycling. Recyclable thermoplastics include polyalkenes (low-density and high-density polyethylene and polypropylene); polyvinylchloride (PVC), used in large quantities to produce pipe, house siding, and other durable materials; polyethylene terephthalate; and polystyrene. Plastic packaging materials are commonly made from thermoplastics and are potentially recyclable. Fortunately, from the viewpoint of recycling, thermoplastics make up most of the quantities of plastics used.

Thermosetting plastics are those that form molecular cross linkages between their polymeric units when they are heated. These bonds set the shape of the plastic, which does not melt when it is heated. Therefore, thermosetting plastics are not very amenable to recycling, and often burning them for their heat content is about the only use to which they may be put. An important class of thermosetting plastics consists of the epoxy resins, characterized by an oxygen atom bonded between adjacent carbons (1,2-epoxide or oxirane). Epoxies are widely used in composite materials combined with fibers of glass or graphite. Other thermosetting plastics include cross-linked phenolic polymers, some kinds of polyesters, and silicones. When recycling is considered, the best use for thermosetting plastics is for the fabrication of entire components that can be recycled.

Contaminants are an important consideration in recycling plastics. A typical kind of contaminant is paint used to color the plastic object. Adhesives and coatings of various kinds may also act as contaminants. Such materials may weaken the recycled material or decompose to produce gases when the plastic is heated for recycling. Toxic cadmium used to enable polymerization of plastics, a "tramp element" in recycling parlance, can hinder recycling of plastics and restrict the use of the recycled products.

Lubricating Oil

Lubricating oils are used in vast quantities and are prime candidates for recycling. The most simplistic means of recycling lubricating oil is to burn it, and large volumes of oil are burned for fuel. This is a very low level of recycling and will not be addressed further here.

For many years the main process for reclaiming waste lubricating oil was the acid-clay process that treated the material with sulfuric acid, then clay. This process generated large quantities of acid sludge and spent clay contaminated with oil, frequent contributors to hazardous waste sites. Current practice of lubricating oil reclamation use solvents, vacuum distillation, and catalytic hydrofinishing to produce a usable material from spent lubricating oil. The first step is dehydration to remove water and stripping to remove contaminant fuel (gasoline) fractions. If solvent treatment is used, the oil is then extracted with a solvent, such as isopropyl or butyl alcohols or methylethyl ketone, then centrifuged to remove impurities that are not soluble in the solvent. The solvent is then stripped from the oil. The next step is a vacuum distillation that removes a light fraction useful for fuel and a heavy residue. The lubricating oil can then be subjected to hydrofinishing over a catalyst to produce a suitable lubricating oil product.

5.8. EFFICIENT USE OF MATERIALS THROUGH INDUSTRIAL ECOLOGY

Strategies for reduced material use may be driven by either technology or by policy (regulation). There are four distinct approaches to reducing material consumption categorized as (1) dematerialization, (2) substitution, (3) recycling, and (4) waste mining.[6]

As the name implies, **dematerialization** is the use of smaller quantities of materials. The most striking example of technology-driven dematerialization has been achieved with solid-state circuitry in the electronics industry. Thus, for example, the circuit components required by the bulky vacuum tube-based radio of the 1940s have been reduced to a circuit so small that individual components cannot be seen at all with the naked eye. A prime example of policy-driven dematerialization is that of lighter automobiles required to meet federally mandated fuel standards in the U.S. during the 1970s. Other examples of dematerialization driven by technology and/or policy include substitution of long-lived steel-belted radial tires for bias-ply tires, thinner tin plating in "tin cans," thinner aluminum in aluminum cans, less silver in photographic film, and substitution of polymers and lighter metals, particularly aluminum, in place of heavier metals, such as lead or copper.

Technology-driven **material substitution** is exemplified by the substitution of glass optical fibers in place of copper wire for telecommunications. Another example is the use of noncorrodable fiber glass composite tanks in place of steel for underground storage of fuel. Federally mandated elimination of tetraethyllead to boost octane in gasoline is a policy-driven material substitution in which a nonrenewable resource that was dispersed to the environment (lead) has been displaced by a much more benign, and even renewable material (octane-boosting methyltertiarybutyl ether). Often reduced material consumption is driven by both technology and policy. This is the case with light-weight plastics now widely used for automobile parts. Policy has mandated use of lighter weight materials and the improved technology for plastics manufacture has made the substitution technologically possible. The substitution of polyvinyl chloride plastic pipe for copper water pipes and cast iron sewer lines; the use of plastic components in place of zinc castings for automobile parts; the replacement of pigments containing toxic cadmium, chromium, lead, and zinc by organic dyes; and the policy-mandated replacement in transformer fluids of highly persistent polychlorinated biphenyls (PCBs) by more environmentally friendly fluids are all examples of material substitution.

One crucial substance for which there is no possible substitute is phosphorus used in fertilizer. In the case of this limited resource, it is essential that it be used as efficiently as possible. This can be accomplished by precision agriculture in which quantities of fertilizers and other materials are carefully tailored to the needs of a particular agricultural production situation. Genetic engineering of crops that require minimum phosphorus and that use it as efficiently as possible also has a significant potential for reduced fertilizer phosphorus use.

The technology is available for **recycling** a wide range of materials, including essentially all metals. Fortunately, some of the more toxic metals, such as mercury and lead, are among the easiest to recycle and among the more costly, so that there is a financial incentive to recycle them. "Can-ban" ordinances and state laws that require charging refundable deposits for containers at the time of purchase have increased the policy-driven recycling of aluminum cans. Another example is that of mandates for recycled content in paper products. Many municipalities now have recycling programs for paper, glass, plastic, and aluminum in conjunction with their municipal refuse collection activities.

Recycling can be divided into three categories. **Reuse** of a material or object includes such things as using brick from a demolished building to construct decorative patios. Water is frequently reused, for example, treated sewage effluent can be discharged to wetlands that support waterfowl. **Repair** enables a product to have an extended lifetime. Automobiles are usually repaired many times during their lifetime. In some cases sophisticated design has made repair more difficult, and easier in others. High costs and difficulties of repair of modern gadgets have been blamed for the "throwaway society." The automobile industry provides the most common example of **remanufacture** in which components, such as entire engines are taken apart and some parts replaced with new ones, whereas others are remachined to give a product that may work as well and last as long as a new one. (Expensive office equipment, such as copy machines, are frequently remanufactured.) Repair and manufacture are especially well adapted to poorer, low-wage societies. As an example, Cuba has many 1950s-vintage American automobiles that have been kept running by the ingenuity of their owners and Cuban mechanics.

Waste mining is the term given to the extraction of useful materials from waste streams. Sulfur is one of the best examples of waste mining. Technology has been developed for the removal of sulfur from flue gas in nonferrous metal smelters. The recovery of sulfur from the smelting of lead, copper, zinc, and other metals also provides an example of policy-driven waste mining because it is mandated by law in most developed countries. Waste mining is employed to recover some metals as part of the production of other metals. Byproduct metals can be recovered from the gangue remaining from beneficiation of ore, slag from smelting, or dust collected from flues in metal refining operations. Arsenic and cadmium are recovered from the production of copper and cadmium, respectively. Coal ash is a huge untapped resource for waste mining of aluminum and ferrosilicon. Factoring in the costs of waste disposal and potential environmental degradation can make the economics of waste mining relatively more attractive. Some caution is suggested in that policy-driven waste mining of some substances creates a need to market them, sometimes to the detriment of the environment. Cadmium and arsenic are both examples of substances recovered from waste mining that should not be used any more than necessary because of their toxicities.

There are two major categories of scrap, such as scrap metal. **New scrap** consists of materials that are reclaimed during the manufacture of an item, such as metal shavings from machining. **Old scrap** consists of material that has been in products used in the consumer market and reclaimed as scrap material. The quality of new scrap can be carefully controlled and it can be reclaimed very quickly and efficiently. However, old scrap has a recycling time that depends upon the life of the product in which it is contained and it is difficult to control its quality. Furthermore, the percentage return of material is lower from old scrap because of the products that are discarded and not recycled.

The anthrosphere provides a large reservoir of materials that eventually can be recycled. A prime example of such a material is that of copper contained in copper wiring and plumbing in buildings, electrical lines, and other anthrospheric structures. The relatively high price of copper metal makes it attractive to reclaim from these sources. Another example is the large amount of iron contained in vehicles, machinery, rail lines, and other artifacts. Landfills constitute another anthrospheric construct that contains large amounts of materials, including metals. Unfortunately, the usable materials in landfills are usually too dispersed and mixed with other materials to be reclaimable.

Materials Price and Concentration

The economics of materials utilization are functions of its concentration from natural and waste sources and the economic value assigned to the material. A commodity that is relatively cheap in the marketplace normally comes from readily available, concentrated sources. Such is the case with elemental nitrogen, for example, which makes up about 79% of dry air and is relatively easy and inexpensive to obtain in the liquid form by liquefaction and distillation of the air. Platinum, on the other hand is so expensive that it can be economically extracted from very dilute sources.

Figure 5.5 illustrates the relationship between material price and the concentration of the source. Various commodities tend to be clustered along the line. Inexpensive liquid nitrogen, for example, would be located in the lower left portion of the plot, copper somewhat higher and to the right, and platinum toward the upper right of the plot.

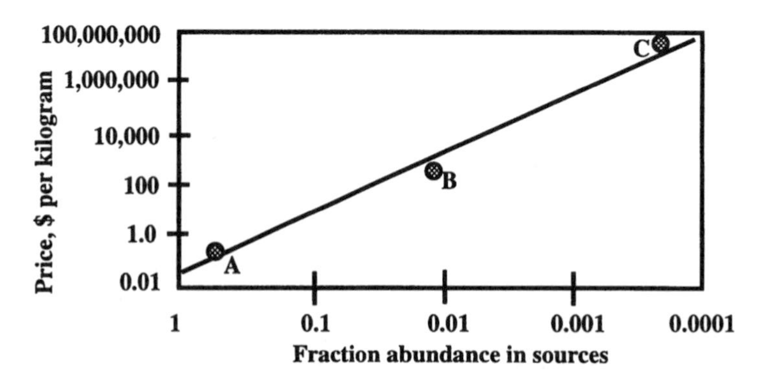

Figure 5.5. Generalized plot of material value as a function of fraction abundance in sources. Point A represents a relatively low value commodity, point B one of intermediate value, and point C one of very high value.

5.9. IMPROVEMENT ANALYSIS

As explained in Section 5.2, the first two stages of a life-cycle assessment are an inventory analysis and an impact analysis. The third stage, which is addressed in this section, is an **improvement analysis** consisting of a systematic approach applied to the entire product, process, or activity life cycle with the objective of minimizing the overall environmental burden by reducing materials and energy consumption and detrimental effects to the environment. It evaluates both needs and opportunities for reducing environmental impact, considering changes in design, processing, and management. It considers raw material use, material processing, manufacturing procedures, consumer use, and waste disposition, all with the objective of increasing positive effects on the environment.

In performing an improvement analysis, it is useful to divide the entire life cycle of the product into five stages.[7]

1. **Materials extraction and processing**. In this stage, materials are extracted from their sources and converted to needed finished materials or components.
2. **Manufacturing**. In this stage, materials may be further processed and the final product is assembled.
3. **Distribution**. This stage involves shipping and the packaging and unpackaging associated with moving the product to the consumer.
4. **Consumer use**. This stage is the one in which the product is used by the consumer.
5. **Product fate**. This final stage is the one in which the product is discarded or, at least in part, refurbished.

A properly performed improvement analysis considers opportunities for environmental improvement at each of these stages. The analysis considers products and various alternatives involving products.

An important choice in improvement analysis is that of materials, such as those going into manufacturing the product or packing it for shipping. When such materials are obtained by extractive industries, such as mining, the environmental impacts can be severe, but the opportunities for improvement are substantial. The use of recycled materials is definitely advantageous to the environment because they do not have to be extracted from the geosphere with the environmental disturbance that entails, they require less energy because of their embedded energy, they produce lower quantities of byproducts requiring disposal, and their use prevents (at least delays) filling landfills that their disposal would require. There are several potential disadvantages of the use of recycled materials including uncertain source of supply and contamination with impurities that might be detrimental to their use.

Energy use is another important issue. Important considerations include how much energy is required for extraction of a particular kind of raw material or the energy consumption involved in consumer use. For example, a refrigerator may take relatively more material and energy to manufacture, but its energy efficiency rating in use might be far superior.

Solid and liquid residues must be considered for each of the five stages. Obviously, any product that generates large volumes of such residues at any stage of its production, use, or ultimate fate is not environmentally desirable.

Potential gaseous and vapor emissions must also be taken into account. As an example, the manufacture of some kinds of lightweight foam plastics once used stratospheric ozone-depleting chlorofluorocarbons as "blowing agents" to impart porosity. This use could have been stopped by a proper improvement analysis.

Improvement Analysis of Processes

The discussion above has emphasized products; of equal, often greater importance are considerations applied to processes. One thing that makes improvement analysis especially important for processes is their relatively long life spans compared to some products. In many cases when a product is changed, often in a way that makes it more environmentally friendly, the process by which it is made remains locked in place, along with its environmentally damaging aspects.

An interesting example of the intricacies of processes insofar as environmental considerations are concerned is an unexpected side effect of the highly promoted just-in-time system for delivering components and assemblies for manufacturing. Although this system is clearly efficient and cost-effective, atmospheric emissions by the trucks used in making the deliveries contribute substantially to pollution in highly industrialized areas such as Tokyo. For the most part, much more enviromentally friendly delivery by train would normally occur in large batches and would not provide the flexibility and scheduling advantages offered by truck delivery.

Before a process can be put into operation, the equipment needed for it must be installed and made operational. This **process implementation** is largely a single step. Nevertheless it can have some definite environmental effects, such as disrupting terrain during construction.

Process operation has a very high potential for improvement analysis. Process operations handle materials. These materials are not confined to those that go into the product, although that would be the ideal, but include such auxiliary materials as solvents, bleaching agents, and sand used in castings. If at all possible the use of highly toxic and hazardous substances should be avoided and they should not be produced as part of the process. A material with particular environmental implications used in most processes is water. Much of the reduction of the detrimental environmental effects of processes in recent years has resulted from improvements in the use of water.

An important consideration in process operation is that of symbiotic and sequential processes. Symbiotic processes are those that produce byproducts needed by each other, which is the essence of a functional system of industrial ecology. Sequential processes are those in which processes work on materials in sequence, in facilities not usually owned or operated by the same firm. As an example, the corn products industry carries out complex, integrated processes involving materials such as corn starch and corn oil and a variety of processes including separations, enzyme processes, and fermentation. Before any of these operations can occur, however, the corn must be planted, tended, and harvested by farmers, not the agribusiness concerns processing the corn. And before corn is even grown, modern biotechnology companies produce hybrid seed corn genetically engineered to resist insects and herbicides that are applied to the cornfields to kill weeds. Each of these processes depends upon the other, and the whole sequence should be considered in an improvement analysis.

Although products usually come to mind in considering disposal and recycling, these considerations must be applied to processes as well. Therefore, the machines, structures, and other components of a process should be designed for reuse, recycling, refurbishment, and, if none of those is possible, disposal in an environmentally benign way. One development that is favorable in that respect is the increased use of computerized, programmable robots that can be adapted to a number of different tasks and until they become totally obsolete, can be used to manufacture a number of different products.

Improvement Analysis of Facilities

Somewhat apart from products and operations is the facility through which processes are carried out. Such a facility may be where a product is made, or it may be where a product is marketed, or a service provided. An automotive service center, for example, may market products, such as motor oil, but also perform service work on the automobile.

A fundamental decision regarding facilities is location. The location of a facility generally depends upon the use of the facility in one of the three following broad categories:

1. **Materials extraction/processing**. Often the location of a facility is determined by the source of the raw materials, leaving relatively lower flexibility in location.
2. **Manufacturing**. Two major determinants of facility location are availability of labor and considerations of transportation of materials to the facility and products to market from it.
3. **Service**. In general, service facilities are located near customers.

Obviously, facilities should be sited and constructed with environmental quality in mind. Ideally, industrial facilities should be located on sites that have been previously developed, not on prime agricultural land, for example. But with current environmental laws in the U.S., this sensible approach is often avoided because of potential liabilities from wastes left by previous owners. With a modern functional system of industrial ecology it has become more important to locate facilities close to others with which they may interact symbiotically.

With enlightened modern practice of industrial ecology, symbiotic relationships among industrial facilities have become important factors in selecting locations for facilities. For example, considering electricity generation alone, a coal-fired power plant might best be located at the site of a coal mine. However, when the opportunities for marketing steam from the plant and using municipal refuse as part of the fuel mix are considered, a location in a populous urban area may be preferable.

The availability of infrastructure in the form of water supplies, sewage systems, transportation systems, and similar elements of the infrastructure can be an important factor in facility location. These considerations usually favor previously developed areas with elements of the infrastructure in place. However, obsolescence and inadequacies of established infrastructures can favor undeveloped sites in which whole new integrated infrastructural systems can be constructed.

The ultimate fates of facilities have important environmental implications. One way that environmental impacts can be minimized is by extending the life of facilities, a process that begins with the initial design and siting of facilities. The importance of extending facility lifetimes is emphasized by the fact that construction materials constitute the greatest single use of materials in a modern industrialized society. The materials in facilities represent significant amounts of embedded energy. Existing facilities also represent embedded environmental value, because by reusing such facilities, most of the environmental disruption caused by building new facilities is avoided.

Facilities/Process/Product Interactions

Strong interactions exist among product characteristics, the nature of the processes by which products are made, and the facilities in which processes are carried out and products are manufactured, distributed, and disposed or recycled. The nature of a product largely determines the process by which it is made. For example, the choice of a plastic component for an automobile part requires a much different manufacturing process compared to an equivalent metal part. The facility in which wood is processed to make a product such as furniture may differ significantly from one in which plastics are molded and assembled to produce an equivalent item of furniture.

LITERATURE CITED

1. "An Introduction to Life-Cycle Assessment," Chapter 8 in *Industrial Ecology*, Thomas E. Graedel and Braden R. Allenby, Prentice Hall, Englewood Cliffs, NJ, 1995, pp. 108–122.

2. "Industrial Ecology Methods and Tools," Chapter 3 in *Discovering Industrial Ecology*, Ernest A Lowe, John L. Warren, and Stephen R. Moran, Battelle Press, Columbus, OH, 1997, pp. 37-74.

3. McCabe, Mark M. and William Newton, "Waste Oil," Section 4.1 in *Standard Handbook of Waste Treatment and Disposal*, 2nd ed., Harry M. Freeman, Ed., McGraw-Hill, New York, NY, 1998, pp. 4.3–4.13.

4. Braungart, Michael, "Product Life-Cycle Management to Replace Waste Management," Chapter 24 in *Industrial Ecology and Global Change*, Robert Socolow, Clinton Andrews, Frans Berkhout, and Valerie Thomas, Eds., Cambridge University Press, Cambridge, U.K., 1994, pp. 335–337.

5. "Design for Recycling," Chapter 19 in *Industrial Ecology*, Thomas E. Graedel and Braden R. Allenby, Prentice Hall, Englewood Cliffs, NJ, 1995, pp. 260-275.

6. "Introduction: Materials Perspective," Chapter 1 in *Industrial Ecology*, Robert U. Ayres and Leslie W. Ayres, Edward Elgar Publishing, Ltd., Cheltenham, U.K., 1996, pp. 1-17.

7. "The Improvement Analysis for Products, Processes, and Facilities," Chapter 20 in *Industrial Ecology*, Thomas E. Graedel and Braden R. Allenby, Prentice Hall, Englewood Cliffs, NJ, 1995, pp. 276-292.

SUPPLEMENTARY REFERENCES

Ayres, Robert U. and Leslie W. Ayres, *Industrial Ecology: Towards Closing the Materials Cycle*, Edward Elgar Publishers, Cheltenham, U.K., 1996.

Curran, Mary Ann, Ed., *Environmental Life-Cycle Assessment*, McGraw-Hill, New York, NY, 1997.

Davis John B., *Product Stewardship and the Coming Age of Takeback: What Your Company Can Learn from The Electronic Industry's Experience*, Cutter Information Corporation, Arlington, MA, 1996.

Fiksel, Joseph, Ed., *Design for Environment: Creating Eco-Efficient Products and Processes*, McGraw-Hill, New York, NY, 1996.

Frosch, Robert A. and Nicholas E. Gallopoulos, "Strategies for Manufacturing," *Scientific American*, **261**, 94–102 (1989).

Graedel, Thomas E. and B. R. Allenby, *Industrial Ecology*, Prentice Hall, Englewood Cliffs, NJ, 1995, pp. 204–230.

Graham, John D. and Jennifer K. Hartwell, *The Greening of Industry*, Harvard University Press, Cambridge, MA, 1997.

Leff, Enrique, *Green Production: Toward an Environmental Rationality*, Guilford Press, New York, NY, 1995.

Lowe, Ernest and John L. Warren, *The Source of Value: An Executive Briefing and Sourcebook on Industrial Ecology*, Battelle, Pacific Northwest National Laboratory, Richland, WA, 1997.

Nemerow, Nelson L., *Zero Pollution for Industry: Waste Minimization Through Industrial Complexes*, John Wiley & Sons, New York, NY, 1995.

Peck, Steven and Elain Hardy, *The Eco-Efficiency Resource Manual*, Fergus, Ontario, Canada, 1997.

von Weizsäcker, Ernst U., Amory B. Lovins, and L. Hunter Lovins, *Factor Four: Doubling Wealth, Halving Resource Use*, Earthscan, London, U.K., 1997.

6 INDUSTRIAL ECOLOGY AND RESOURCES

6.1. INTRODUCTION

Modern civilization depends upon a wide variety of resources consisting largely of minerals that are processed to recover the materials needed for industrial activities. The most common type of mineral material so used, and one that all people depend upon for their existence, is soil, used to grow plants for food. Also of crucial importance are metal ores. Some of these metal sources are common and abundant, such as iron ore; others, such as sources of chromium, are rare and will not last long at current rates of consumption. There are also some crucial sources of nonmetals. Sulfur, for example, is abundant and extracted in large quantities as a byproduct of sulfur-rich fuels. Phosphorus, a key fertilizer element, will last only for several generations at current rates of consumption.

The materials needed for modern societies can be provided from either **extractive** (nonrenewable) and **renewable** sources. Extractive industries remove irreplaceable mineral resources from the Earth's crust. The utilization of mineral resources is strongly tied with technology, energy, and the environment. Perturbations in one usually cause perturbations in the others. For example, reductions in automotive exhaust pollutant levels to reduce air pollution have made use of catalytic devices that require platinum-group metals, a valuable and irreplaceable natural resource. Furthermore, automotive pollution control devices result in greater gasoline consumption than would be the case if exhaust emissions were not a consideration (a particularly pronounced effect in the earlier years of emissions control). The availability of many metals depends upon the quantity of energy used and the amount of environmental damage tolerated in the extraction of low-grade ores. Many other such examples could be cited. Because of these intimate interrelationships, technology, resources, and energy must all be considered together. The practice of industrial ecology has a significant potential to improve environmental quality with reduced consumption of nonrenewable resources and energy.

In discussing nonrenewable sources of minerals and energy, it is useful to define two terms related to available quantities. The first of these is **resources**, defined as quantities that are estimated to be *ultimately* available. The second term is **reserves**, which refers to well-identified resources that can be profitably utilized with existing technology.

6.2. MINERALS IN THE GEOSPHERE

There are numerous kinds of mineral deposits that are used in various ways. These are, for the most part, sources of metals which occur in **batholiths** composed of masses of igneous rock that have been extruded in a solid or molten state into the surrounding rock strata. In addition to deposits formed directly from solidifying magma, associated deposits are produced by water interacting with magma. Hot aqueous solutions associated with magma can form rich **hydrothermal** deposits of minerals. Several important metals, including lead, zinc, and copper, are often associated with hydrothermal deposits.

Some useful mineral deposits are formed as **sedimentary deposits** along with the formation of sedimentary rocks. **Evaporites** are produced when seawater is evaporated. Common mineral evaporites are halite (NaCl), sodium carbonates, potassium chloride, gypsum ($CaSO_4 \cdot 2H_2O$), and magnesium salts. Many significant iron deposits consisting of hematite (Fe_2O_3) and magnetite (Fe_3O_4) were formed as sedimentary bands when earth's atmosphere was changed from reducing to oxidizing as photosynthetic organisms produced oxygen, precipitating the oxides from the oxidation of soluble Fe^{2+} ion.

Deposition of suspended rock solids by flowing water can cause segregation of the rocks according to differences in size and density. This can result in the formation of useful **placer** deposits that are enriched in desired minerals. Gravel, sand, and some other minerals, such as gold, often occur in placer deposits.

Some mineral deposits are formed by the enrichment of desired constituents when other fractions are weathered or leached away. The most common example of such a deposit is bauxite, Al_2O_3, remaining after silicates and other more soluble constituents have been dissolved by the weathering action of water under the severe conditions of hot tropical climates with very high levels of rainfall. This kind of material is called a **laterite.**

Evaluation of Mineral Resources

In order to make its extraction worthwhile, a mineral must be enriched at a particular location in Earth's crust relative to the average crustal abundance. Normally applied to metals, such an enriched deposit is called an **ore**. The value of an ore is expressed in terms of a **concentration factor**:

$$\text{Concentration factor} = \frac{\text{Concentration of material in ore}}{\text{Average crustal concentration}} \qquad (6.2.1)$$

Obviously, higher concentration factors are always desirable. Required concentration factors decrease with average crustal concentrations and with the value of the commodity extracted. A concentration factor of 4 might be adequate for iron, which makes up a relatively high percentage of Earth's crust. Concentration factors must be several hundred or even several thousand for less expensive metals that are not present at very high percentages in Earth's crust. However, for an extremely valuable metal, such as platinum, a relatively low concentration factor is acceptable because of the high financial return obtained from extracting the metal.

Acceptable concentration factors are a sensitive function of the price of a metal. Shifts in price can cause significant changes in which deposits are mined. If the price of a metal increases by, for example, 50%, and the increase appears to be long term, it becomes profitable to mine deposits that had not been mined previously. The opposite can happen, as is often the case when substitute materials are found, or newly discovered, richer sources go into production.

In addition to large variations in the concentration factors of various ores, there are extremes in the geographic distribution of mineral resources. The United States is perhaps about average for all nations in terms of its mineral resources, possessing significant resources of copper, lead, iron, gold, and molybdenum, but virtually without resources of some important strategic metals, including chromium, tin, and platinum-group metals. For its size and population, South Africa is particularly blessed with some important metal mineral resources.

6.3. EXTRACTION AND MINING

Minerals are usually extracted from Earth's crust by various kinds of mining procedures, but other techniques may be employed as well. The raw materials so obtained include inorganic compounds, such as phosphate rock; sources of metal, such as lead sulfide ore; clay used for firebrick; and structural materials, such as sand and gravel.

Surface mining, which can consist of digging large holes in the ground to remove copper ore, or strip mining, is used to extract minerals that occur near the surface. A common example of surface mining is quarrying of rock. Vast areas have been dug up to extract coal. Because of past mining practices, surface mining got a well-deserved bad name. With modern reclamation practices, however, topsoil is first removed and stored. After the mining is complete, the topsoil is spread on top of overburden that has been replaced such that the soil surface has gentle slopes and proper drainage. Topsoil spread over the top of the replaced spoil, often carefully terraced to prevent erosion, is seeded with indigenous grass and other plants, fertilized, and watered, if necessary, to provide vegetation. The end result of carefully done **mine reclamation** projects is a well-vegetated area suitable for wildlife habitat, recreation, forestry, grazing and other beneficial purposes.

Extraction of minerals from placer deposits formed by deposition from water has obvious environmental implications. Mining of placer deposits can be accomplished by dredging from a boom-equipped barge. Another means that can be used is hydraulic mining with large streams of water. One interesting approach for more coherent deposits is to cut the ore with intense water jets, then suck up the resulting small particles with a pumping system. These techniques have a high potential to pollute water and disrupt waterways.

For many minerals, underground mining is the only practical means of extraction. An underground mine can be very complex and sophisticated. The structure of the mine depends upon the nature of the deposit. It is of course necessary to have a shaft that reaches to the ore deposit. Horizontal tunnels extend out into the deposit, and provision must be made for sumps to remove water and for ventilation. Factors that must be considered in designing an underground mine include the depth, shape, and orientation of the ore body, as well as the nature and strength of the rock in and around it; thickness of overburden; and depth below the surface.

Usually, significant amounts of processing are required before a mined product is used or even moved from the mine site. Such processing, and the byproducts of it, can have significant environmental effects. Even rock to be used for aggregate and for road construction must be crushed and sized, a process that has the potential to emit air-polluting dust particles to the atmosphere. Crushing is also a necessary first step for further processing of ore. Some minerals occur to an extent of a few percent or even less in the rock taken from the mine and must be concentrated on site so that the residue does not have to be hauled far. For metals mining, these processes, as well as roasting, extraction, and similar operations are covered under the category of **extractive metallurgy**.

One of the more environmentally troublesome byproducts of mineral refining consists of waste **tailings**. By the nature of the mineral processing operations employed, tailings are usually finely divided and as a result subject to chemical weathering processes. Heavy metals associated with metal ores can be leached from tailings, producing water runoff contaminated with cadmium, lead, and other pollutants. Adding to the problem are some of the processes used to refine ore. Large quantities of cyanide solution are used in some processes to extract low levels of gold from ore, posing obvious toxicological hazards.

Environmental problems resulting from exploitation of extractive resources, including disturbance of land, air pollution from dust and smelter emissions, and water pollution from disrupted aquifers, are aggravated by the fact that the general trend in mining involves utilization of less rich ore. This is illustrated in Figure 6.1, showing the average percentage of copper in copper ore mined since 1900. The average percentage of copper in ore mined in 1900 was about 4%, but by 1982

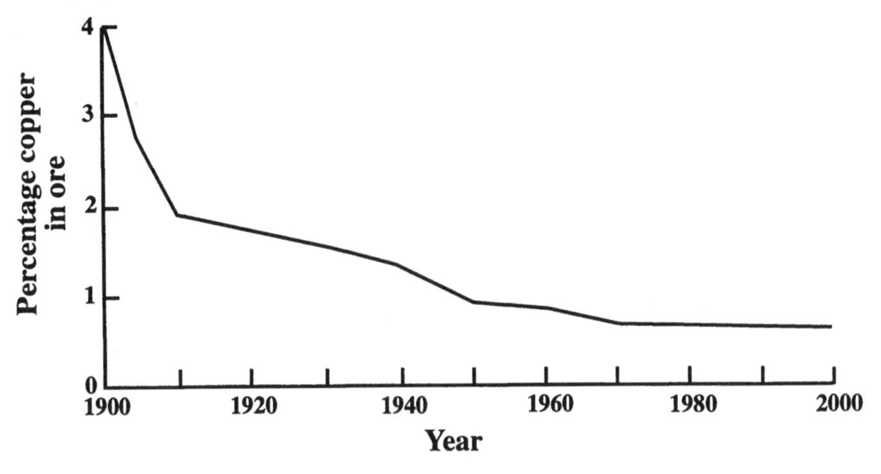

Figure 6.1. Average percentage of copper in ore that has been mined.

it was about 0.6% in domestic ore and 1.4% in richer foreign ore. Ore as low as 0.1% copper may eventually be processed. Increased demand for a particular metal, coupled with the necessity to utilize lower grade ore, has a vicious multiplying effect upon the amount of ore that must be mined and processed, and accompanying environmental consequences.

The proper practice of industrial ecology can be used to significantly reduce the effects of mining and mining byproducts. One way in which this can be done is to entirely eliminate the need for mining, utilizing alternate sources of materials. An example of such utilization, widely hypothesized, but not yet put into practice to a large extent is the extraction of aluminum from coal ash. This would have the double advantage of reducing amounts of waste ash and preventing the mining of scarce aluminum ore.

6.4. METALS

The majority of elements are metals, most of which are of crucial importance as resources. The availability and annual usage of metals vary widely with the kind of metal. Some metals are abundant and widely used in structural applications; iron and aluminum are prime examples. Other metals, especially those of the platinum group (platinum, palladium, iridium, rhodium) are very precious and their use is confined to applications such as catalysts, filaments, or electrodes for which only small quantities are required. Some metals are considered to be "crucial" because of their applications for which no substitutes are available and shortages or uneven distribution in supply that occur. Such a metal is chromium, used to manufacture stainless steel (especially for parts exposed to high temperatures and corrosive gases), jet aircraft, automobiles, hospital equipment, and mining equipment. The platinum-group metals are used as catalysts in the chemical industry, in petroleum refining, and in automobile exhaust antipollution devices.

Metals exhibit a wide variety of properties and uses. They come from a number of different compounds; in some cases two or more compounds are significant mineral sources of the same metal. Usually these compounds are oxides or sulfides. However, other kinds of compounds and, in the cases of gold and platinum-group metals, the elemental (native) metals themselves serve as metal ore. Table 6.1 lists the important metals, their properties, major uses, and sources.

6.5. METAL RESOURCES AND INDUSTRIAL ECOLOGY

Considerations of industrial ecology are very important in extending and efficiently utilizing metal resources.[1] More than any other kind of resource, metals lend themselves to recycling and to the practice of industrial ecology. This section briefly addresses the industrial ecology of metals.

Aluminum

Aluminum metal has a remarkably wide range of uses resulting from its properties of low density, high strength, ready workability, corrosion resistance, and high electrical conductivity. Unlike some metals, such as toxic cadmium or lead,

Table 6.1. Worldwide and Domestic Metal Resources

Metals	Properties[a]	Major uses	Ore, aspects of resources[b]
Aluminum	mp 660°C, bp 2467°C, sg 2.70, malleable, ductile	Metal products, including autos aircraft, electrical equipment. Conducts electricity better than copper per unit mass and is used in electrical transmission lines.	From bauxite ore containing 35-55% Al_2O_3. About 60 million metric tons of bauxite produced worldwide per year. U. S. resources of bauxite are 40 million metric tons, world resources about 15 billion metric tons.
Cadmium	Soft, ductile, silvery-white	Corrosion-resistant plating on steel and iron, alloys, bearings, pigments, rechargeable batteries.	Byproduct of zinc production, so annual production of cadmium parallels that of zinc. Abundant supply of toxic cadmium from this source has resulted in excessive dissipative uses.
Chromum	mp 1903°C, bp 2642°C, sg 7.14, hard, silvery color	Metal plating, stainless steel, wear-resistant and cutting tool alloys, chromium chemicals, including chromates.	From chromite, an oxide mineral containing Cr, Mg, Fe, Al. Resources of 1 billion metric tons in South Africa and Zimbabwe, large deposits in Russia, virtually none in the U.S.
Cobalt	mp 1495°C, bp 2880°C, sg 8.71, bright, silvery	Manufacture of hard, heat-resistant alloys, permanent magnet alloys, driers, pigments, animal feed additive.	From a variety of minerals, such as linnaeite, Co_3S_4, and as a byproduct of other metals. Abundant global and U.S. resources.

Table 6.1. (Cont.)

Copper	mp 1083°C, bp 2582°C, sg 8.96, ductile, maleable	Electrical conductors, alloys, chemicals. Many uses.	Occurs in low percentages as sulfides, oxides, and carbonates. U.S. consumption 1.5 million metric tons per year. World resources of 344 million metric tons, including 78 million in U.S.
Gold	mp 1063°C, bp 2660°C, sg 19.3	Jewelry, basis of currency, electronics, increasing industrial uses.	In various minerals at only around 10 ppm for ore currently processed in the U.S.; byproduct of copper refining. World resources of 1 billion oz., 80 million in U.S.
Iron	mp 1535°C, bp 2885°C, sg 7.86, silvery metal, in (rare) pure form	Most widely produced metal, usually as steel, a high-tensile-strength material containing 0.3-1.7% C. Made into many specialized alloys.	Occurs as hematite (Fe_2O_3), goethite ($Fe_2O_3 \cdot H_2O$), and magnetite (Fe_3O_4), abundant global and U.S. resources.
Lead	mp 327°C, bp 1750°C, sg 11.35, silvery color	Fifth most widely used metal, storage batteries, chemicals; uses in gasoline, pigments, and ammunition decreasing for environmental reasons.	Major source is galena, PbS. Worldwide consumption about 3.5 million metric tons, 1/3 in U.S. Global reserves about 140 million metric tons, 39 million metric tons U.S.
Manganese	mp 1244°C, bp 2040°C, sg 7.3, hard, brittle, gray-white	Sulfur and oxygen scavenger in steel, manufacture of alloys, dry cells, gasoline additive, chemicals	Found in several oxide minerals. About 20 million metric tons per year produced globally, 2 million consumed in U.S., no U.S. production, world reserves 6.5 billion metric tons.

Table 6.1. (Cont.)

Mercury	mp -38°C, bp 357°C, sg 13.6, shiny, liquid metal	Instruments, electronic apparatus electrodes, chemicals.	From cinnabar, HgS. Annual world production 11,500 metric tons, 1/3 used in U.S. World resources 275,000 metric tons, 6,600 U.S.
Molybdenum	mp 2620°C, bp 4825°C, sg 9.01, ductile, silvery-gray	Alloys, pigments, catalysts, chemicals, lubricants.	Molybdenite (MoS_2) and wulfenite ($PbMoO_4$) are major kinds or ore. About 2/3global Mo production in U.S., large global resources.
Nickel	mp 1455°C, bp 2835°C, sg 8.90, silvery color	Alloys, coins, storage batteries, catalysts (such as for hydrogenation of vegetable oil).	Found in ore associated with iron. U.S. consumes 150,000 metric tons per year, 10% from domestic production, large domestic reserves of low-grade ore.
Platinum-group[c]	Resist chemical attack, perform well at high temperatures, good electrical properties, catalytic properties	Jewelry, alloys, catalysts, electrodes, filaments	In alluvial deposits produced by weathering and gravity separation. Most resources in Russia, South Africa, and Canada.
Silver	mp 961°C, bp 2193°C, sg 10.5, shiny metal	Photographic film, electronics, sterling ware, jewelry, bearings, dentistry.	Found with sulfide minerals, a by-product of Cu, Pb, Zn. Annual U.S. consumption of 150 million troy ounces, short supply.

Table 6.1. (Cont.)

Tin	mp 232°C, bp 2687°C, sg 7.31	Coatings, solders, bearing alloys, bronze, chemicals, organometallic biocides.	Many forms associated with granitic rocks and chrysolites. Global consumption 190,000 metric tons/year, U.S. 60,000 metric tons/year, world resources 10 million metric tons.
Titanium	mp 1677°C, bp 3277°C, sg 4.5, silvery color	Strong, corrosion-resistant, used in aircraft, valves, pumps, paint pigments.	Commonly as TiO_2, ninth in elemental abundance, no shortages.
Tungsten	mp 3380°C, bp 5530°C, sg 19.3, gray	Very strong, high boiling point, used in alloys, tungsten carbide, drill bits, turbines, nuclear reactors.	Found as tungstates, such as scheelite ($CaWO_4$); U.S. has 7% world reserves, China 60%.
Vanadium	mp 1917°C, bp 3375°C, sg 5.87, gray	Used to make strong steel alloys.	In igneous rocks, primarily a byproduct of other metals. U.S. consumption of 5,000 metric tons per year equals production.
Zinc	mp 420°C, bp 907°C, sg 7.14, bluish-white	Widely used in alloys (brass), galvanized steel, paint pigments, chemicals. Fourth in world metal production.	Found in many ore minerals. World production is 5 million metric tons per year, U.S. consumes 1.5 million metric tons per year. World resources 235 million metric tons, 20% in U.S.

[a] Abbreviations: mp, melting point; bp, boiling point; sg, specific gravity.

[b] All figures are approximate; quantities of minerals considered available depend upon price, technology, recent discoveries, and other factors, so that quantities quoted are subject to fluctuation.

[c] The platinum-group metals consist of platinum, palladium, osmium, iridium, ruthenium, and osmium, all of which are very valuable.

the use and disposal of aluminum presents no environmental problems. Furthermore, it is about the most readily recycled of all metals.

The environmental problems associated with aluminum result from the mining and processing of aluminum ore. It occurs as a mineral called **bauxite**, which contains 40–60% alumina, Al_2O_3, associated with water molecules. Hydrated alumina is concentrated in bauxite, particularly in high-rainfall regions of the tropics, by the weathering away of more water-soluble constituents of soil. Bauxite ore is commonly strip mined from thin seams, so its mining causes significant disturbance to the geosphere. The commonly used Bayer process for aluminum refining dissolves alumina, shown below as the hydroxide $Al(OH)_3$, from bauxite at high temperatures with sodium hydroxide as sodium aluminate,

$$Al(OH)_3 + NaOH \rightarrow NaAlO_2 + 2H_2O \qquad (6.5.1)$$

leaving behind large quantities of caustic "red mud." This residue, which is rich in oxides of iron, silicon, and titanium, has virtually no uses and a high potential to produce pollution. Aluminum hydroxide is then precipitated in the pure form at lower temperatures and calcined at about 1200°C to produce pure anhydrous Al_2O_3. The anhydrous alumina is then electrolyzed in molten cyrolite, Na_3AlF_6, at carbon electrodes to produce aluminum metal.

All aspects of aluminum production from bauxite are energy intensive. Large amounts of heat energy are required to heat the bauxite treated with caustic to extract sodium aluminate, and heat is required to calcine the hydrated alumina before it can be electrolyzed. Very large amounts of electrical energy are required to reduce aluminum to the metal in the electrolytic process for aluminum production.

An interesting possibility that could avoid many of the environmental problems associated with aluminum production is the use of coal fly ash as a source of the metal. Fly ash is produced in large quantities as a byproduct of electricity generation, so it is essentially a free resource. As a raw material, coal fly ash is very attractive because it is anhydrous, thus avoiding the expense of removing water, it is finely divided, and it is homogeneous. Aluminum, along with iron, manganese, and titanium, can be extracted from coal fly ash with acid. If aluminum is extracted as the chloride salt, $AlCl_3$, it can be electrolyzed as the chloride by the ALCOA process. Although this process has not yet been proven to be competitive with the Bayer process, it may become so in the future.

Gallium is a metal that commonly occurs with aluminum ore and may be produced as a byproduct of aluminum manufacture. Gallium combined with arsenic or with indium and arsenic is useful in semiconducter applications, including integrated circuits, photoelectric devices, and lasers. Although important, these applications require only miniscule amounts of gallium compared to major metals.

Chromium

Chromium is of crucial importance because of its use in stainless steel and superalloys. These materials are of crucial importance in industrialized societies because of their applications in jet engines, nuclear power plants, chemical-resistant valves, and other applications in which a material that resists heat and chemical attack is required.

As shown in Table 6.1, supplies of chromium are poorly distributed around the Earth. It is important that chromium be handled according to good practices of industrial ecology. Several measures may be taken in this respect. Chromium is almost impossible to recover from chrome plated objects, and this use should be eliminated insofar as possible, as has been done with much of the decorative chrome plated adornments formerly put on automobiles. Chromium(VI) (chromate) is a toxic form of the metal and its uses should be eliminated wherever possible. The use of chromium in leather tanning and miscellaneous chemical applications should be curtailed. One important use of chromium is in the preparation of treated CCA lumber, which resists fungal decay and termites. The widespread use of this lumber has greatly extended the lives of wood products, which is in keeping with the practice of industrial ecology. However, its use of toxic arsenic, scarce copper, and even more scarce chromium are negatives, and alternative means of preserving lumber still need to be found.

Copper

Copper is a low-toxicity, corrosion-resistant metal widely used because of its workability (ductility and malleability), electrical conductivity, and ability to conduct heat. In addition to its use in electrical wire, where in some applications it is now challenged by aluminum, copper is also used in tubing, copper pipe, shims, gaskets, and other applications.

There are at least two major environmental problems associated with the extraction and refining of copper. The first of these is the dilute form in which copper ore now occurs (see Figure 6.1), such that in the U.S. 150–175 tons of inert material (not counting overburden removed in strip mining) must be processed and discarded to produce a ton of copper metal. The second problem is the occurrence of copper as the sulfide, so that in the production of copper large amounts of sulfur must be recovered as a byproduct or, unfortunately in some less developed countries, released to the atmosphere as pollutant SO_2.

An advantage to copper for recycling is that it is used primarily as the metal. Metallic copper represents "stored energy" in that copper from this source does not require energy for reduction to the metal. Recycling rates of scrap copper appear low in part because so much of the inventory of copper metal is tied up in long-lasting electrical wire, in structures, and other places where the lifetime of the metal is long. (This is in contrast to lead, where the main source of recycled metal is storage batteries, which last only 2–4 years.) An impediment to copper recycling is the difficulty of recovering copper components from circuits, plumbing, and other applications.

Cobalt

Cobalt is a "strategic" metal with very important applications in alloys, particularly in heat-resistant applications, such as jet engines. The major source of cobalt is as a byproduct of copper refining, although it can also be obtained as a byproduct of nickel and lead. As much as 50% of the cobalt in these sources is lost to tailings, slag, or other wastes, so there is a significant potential to improve the recovery of cobalt. Relatively low percentages of cobalt are recycled as scrap.

Lead

The industrial ecology of lead is very important because of the widespread use of this metal and its toxicity. Global fluxes of lead from the anthrosphere are shown in Figure 6.2.

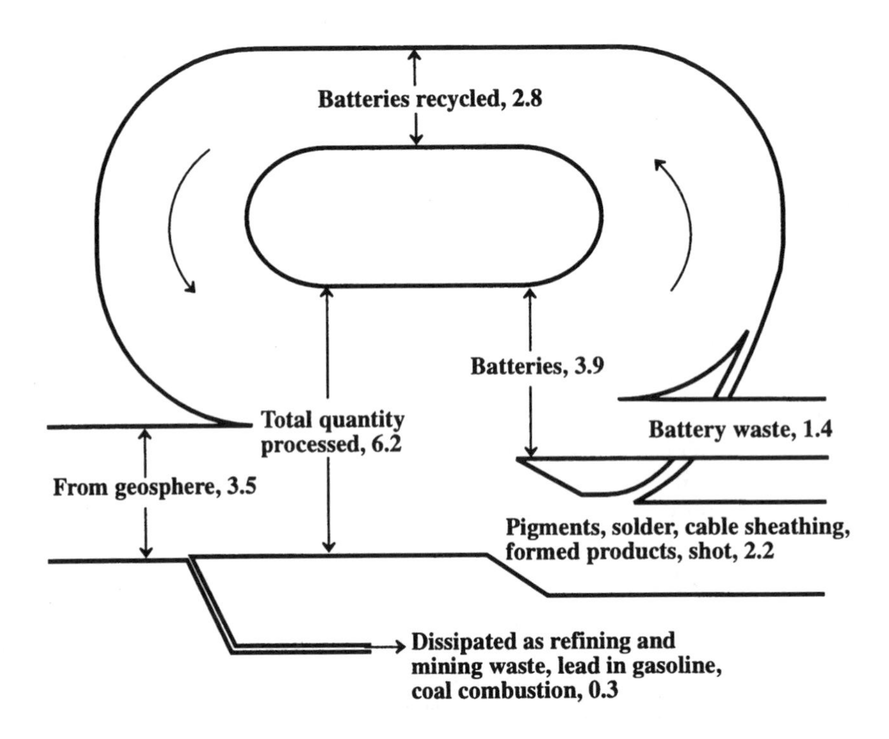

Figure 6.2. Flux of lead in the anthrosphere, globally, on an annual basis in millions of tons per year. Lead from the geosphere includes metal mined and a small quantity dissipated by coal combustion.

Somewhat more than half of the lead processed by humans comes from the geosphere, mostly as lead mined for the metal, and with a very small fraction contained in coal that is burned. By far the greatest use of lead is in batteries, and the amount of battery lead recycled each year approaches that taken from the geosphere. A small fraction of lead is dissipated as wastes associated with the mining and refining of the metal and as lead in gasoline, an amount that is decreasing as use of unleaded gasoline becomes prevalent around the world. A significant quantity of lead goes into various uses other than batteries, including pigments, solder, cable sheathing, formed products, and shot in ammunition. Only a small fraction of lead from these uses is recycled, and this represents a potential improvement in the conservation of lead. Another area in which improvements can be made is to eliminate or greatly reduce nonbattery uses of lead, as has been done in the case of lead shot and pigments. Although a large fraction of lead in batteries is recycled, about 1/3 of the lead used in batteries is lost; this represents another area of potential improvement in the utilization of lead.

Zinc

Zinc is relatively abundant and not particularly toxic, so its industrial ecology is of less concern than that of toxic lead or scarce chromium. As with other metals, the mining and processing of zinc can pose some environmental concerns. Zinc occurs as ZnS (a mineral called sphalerite), and the sulfur must be reclaimed in the smelting of zinc. Zinc minerals often contain significant fractions of lead and copper, as well as significant amounts of toxic arsenic and cadmium.

Zinc is widely used as the metal, and lesser amounts are used to make zinc chemicals. One of the larger uses for zinc is as a corrosion-resistant coating on steel. This application, refined to a high degree in the automotive industry in recent years, has significantly lengthened the life span of automotive bodies and frames. It is difficult to reclaim zinc from zinc plating. However, zinc is a volatile element and it can be recovered in baghouse dust from electric arc furnaces used to reprocess scrap steel.

Zinc is used along with copper to make the alloy called brass. Brass is very well adapted to the production of various parts and objectives. It is recyclable, and significant quantities of brass are recycled as wastes from casting, machining, and as postconsumer waste.

Although a number of zinc compounds are synthesized and used, by far the most important of them is zinc oxide, ZnO. Formerly widely used as a paint pigment, this white substance is now employed as an accelerating and activating agent for hardening rubber products, particularly tires. Tire wear is a major vector for the transfer of zinc to the environment (and, since it occurs with zinc) toxic cadmium is also dissipated to the environment by tire wear. The other two major compounds of zinc employed commercially are zinc chloride used in dry cells, as a disinfectant, and to vulcanize rubber, and zinc sulfide, used in zinc electroplating baths and to manufacture zinc-containing insecticides, particularly Zineb.®

Two aspects of zinc may be addressed in respect to its industrial ecology. The first of these is that, although it is not very toxic to animals, zinc is phytotoxic (toxic to plants) and soil can be "poisoned" by exposure to zinc from zinc smelting or from application of zinc-rich sewage sludge. The second of these is that the recycling of zinc is complicated by its dispersal as a plating on other metals. However, means do exist to reclaim significant fractions of such zinc, such as from electric arc furnaces mentioned above.

Potassium

Potassium deserves special mention as a metal because the potassium ion, K^+, is an essential element required for plant growth. It is mined as potassium minerals and applied to soil as plant fertilizer. Potassium minerals consist of potassium salts, generally KCl. Such salts are found as deposits in the ground or may be obtained from some brines. Very large deposits are found in Saskatchewan, Canada. These salts are all quite soluble in water.

6.6. HAZARDOUS WASTES AND METALS

Given the toxicities of many metal compounds, it is not surprising that the processing and smelting of metals produces hazardous waste materials. Some of these are from metals, particularly lead, noted for their toxicities. Others are from metals, such as iron, which are not notably toxic, but whose widespread production presents hazards. The listed wastes associated with metals production include the following, where the "K number" preceding the waste is a U.S. Environmental Protection Agency (EPA) hazardous waste number:

Iron and Steel

- K061, emission control dust/sludge from the primary production of steel in electric furnaces
- K062, spent pickle liquor generated by steel finishing operations of facilities within the iron and steel industry (SIC codes 331 and 332)

Primary Copper

- K064, acid plant blowdown slurry/sludge resulting from the thickening of blowdown slurry from primary copper production

Primary Lead

- K065, surface impoundment solids contained in and dredged from surface impoundments at primary lead smelting facilities

Primary Zinc

- K066, sludge from treatment of process wastewater and/or acid plant blowdown from primary zinc production

Primary Aluminum

- K088, spent potliners from primary aluminum reduction

Ferroalloys

- K090, emission control dust or sludge from ferrochromiumsilicon production
- K091, emission control dust or sludge from ferrochromium production

Secondary Lead

- K069, emission control dust/sludge from secondary lead smelting
- K100, waste leaching solution from acid leaching of emission control dust/sludge from secondary lead smelting

6.7. NONMETAL MINERAL RESOURCES

A number of minerals other than those used to produce metals are important resources. There are so many of these that it is impossible to discuss them all in this chapter; however, mention will be made of the major ones. As with metals, the environmental aspects of mining many of these minerals are quite important. Typically, even the extraction of ordinary rock and gravel can have important environmental effects.

Clays are secondary minerals formed by weathering processes on parent minerals. Clays have a variety of uses. About 70% of the clays used are miscellaneous clays of variable composition that have uses for a number of applications including filler (such as in paper), brick manufacture, tile manufacture, and portland cement production. Somewhat more than 10% of the clay used is fireclay, which has the characteristic of being able to withstand firing at high temperatures without warping. This clay is used to make a variety of refractories, pottery, sewer pipe, tile, and brick. Somewhat less than 10% of the clay that is used is kaolin, which has the general formula $Al_2(OH)_4Si_2O_5$. Kaolin is a white mineral that can be fired without losing shape or color. It is employed to make paper filler, refractories, pottery, dinnerware, and as a petroleum-cracking catalyst. About 7% of clay mined consists of bentonite and fuller's earth, a clay of variable composition used to make drilling muds, petroleum catalyst, carriers for pesticides, sealers, and clarifying oils. Very small quantities of a highly plastic clay called ball clay are used to make refractories, tile, and whiteware. U.S. production of clay is about 60 million metric tons per year, and global and domestic resources are abundant.

Fluorine compounds are widely used in industry. Large quantities of fluorspar, CaF_2, are required as a flux in steel manufacture. Synthetic and natural cryolite, Na_3AlF_6, is used as a solvent for aluminum oxide in the electrolytic preparation of aluminum metal. Sodium fluoride is added to water to help prevent tooth decay, a measure commonly called water fluoridation. World reserves of high-grade fluorspar are around 190 million metric tons, about 13% of which is in the United States. This is sufficient for several decades at projected rates of use. A great deal of byproduct fluorine is recovered from the processing of fluorapatite, $Ca_5(PO_4)_3F$, used as a source of phosphorus (see below).

Micas are complex aluminum silicate minerals, which are transparent, tough, flexible, and elastic. Muscovite, $K_2O{\cdot}3Al_2O_3{\cdot}6\ SiO_2{\cdot}2H_2O$, is a major type of mica. Better grades of mica are cut into sheets and used in electronic apparatus, capacitors, generators, transformers, and motors. Finely divided mica is widely used in roofing, paint, welding rods, and many other applications. Sheet mica is imported into the United States and finely divided "scrap" mica is recycled domestically. Shortages of this mineral are unlikely.

Pigments and **fillers** of various kinds are used in large quantities. The only naturally occurring pigments still in wide use are those containing iron. These minerals are colored by limonite, an amorphous brown-yellow compound with the formula $2Fe_2O_3 \cdot 3H_2O$, and hematite, composed of gray-black Fe_2O_3. Along with varying quantities of clay and manganese oxides, these compounds are found in ocher, sienna, and umber. Manufactured pigments include carbon black, titanium dioxide, and zinc pigments. About 1.5 million metric tons of carbon black, manufactured by the partial combustion of natural gas, are used in the U.S. each year, primarily as a reinforcing agent in tire rubber.

Over 7 million metric tons of minerals are used in the U.S. each year as fillers for paper, rubber, roofing, battery boxes, and many other products. Among the minerals used as fillers are carbon black, diatomite, barite, fuller's earth, kaolin (see clays, above), mica, limestone, pyrophyllite, and wollastonite ($CaSiO_3$).

Although **sand** and **gravel** are the cheapest of mineral commodities per ton, the average annual dollar value of these materials is greater than all but a few mineral products because of the huge quantities involved. In tonnage, sand and gravel production is by far the greatest of nonfuel minerals. Almost 1 billion tons of sand and gravel are employed in construction in the U.S. each year, largely to make concrete structures, road paving, and dams. Slightly more than that amount is used to manufacture portland cement and as construction fill. Although ordinary sand is predominantly silica, SiO_2, about 30 million tons of a more pure grade of silica are consumed in the U.S. each year to make glass, high-purity silica, silicon semiconductors, and abrasives.

At present, old river channels and glacial deposits are used as sources of sand and gravel. Many valuable deposits of sand and gravel are covered by construction and lost to development. Transportation and distance from source to use are especially crucial for this resource. Environmental problems involved with defacing land can be severe, although bodies of water used for fishing and other recreational activities frequently are formed by removal of sand and gravel.

6.8. PHOSPHATES

Phosphate minerals are of particular importance because of their essential use in the manufacture of fertilizers applied to land to increase crop productivity. In addition, phosphorus is used for supplementation of animal feeds, synthesis of detergent builders, and preparation of chemicals such as pesticides and medicines. The most common phosphate minerals are hydroxyapatite, $Ca_5(PO_4)_3(OH)$, and fluorapatite, $Ca_5(PO_4)_3F$. Ions of Na, Sr, Th, and U are found substituted for calcium in apatite minerals. Small amounts of PO_4^{3-} may be replaced by AsO_4^{3-} and the arsenic must be removed for food applications. Approximately 17% of world phosphate production is from igneous minerals, primarily fluorapatites. About three-fourths of world phosphate production is from sedimentary deposits, generally of marine origin. Vast deposits of phosphate, accounting for approximately 5% of world phosphate production, are derived from guano droppings of seabirds and bats. Current U.S. production of phosphate rock is around 40 million metric tons per year, most of it from Florida. Idaho, Montana, Utah, Wyoming, North Carolina, South Carolina, and Tennessee also have sources of phosphate.

Reserves of phosphate minerals in the United States amount to 10.5 billion metric tons, containing approximately 1.4 billion metric tons of phosphorus.

Phosphate in the naturally occurring minerals is not sufficiently available to be used as fertilizer. For commercial phosphate fertilizer production, these minerals are treated with phosphoric or sulfuric acids to produce more soluble superphosphates.

$$2Ca_5(PO_4)_3F(s) + 14H_3PO_4 + 10H_2O \rightarrow 2HF(g) + 10Ca(H_2PO_4)_2 \cdot H_2O \quad (6.8.1)$$

$$2Ca_5(PO_4)_3F(s) + 7H_2SO_4 + 3H_2O \rightarrow 2HF(g) + 3Ca(H_2PO_4)_2 \cdot H_2O + 7CaSO_4 \quad (6.8.2)$$

The HF produced as a byproduct of superphosphate production can create air pollution problems, and the recovery of fluorides is an important aspect of the industrial ecology of phosphate production.

Phosphate minerals are rich in trace elements required for plant growth, such as boron, copper, manganese, molybdenum, and zinc. Ironically, these elements are lost in processing phosphate for fertilizers and are sometimes added later for fertilizers.

Ammonium phosphates are excellent, highly soluble phosphate fertilizers. Liquid ammonium polyphosphate fertilizers consisting of ammonium salts of pyrophosphate, triphosphate, and small quantities of higher polymeric phosphate anions in aqueous solution are becoming more popular as phosphate fertilizers. The polyphosphates are believed to have the additional advantage of chelating iron and other micronutrient metal ions, thus making the metals more available to plants.

There are at least two major reasons that the industrial ecology of phosphorus is particularly important. The first of these is that current rates of phosphate use would exhaust known reserves of phosphate within two or three generations.[2] Although additional sources of phosphorus will be found and exploited, it is clear that this essential mineral is in distressingly short supply relative to human consumption and shortages, along with sharply higher prices, will eventually cause a crisis in food production. The second significant aspect of the industrial ecology of phosphorus is the pollution of waterways by waste phosphate, a plant and algal nutrient. This results in excessive growth of algae in the water, followed by decay of the plant biomass, consumption of dissolved oxygen, and an undesirable condition of eutrophication.

Excessive use of phosphate coupled with phosphate pollution suggests that phosphate wastes, such as from sewage treatment, should be substituted as sources of plant fertilizer. Several other partial solutions to the problem of phosphate shortages are the following:

- Development and implementation of methods of fertilizer application that maximize efficient utilization of phosphate.
- Genetic engineering of plants that have minimal phosphate requirements and that utilize phosphorus with maximum efficiency.

- Development of systems to maximize the utilization of phosphorus-rich animal wastes.

6.9. SULFUR

Sulfur is an important nonmetal; its greatest single use is in the manufacture of sulfuric acid. However, the element is employed in a wide variety of other industrial and agricultural products. Current consumption of sulfur amounts to approximately 10 million metric tons per year in the United States. The four most important sources of sulfur are (in decreasing order) deposits of elemental sulfur; H_2S recovered from sour natural gas; organic sulfur recovered from petroleum; and pyrite (FeS_2). Recovery of sulfur from coal used as a fuel is a huge potential, largely untapped source of this important nonmetal.

The resource situation for sulfur differs from that of phosphorus in several significant respects. Although sulfur is an essential nutrient like phosphorus, most soils contain sufficient amounts of nutrient sulfur, and the major uses of sulfur are in the industrial sector. The sources of sulfur are varied and abundant and supply is no problem either in the United States or worldwide; sulfur recovery from fossil fuels as a pollution control measure could even result in surpluses of this element.

About 90% of the use of sulfur in the world is for the manufacture of sulfuric acid. Almost 2/3 of the sulfuric acid consumed is used to make phosphate fertilizers as discussed in Section 6.7, in which case the phosphorus ends up as waste "phospho-gypsum," $CaSO_4 \cdot xH_2O$. Other uses of sulfur include lead storage batteries, steel pickling, petroleum refining, extraction of copper from copper ore, and the chemical industry.

The industrial ecology of sulfur needs to emphasize reduction of wastes and sulfur pollution, rather than supply of this element. Unlike many resources, such as most common metals, the uses of sulfur are for the most part dissipative, and the sulfur is "lost" to agricultural land, paper products, petroleum products, or other environmental sinks. There are two major environmental concerns with sulfur. One of these is the emission of sulfur to the atmosphere, which occurs largely as pollutant sulfur dioxide and is largely manifested by production of acidic precipitation and dry deposition. The second major environmental concern with sulfur is that it is used mostly as sulfuric acid and is not incorporated into products, thus posing the potential to pollute water and create acidic wastes. Acid purification units are available to remove significant amounts of sulfuric acid from waste acid solutions for recycling.[3]

Gypsum

Calcium sulfate in the form of the dihydrate $CaSO_4 \cdot 2H_2O$ is the mineral **gypsum**, one of the most common forms in which waste sulfur is produced. As noted, large quantities of this material are produced as a byproduct of phosphate fertilizer manufacture. Another major source of gypsum is its production when lime is used to remove sulfur dioxide from power plant stack gas,

$$Ca(OH)_2 + SO_2 \rightarrow CaSO_3 + H_2O \qquad (6.9.1)$$

to produce a calcium sulfite product that can be oxidized to calcium sulfate. About 100 million metric tons of gypsum are mined each year for a variety of uses, including production of Portland cement, to produce wallboard, as a soil conditioner to loosen tight clay soils, and numerous other applications.

Calcium sulfate from industrial or natural (gypsum) sources can be calcined at a very low temperature of only 159°C to produce $CaSO_4 \cdot {}^1/_2H_2O$, a material known as plaster of Paris, which was once commonly used for the manual application of plaster to walls. Plaster of Paris mixed with water forms a plastic material which sets up as the solid dihydrate,

$$CaSO_4 \cdot {}^1/_2H_2O + {}^3/_2H_2O \rightarrow CaSO_4 \cdot 2H_2O \qquad (6.9.2)$$

Cast into sheets coated with paper, this material produces plasterboard commonly used for the interior walls of homes and other buildings. Historically, plaster of Paris was used for mortar and other structural applications, and it has the potential for similar applications today.

The very large quantities of gypsum that are mined suggest that byproduct calcium sulfate, especially that produced with phosphate fertilizers and from flue gas desulfurization, should be a good candidate for reclamation through the practice of industrial ecology. The low temperature (see above) required to convert hydrated calcium sulfate to $CaSO_4 \cdot {}^1/_2H_2O$, which can be set up as a solid by mixing with water, suggest that the energy requirements for a gypsum-based byproducts industry should be modest. Low-density gypsum blown as a foam and used as a filler in composites along with sturdy reinforcing materials should have good insulating, fire-resistant, and structural properties for building construction.

6.10. SOIL RESOURCES

Soil (Figure 6.3) is the medium in which crops and other plants grow and is the essential support medium for the production of food required by most of Earth's organisms, including humans. Though clearly a part of the geosphere, soil interacts strongly with the other four environmental spheres. It is required to support the biosphere and a large fraction of Earth's biomass exists within soil in the form of plant roots, soil microorganisms, worms, and other life forms. A significant fraction of the volume of a healthy soil consists of air spaces. Soil takes up gases from the atmosphere, such as carbon dioxide fixed as biomass from plants and nitrogen fixed as organic nitrogen by *Rhizobium* bacteria growing synergistically with plant roots, and it is instrumental in releasing gases to the atmosphere, such as greenhouse-gas methane released by anaerobic bacteria growing in waterlogged rice paddies. The productivity of soil depends upon water associated with the hydrosphere, and soil releases solutes to the hydrosphere, such as inorganic nitrogen produced by the decay of nitrogeneous biomass and herbicides applied to soil for weed control. The interaction of soil with the anthrosphere is enormous, with intense cultivation by humans, chemical modification of soil by agricultural practices, and paving of large areas of soil by highways and urban developments.

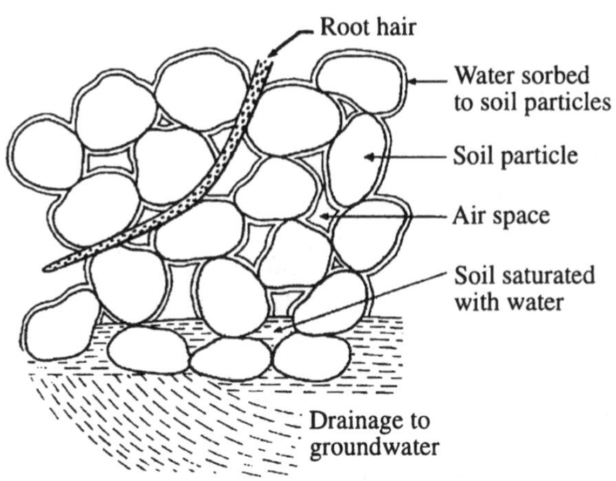

Figure 6.3. Soil is a complex material consisting of inorganic and organic solids, water held to various degrees by the soil solids, air spaces, and living matter.

Soil is an essential resource and a vulnerable environmental system.[4] Two broad categories of threats to soil are **soil degradation** and **soil pollution**.

Soil degradation is largely associated with human misuse of soil in extracting resources from it. These practices include overgrazing by animals, deforestation to produce more agricultural land, and cultivation practices that lead to erosion and loss of essential nutrients from soil. Improper irrigation and cultivation practices can cause soil salinization in which levels of salt become so high that the soil will no longer support plant growth. In arid regions soil salinization, overgrazing, and other insults can result in the soil becoming barren and unable to support crops or grazing, a condition called **desertification.**

Despite a great deal of industrial activity leading to the potential release of pollutants to soil, soil pollution is relatively less important than soil degradation. Soil pollution overlaps soil degradation in that a soil that is too polluted to grow crops is obviously degraded. There are numerous ways in which soil can be polluted by human activities. Examples include acid and heavy metal deposition from the smelting of nonferrous metal sulfides, contamination by saltwater from petroleum production, acidification from acid precipitation, and contamination by pesticides that have been applied improperly or at excessive levels.

Soil and Industrial Ecology

The proper practice of industrial ecology must consider soil as part of industrial ecosystems as a whole. Some industrial byproducts can be applied to soil to enhance its productivity. Examples include ammonium sulfate fertilizer as a source of nitrogen and sulfur, waste calcium sulfate from power plant flue gas desulfurization applied as a soil conditioner, and sewage sludge applied as a source of fertilizer nitrogen and phosphorus. Soil in turn is a source of products and byproducts that can be used in manufacturing. The prime example is wood used for construction and to make paper. Huge amounts of agricultural byproducts, such as straw, are produced each year. Although a certain fraction of such materials must be returned to the soil to keep it in proper condition, the potential exists to utilize significant amounts in applications such as pressed board for structural purposes. At a lower level of utilization, excess biomass grown on soil can be used as a fuel to generate electricity.

6.11. WATER RESOURCES

In parts of the world, water is a resource that is in short supply; in many cases, critically so. Although the United States has abundant water resources in comparison to much of the world, it is a voracious user of water, which is resulting in some acute water resources problems. In the continental United States, an average of approximately 1.48×10^{13} liters of water fall as precipitation each day, which translates to 76 cm per year. Of that amount, approximately 1.02×10^{13} liters per day, or 53 cm per year, are lost by evaporation and transpiration. Thus, the water theoretically available for use is approximately 4.6×10^{12} liters per day, or only 23 centimeters per year. At present, the U.S. uses 1.6×10^{12} liters per day, or 8 centimeters of the average annual precipitation. This amounts to an almost 10-fold increase from a usage of 1.66×10^{11} liters per day around the year 1900. Even more striking is the per capita increase from about 40 liters per day in 1900 to around 600 liters per day now. Much of this increase is accounted for by high agricultural and industrial use, each of which accounts for somewhat more than 40% of total consumption. Municipal and miscellaneous industrial uses account for the remaining water use.

Water use in the U.S. has shown an encouraging trend in recent years with total consumption down by about 9% during a time in which population grew 16% according to figures compiled by the U.S. Geological Survey.[5] This trend, which is illustrated in Figure 6.4, has been attributed to the success of efforts to conserve water, especially in the industrial (including power generation) and agricultural sectors. Conservation and recycling have accounted for much of the decreased use in the industrial sector. Irrigation water has been used much more efficiently by replacing spray irrigators, which lose large quantities of water to the action of wind and to evaporation, with irrigation systems that apply water directly to soil. Trickle irrigation systems that apply just the amount of water needed directly to plant roots are especially efficient.

A major problem with water supply is its nonuniform distribution with location and time; precipitation falls unevenly in the continental U.S. This causes difficulties because people in areas with low precipitation often consume more water than people in regions with more rainfall. Rapid population growth in the more arid southwestern states of the U.S. since World War II has further aggravated the problem. Water shortages are becoming more acute in the southwestern U.S., which contains six of the nation's eleven largest cities (Los Angeles, Houston, Dallas, San Diego, Phoenix, and San Antonio). Other problem areas include Florida, where overdevelopment of coastal areas threatens Lake Okeechobee; the Northeast, plagued by deteriorating water systems; and the High Plains, ranging from the Texas panhandle to Nebraska, where irrigation demands on the Ogallala aquifer are dropping the water table steadily, with no hope of recharge. These problems are minor, however, in comparison to those in some parts of Africa where water shortages are contributing to real famine conditions.

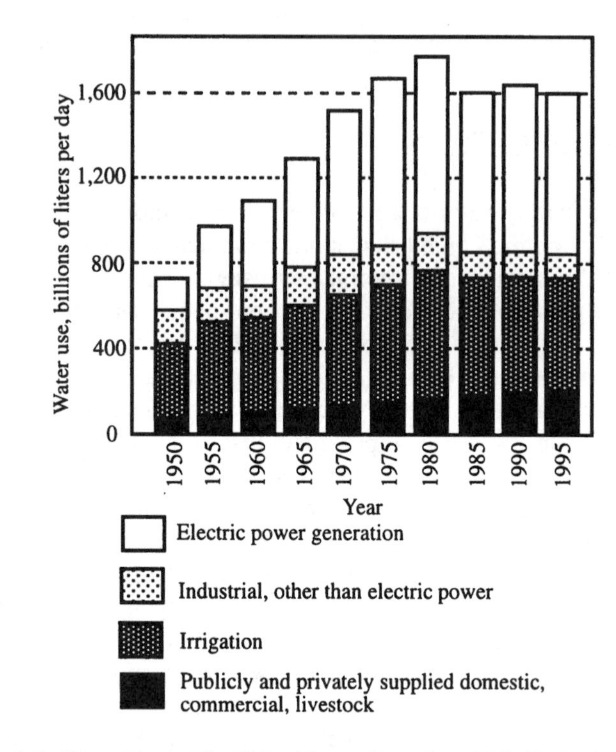

Figure 6.4. Trends in Water Use in The United States (Data from U.S. Geological Survey).

Ultrapure Water

Modern industrial processes have created a market for specialty materials and chemicals. This demand has been particularly notable in the electronics industries where ultrapure chemicals are an absolute requirement. One of the most widely used "electronic chemicals" is ultrapure water, which can tolerate impurities only in the parts-per-trillion or even only parts-per-quadrillion range, with a dollar value of about $500 million per year worldwide. The manufacture of a typical 20-cm silicon wafer requires almost 1500 liters of ultrapure water!

Industrial Ecology and Water Resources

Fortunately, water resources are highly amenable to the practice of industrial ecology. With water more than any other resource, industrial ecology (though not so called for the most part) has been applied for decades. Nature provides the ultimate water recycling mechanism through the hydrologic cycle in which water is evaporated by solar heat from the oceans, carried inland through the atmosphere, and falls as precipitation utilized by humans. Sequential uses of water in industrial and municipal systems occur through, as examples, use of industrial process water as cooling water or sewage effluent for irrigation. Careful planning for water use and efficient technologies for water purification and recycle can enable complete utilization of water resources through a system of industrial ecology.

6.12. EXTENDING RESOURCES WITH INDUSTRIAL ECOLOGY

Increasing populations and demands for higher living standards continue to put pressure on mineral and other resources. The emergence of newly developing economies, particularly those in the highly populous countries of China and India, will continue to put pressures on resource availability.

To a degree, the economic demand for and price of a resource determine its availability. Higher prices lead to greater exploration, exploitation of less available resources, and often spectacular increases in supply. This phenomenon has led to misinterpretation of the resource supply-and-demand equation by authorities whose understanding of economics is limited to conventional monetary supply-and-demand models, without due consideration of environmental aspects. The fact is that the total available amounts of most resources are limited, painfully so in terms of the time span over which they will be needed by humankind. Although higher prices and improved technologies can significantly increase supplies of critical resources, the ultimate result will be the same — the resource will run out. Furthermore, exploitation of lower and lower grades of resources results in ever-increasing environmental disruption, adding significantly to the cost, considering environmental economics.

Mineral resources may be divided into several categories based upon current production and consumption and known reserves. In the first category are those that are in relatively comfortable supply, with supply of at least 100 to several hundred years. Minerals in this category include bauxite, the source of aluminum, iron ore, platinum-group metals, and potassium salts. In an intermediate category are minerals with a current projected lifetime supply of 25-100 years. These include chromium, cobalt, copper, manganese, nickel, gypsum, phosphate minerals, and sulfur. The most critical group consists of minerals for which the supply, based upon current rates of consumption and known reserves, is 25 years or less. Among these minerals are sources of lead, tin, zinc, gold, silver, and mercury.

The world economy will never totally run out of any of the minerals listed above. However, severely constrained supplies of any one or several of them will have some marked effects. For example, world food production now depends on fertilizers, which require phosphorus, of which resources are limited. Within the next century a food crisis related to phosphate shortages may be anticipated.

Modern technology and human ingenuity are very effective in alleviating shortages of important minerals. Applications of materials science continue to produce substances made from readily available materials that provide good substitutes for more scarce resources. For example, concrete covered by strong layers of composite materials can readily substitute for iron in construction. Ceramics with special heat- and abrasion-resistant qualities are being used where high-temperature alloys were formerly required.

Modern technology enables exploitation of lower grade ore, thus significally increasing supplies. A striking example of this phenomenon has been provided by copper. During the 1800s, the average copper content of ore mined in the United States reached levels around 5%; now it is only about 1/10 that figure for copper ore mined globally. Despite the decline in copper ore quality, during the last 50 years known copper reserves have increased about 5-fold. Perhaps even more surprising is the fact that after inflation is taken into account, the price of copper is now less than it was a century ago.

The ability to exploit much less rich sources of ore has resulted from improved technologies. Of particular importance have been advances in the means of moving huge quantities of rock, essential for the exploitation of lower grade ore. Earth-moving equipment has greatly increased in size and versatility during the last several decades. There has been an environmental cost, of course, for these advances. As an approximation, for each 10-fold decrease in mineral content in an ore, it is necessary to move 10 times as much material to obtain the same amount of metal. The amount of waste byproduct produced increases proportionately. In addition to disruption of land, disturbed material is more prone to erosion, landslides, and water pollution. Much more energy is required, as is more water for those mining operations that use large quantities of water.

Another potentially exciting source of minerals is on ocean floors, which remain today largely unexplored. Here, relatively new technologies, such as remote-controlled submarines capable of withstanding crushing pressure, have opened new possibilities for exploration and resource utilization. Large areas of the ocean are covered with manganese-rich lumps called **manganese nodules**. Obvious sources of manganese, these lumps also contain other metals, including valuable platinum, copper, and nickel. Extraction of these metals as byproducts adds to the economic attractiveness of mining manganese nodules.

The least efficient use of materials occurs when an industry acquires all of its raw materials from the outside and discards all of its waste materials. There is a balance between materials efficiency and energy efficiency. In general, more energy is required to use materials efficiently, so a tradeoff is involved between material recovery and energy use.

The practice of industrial ecology, which ideally requires no raw materials and produces no waste, automatically conserves resources. If industrial ecology could become the norm throughout the world, many of the problems of resources discussed above would disappear. Partial implementation of the practices of industrial ecology, particularly in those industrial sectors that use scarce resources, can substantially extend resources of materials for which shortages exist or are now projected.

LITERATURE CITED

1. Chapters 3–6 in *Industrial Ecology: Towards Closing the Materials Cycle*, Robert U. Ayres and Leslie W. Ayres, Edward Elgar, Cheltenham, U.K., 1996, pp. 32–96.

2. "Phosphorus, Fluorine and Gypsum," Chapter 8 in *Industrial Ecology: Towards Closing the Materials Cycle*, Robert U. Ayres and Leslie W. Ayres, Edward Elgar, Cheltenham, U.K., 1996, pp. 116–130.

3. Freeman, Harry M., Ed., *Standard Handbook of Hazardous Waste Treatment and Disposal*, 2nd edition, McGraw-Hill, New York, NY, 1998, pp. 6.67–6.68.

4. Schnoor, Jerald and Valerie Thomas, "Soil as a Vulnerable Environmental System, Chapter 16 in *Industrial Ecology and Global Change*, Robert Socolow, Clinton Andrews, Frans Berkhout, and Valerie Thomas, Eds., Cambridge University Press, New York, NY, pp. 233–244.

5. Stevens, Willliam K., "Expectation Aside, Water Use in U.S. is Showing Decline," *New York Times*, November 10, 1998, p. 1.

SUPPLEMENTARY REFERENCES

Ayres, Robert U. and Leslie W. Ayres, *Industrial Ecology: Towards Closing the Materials Cycle*, Edward Elgar Publishers, Cheltenahm, U.K., 1996.

Fiksel, Joseph, Ed., *Design for Environment: Creating Eco-Efficient Products and Processes*, McGraw-Hill, New York, NY, 1996.

Peck, Steven and Elain Hardy, *The Eco-Efficiency Resource Manual*, Fergus, Ontario, Canada, 1997.

von Weizsäcker, Ernst U., Amory B. Lovins, and L. Hunter Lovins, *Factor Four: Doubling Wealth, Halving Resource Use*, Earthscan, London, U.K., 1997.

7 ENVIRONMENTAL CHEMISTRY AND INDUSTRIAL ECOLOGY

7.1. ENVIRONMENTAL CHEMISTRY

Environmental chemistry is *the study of the sources, reactions, transport, effects, and fates of chemical species in the water, air, terrestrial, and living environments and the effects of human activities thereon.*[1] Some idea of the complexity of environmental chemistry as a discipline may be realized by examining Figure 7.1, which shows the interchange of chemical species among various environmental spheres. Throughout an environmental system there are variations in temperature, mixing, intensity of solar radiation, input of materials, and various other factors that strongly influence chemical conditions and behavior. Because of its complexity, environmental chemistry must be approached with simplified models. This chapter presents an overview of environmental chemistry and how it relates to industrial ecology. The discussion of environmental chemistry in this chapter is organized into the four major categories of aquatic chemistry, atmospheric chemistry, geospheric chemistry, and the chemistry of life (toxicological chemistry).

Potentially, environmental chemistry and industrial ecology have many strong connections. The design of an integrated system of industrial ecology must consider the principles and processes of environmental chemistry. Environmental chemistry must be considered in the extraction of materials from the geosphere and other environmental spheres to provide the materials required by industrial systems in a manner consistent with minimum environmental impact. The facilities and processes of an industrial ecology system can be sited and operated for minimal adverse environmental impact if environmental chemistry is considered in their planning and operation. Environmental chemistry clearly points the way to minimize the environmental impacts of the emissions and byproducts of industrial systems, and is very helpful in reaching the ultimate goal of a system of industrial ecology, which is to reduce these emissions and byproducts to zero.

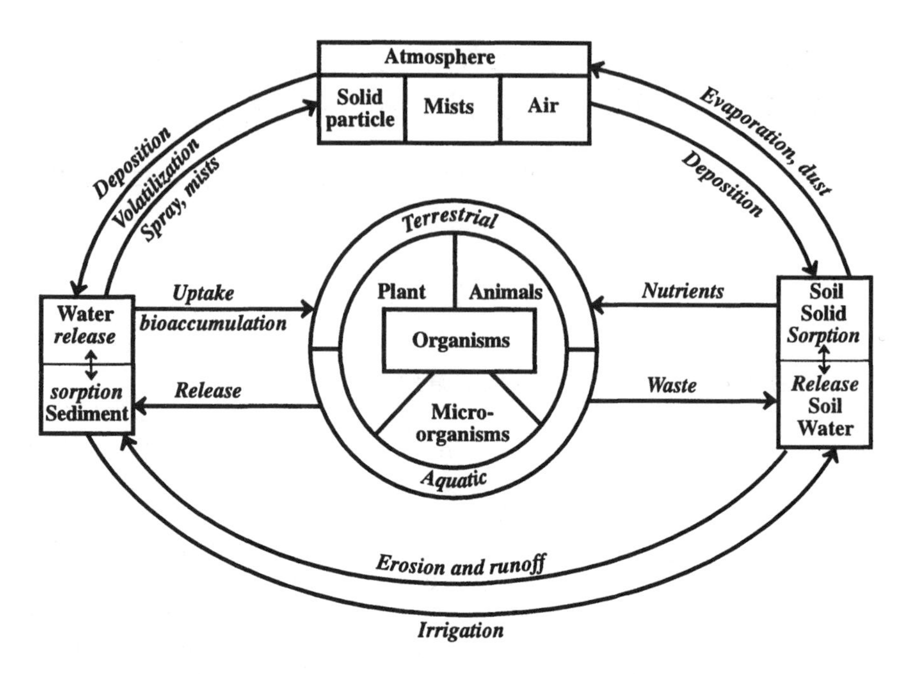

Figure 7.1. Interchange of environmental chemical species among the atmosphere, hydrosphere, geosphere, and biosphere. Human activities (the anthrosphere) have a strong influence on the various processes shown.

7.2. WATER

Although water is part of all environmental spheres, it is convenient to regard portions of the environment as constituting the **hydrosphere**. Water circulates throughout the hydrosphere in one of nature's great cycles, the **hydrologic cycle**. As shown in Figure 7.2, there are several important parts of the hydrosphere and these are, in turn, closely related to the other environmental spheres. Earth's water is contained in several **compartments** of the hydrologic cycle. The amounts of water and the **residence times** of water in these compartments vary greatly. By far the largest of these compartments consists of the **oceans**, containing about 97% of all Earth's chemically unbound water with a residence time of about 3000 years. Oceans serve as a huge reservoir for water and as the source of most water vapor that enters the hydrologic cycle. As vast heat sinks, oceans have a tremendous moderating effect on climate. A relatively large amount of water is also contained in the solid state as ice, snowpacks, glaciers, and the polar ice caps. Surface water is found in lakes, streams, and reservoirs. Groundwater is located in aquifers underground. Another fraction of water is present as water vapor in the atmosphere (clouds). There is a strong connection between the *hydrosphere,* where water is found, and the *geosphere*, or land; human activities affect both.

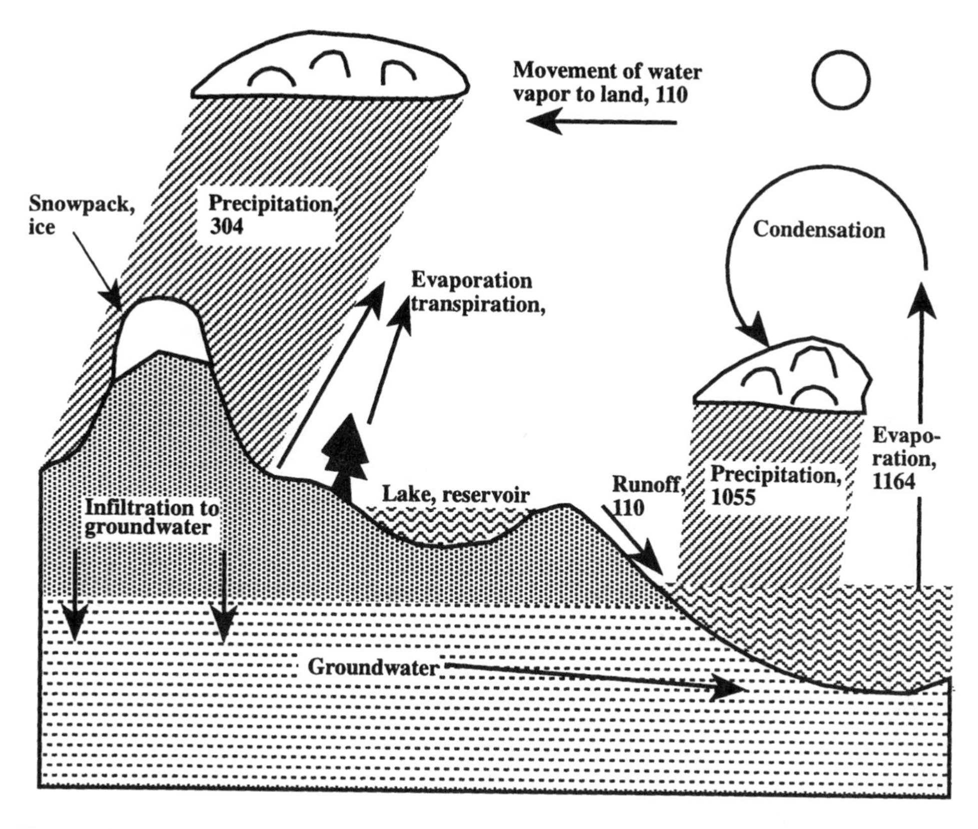

Figure 7.2. The hydrologic cycle, quantities of water in trillions of liters per day.

Water has a number of unique properties that are essential to life, due largely to its molecular structure and bonding properties. Among the special characteristics of water are the fact that it is an excellent solvent, it has a temperature/density relationship that results in bodies of water becoming stratified in layers, it is transparent, and it has extraordinary capacity to absorb, retain, and release heat per unit mass of ice, liquid water, or water vapor.

Water's unique temperature-density relationship results in the formation of distinct layers within nonflowing bodies of water, as shown in Figure 7.3. During the summer a surface layer (**epilimnion**) is heated by solar radiation and, because of its lower density, floats upon the bottom layer, or **hypolimnion**. This phenomenon is called **thermal stratification.** When an appreciable temperature difference exists between the two layers, they do not mix, but behave independently and have very different chemical and biological properties. The epilimnion, which is exposed to light, may have a heavy growth of algae. As a result of exposure to the atmosphere and (during daylight hours) because of the photosynthetic activity of algae, the epilimnion contains relatively higher levels of dissolved oxygen, and is said to be *aerobic*. In the hypolimnion, bacterial action on biodegradable organic material consumes oxygen and may cause the water to become *anaerobic*, that is, essentially free of oxygen. As a consequence, chemical species in a relatively reduced form tend to predominate in the hypolimnion.

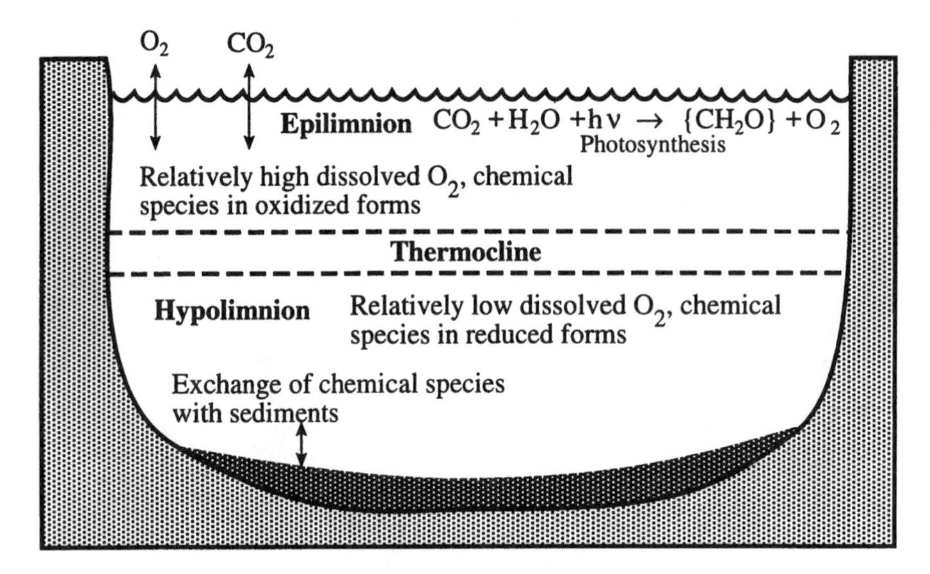

Figure 7.3. Stratification of a body of water.

7.3. AQUATIC CHEMISTRY

Figure 7.4 summarizes the more important aspects of **aquatic chemistry** as it applies to environmental chemistry. As shown in this figure, a number of chemical phenomena occur in water. Many aquatic chemical processes are influenced by the action of algae and bacteria in water. For example, it is shown that algal photosyn-

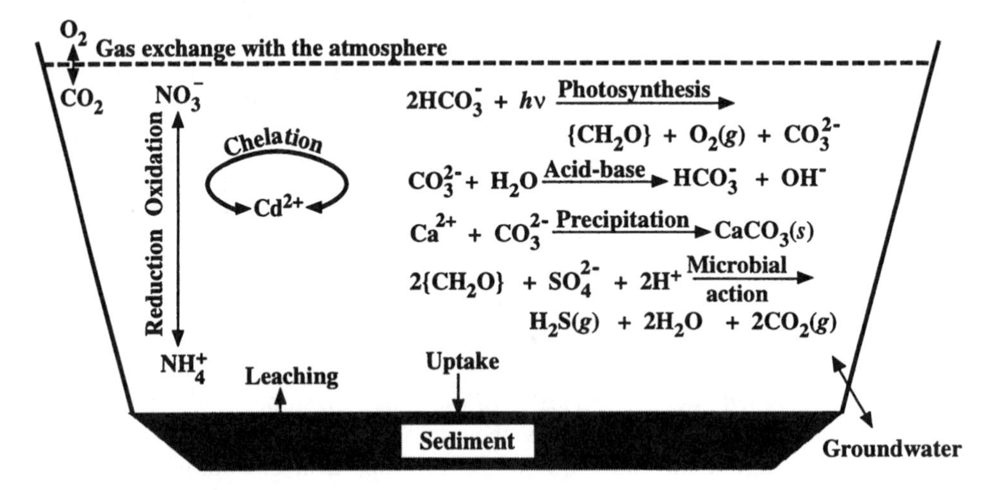

Figure 7.4. Major aquatic chemical processes.

thesis fixes inorganic carbon from HCO_3^- ion in the form of biomass (represented as $\{CH_2O\}$), in a process that also produces carbonate ion, CO_3^{2-}. Carbonate undergoes an acid-base reaction to produce OH^- ion and raise the pH, or it reacts with Ca^{2+} ion to precipitate solid $CaCO_3$. Most of the many oxidation-reduction reactions that occur in water are mediated (catalyzed) by bacteria. For example, bacteria convert inorganic nitrogen largely to ammonium ion, NH_4^+, in the oxygen-deficient (anaerobic) lower layers of a body of water. Near the surface, which is aerobic because O_2 is available from the atmosphere, bacteria convert inorganic nitrogen to nitrate ion, NO_3^-. Metals in water may be bound to organic metal-binding species called chelating agents, such as pollutant nitrilotriacetic acid (NTA) or naturally occurring fulvic acids produced by decay of plant matter. Gases are exchanged with the atmosphere, and various solutes are exchanged between water and sediments in bodies of water.

Oxidation-Reduction

Oxidation-reduction (redox) reactions in water involve the transfer of electrons between chemical species, usually through the action of bacteria. The relative oxidation-reduction tendencies of a chemical system depend upon the **activity of the electron** e^-. When the electron activity is relatively high, chemical species, including water, tend to accept electrons,

$$2H_2O + 2e^- \leftrightarrow H_2(g) + 2OH^- \quad (7.3.1)$$

and are said to be **reduced**. When the electron activity is relatively low, the medium is **oxidizing**, and chemical species such as H_2O may be **oxidized** by the loss of electrons,

$$2H_2O \leftrightarrow O_2(g) + 4H^+ + 4e^- \quad (7.3.2)$$

The relative tendency toward oxidation or reduction may be expressed by the electrode potential, E, which is more positive in an oxidizing medium and more negative in a reducing medium.

Complexation and Chelation

Metal ions in water are always bonded to water molecules in the form of hydrated ions represented by the general formula, $M(H_2O)_x^{n+}$, from which the H_2O is often omitted for simplicity. Other species may be present that bond to the metal ion more strongly than does water. For example, cadmium ion dissolved in water, Cd^{2+}, reacts with cyanide ion, CN^-, as follows:

$$Cd^{2+} + CN^- \rightarrow CdCN^+ \quad (7.3.3)$$

The product of the reaction is called a **complex** (complex ion) and the cyanide ion is called a **ligand**. Some (usually organic) ligands, called chelating agents, can bond with a metal ion in two or more places, forming particularly stable complexes.

In addition to metal complexes and chelates, another major type of environmentally important metal species consists of **organometallic compounds**. These differ from complexes and chelates in that the organic portion is bonded to the metal by a carbon-metal bond and the organic ligand is frequently not capable of existing as a stable separate species.

Complexation, chelation, and organometallic compound formation have strong effects upon metals in the environment. For example, complexation with negatively charged ligands may convert a soluble metal species from a cation, which is readily bound and immobilized by ion exchange processes in soil, to an anion, such as $Ni(CN)_4^{2-}$, that is not strongly held by soil. On the other hand, some chelating agents are used for the treatment of heavy metal poisoning and insoluble chelating agents, such as chelating resins, can be used to remove metals from waste streams.

Water Interactions with Other Phases

Most of the important chemical phenomena associated with water do not occur in solution, but rather through interaction of solutes in water with other phases. Such interactions may involve exchange of solute species between water and sediments, gas exchange between water and the atmosphere, and effects of organic surface films. Substances dissolve in water from other phases, and gases are evolved and solids precipitated as the result of chemical and biochemical phenomena in water.

Sediments are repositories of a wide variety of chemical species and are the media in which many chemical and biochemical processes occur. Sediments are sinks for many hazardous organic compounds and heavy metal salts that have entered into water.

Colloids, which consist of very small particles ranging from 0.001 micrometer (μm) to 1 μm in diameter, have a strong influence on aquatic chemistry. Colloids have very high surface-to-volume ratios, so that they can be very physically, chemically, and biologically active. Colloids may be very difficult to remove from water during water treatment.

Water Pollutants

Natural waters are afflicted with a wide variety of inorganic, organic, and biological pollutants. In some cases, such as that of highly toxic cadmium, a pollutant is directly toxic at a relatively low level. In other cases the pollutant itself is not toxic, but its presence results in conditions detrimental to water quality. For example, biodegradable organic matter in water is often not toxic, but the consumption of oxygen during its degradation prevents the water from supporting fish life. Some contaminants, such as NaCl, are normal constituents of water at low levels, but harmful pollutants at higher levels. The proper design of industrial ecosystems minimizes the release of water pollutants.

Water Treatment

The treatment of water can be considered under the two major categories of (1) treatment before use and (2) treatment of contaminated water after it has passed through a municipal water system or industrial process. In both cases, consideration must be given to potential contamination by pollutants.

Several operations may be employed to treat water prior to use. Aeration is used to drive off odorous gases, such as H_2S, and to oxidize soluble Fe^{2+} and Mn^{2+}

ions to insoluble forms. Lime is added to remove dissolved calcium (water hardness). $Al_2(SO_4)_3$ forms a sticky precipitate of $Al(OH)_3$, which causes very fine particles to settle. Various filtration and settling processes are employed to treat water. Chlorine, Cl_2, is added to kill bacteria.

Municipal wastewater may be subjected to primary, secondary, or advanced water treatment. **Primary** water treatment consists of settling and skimming operations that remove grit, grease, and physical objects from water. **Secondary** water treatment is designed to take out biochemical oxygen demand, BOD. This is normally accomplished by introducing air and microorganisms such that waste biomass in the water, $\{CH_2O\}$, is removed by aerobic respiration.

$$\{CH_2O\} + O_2 \rightarrow CO_2 + H_2O \quad \text{(Aerobic respiration)} \qquad (7.3.4)$$

7.4. THE GEOSPHERE

The **geosphere**, or solid Earth, is that part of the Earth upon which humans live and from which they extract most of their food, minerals, and fuels. Once thought to have an almost unlimited buffering capacity against the perturbations of humankind, the geosphere is now known to be rather fragile and subject to harm by human activities, such as mining, acid rain, erosion from poor cultivation practices, and disposal of hazardous wastes. It may be readily seen that the preservation of the geosphere in a form suitable for human habitation is one of the greatest challenges facing humankind. Because anthrospheric structures are almost always located on the geosphere and because most of the raw materials used in manufacturing are extracted from it, the geosphere is especially important in the design of systems of industrial ecology.

Solids in the Geosphere

The part of Earth's geosphere that is accessible to humans is the **crust**, which is extremely thin compared to the diameter of the earth, ranging from 5 to 40 km thick. Most of the solid earth crust consists of rocks. Rocks are composed of minerals, where a **mineral** is a naturally occurring inorganic solid with a definite internal crystal structure and chemical composition. A **rock** is a solid, cohesive mass of pure mineral or an aggregate of two or more minerals.

At elevated temperatures deep beneath Earth's surface, rocks and mineral matter melt to produce a molten substance called **magma**. Cooling and solidification of magma produces **igneous rock**. Common igneous rocks are granite, basalt, quartz (SiO_2), feldspar ($(Ca,Na,K)AlSi_3O_8$), and magnetite (Fe_3O_4). Exposure of igneous rocks formed under water-deficient, chemically reducing conditions of high temperature and high pressure to wet, oxidizing, low-temperature and low-pressure conditions at the surface causes the rocks to disintegrate by a process called **weathering**.

Erosion from wind, water, or glaciers picks up materials from weathering rocks and deposits it as **sediments** or **soil**. A process called **lithification** describes the conversion of sediments to **sedimentary rocks**. In contrast to the parent igneous rocks, sediments and sedimentary rocks are porous, soft, and chemically reactive. **Metamorphic rock** is formed by the action of heat and pressure on sedimentary, igneous, or other kinds of metamorphic rock that are not in a molten state.

Because rocks are the sources of required raw materials, the practice of industrial ecology obviously requires a detailed knowledge of the physical and chemical properties of rocks and the minerals that comprise them. Although over two thousand minerals are known, only about 25 **rock-forming minerals** make up most of the earth's crust. Oxygen and silicon comprise 49.5% and 25.7% by mass of the earth's crust, respectively; therefore, most minerals are **silicates** such as quartz, SiO_2, or potassium feldspar, $KAlSi_3O_8$. In descending order of abundance, the other elements in the earth's crust are aluminum (7.4%), iron (4.7%), calcium (3.6%), sodium (2.8%), potassium (2.6%), magnesium (2.1%), and other (1.6%).

7.5. SOIL

Insofar as the human environment and life on Earth are concerned, the most important part of the Earth's crust is **soil,** which consists of particles that make up a variable mixture of minerals, organic matter, and water, capable of supporting plant life on Earth's surface. It is the final product of the weathering action of physical, chemical, and biological processes on rocks. The organic portion of soil consists of plant biomass in various stages of decay. High populations of bacteria, fungi, and animals such as earthworms may be found in soil. Soil contains air spaces and generally has a loose texture. These aspects of soil are shown in Figure 6.3 of Chapter 6.

Soils usually exhibit distinctive layers with increasing depth (Figure 7.5). These layers, called **horizons,** form as the result of complex interactions among processes that occur during weathering. Rainwater percolating through soil carries dissolved and colloidal solids to lower horizons where they are deposited. Biolog-

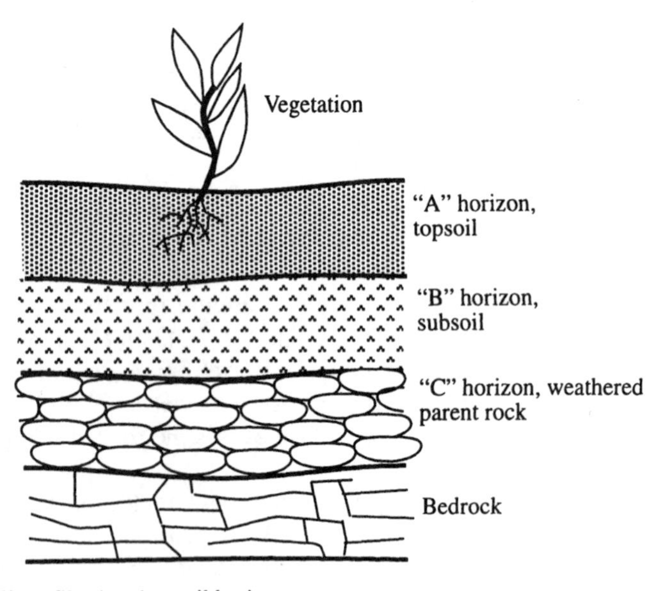

Figure 7.5. Soil profile showing soil horizons.

ical processes, such as bacterial decay of residual plant biomass, produce slightly acidic CO_2, organic acids, and complexing compounds that are carried by rainwater to lower horizons where they interact with clays and other minerals, altering the properties of the minerals. The top layer of soil, typically several inches in thickness, is known as the A horizon, or **topsoil**. This is the layer of maximum biological activity in the soil and contains most of the soil's organic matter. Metal ions and clay particles in the A horizon are subject to considerable leaching, such that it is sometimes called the zone of leaching. The next layer is the B horizon, or **subsoil**. It receives material such as organic matter, salts, and clay particles leached from the topsoil, so it is called the zone of accumulation. The C horizon is composed of fractured and weathered parent rocks from which the soil originated.

Soils exhibit a large variety of characteristics that are used to classify them for various purposes, including crop production, road construction, and waste disposal. The parent rocks from which soils are formed obviously play a strong role in determining the composition of soils. Other soil characteristics include strength, workability, soil particle size, permeability, and degree of maturity.

Water is a crucial part of the three-phase, solid-liquid-gas system making up soil. It is the solvent of the soil solution (see Section 7.6) and is the basic transport medium for carrying essential plant nutrients from solid soil particles into plant roots and to the farthest reaches of the plant's leaf structure (Figure 7.6). The water enters the atmosphere from the plant's leaves, a process called **transpiration**. Large quantities of water are required for the production of most plant materials.

Normally, because of the small size of soil particles and the presence of small capillaries and pores in the soil, the water phase is not totally independent of soil solid matter. Water present in larger spaces in soil is relatively more available to plants and readily drains away. Water held in smaller pores or between the unit layers of clay particles is held much more firmly. Water in soil interacts strongly with organic matter and with clay minerals.

7.6. GEOCHEMISTRY AND SOIL CHEMISTRY

Geochemistry deals with chemical species, reactions, and processes in the lithosphere and their interactions with the atmosphere and hydrosphere. The branch of geochemistry that explores the complex interactions among the rock/water/air/life (and human) systems that determine the chemical characteristics of the surface environment is **environmental geochemistry**. Obviously, geochemistry and its environmental subdiscipline are very important in environmental science and in considerations of industrial ecology.

Geochemistry addresses a large number of chemical and related physical phenomena. Some of the major areas of geochemistry are the following:

- The chemical composition of major components of the geosphere, including magma and various kinds of solid rocks.
- Processes by which elements are mobilized, moved, and deposited in the geosphere through a cycle known as the **geochemical cycle**.

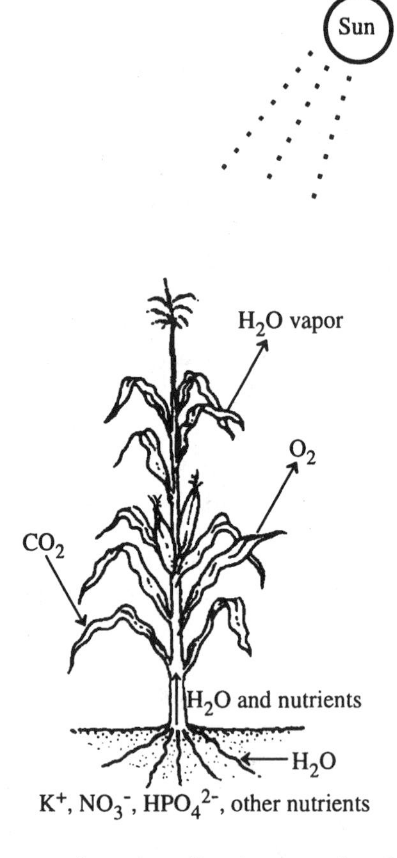

Figure 7.6. Plants transport water from the soil to the atmosphere by transpiration. Nutrients are also carried from the soil to the plant extremities by this process. Plants remove CO_2 from the atmosphere and add O_2 by photosynthesis. The reverse occurs during plant respiration.

- Chemical processes that occur during the formation of igneous rocks from magma.
- Chemical processes that occur during the formation of sedimentary rocks.
- Chemistry of rock weathering.
- Chemistry of volcanic phenomena.
- Role of water and solutions in geological phenomena, such as deposition of minerals from hot brine solutions.
- The behavior of dissolved substances in concentrated brines.

An important consideration in geochemistry is that of the interaction of life forms with geochemical processes addressed as biogeochemistry or organic geochemistry. The deposition of biomass and the subsequent changes that it undergoes have caused huge deposits of petroleum, coal, and oil shale. Chemical changes induced by photosynthesis have resulted in massive deposits of calcium carbonate (limestone). Deposition of the biochemically synthesized shells of microscopic animals have led to the formation of large formations of calcium carbonate and silica. Biogeochemistry is closely involved with elemental cycles, such as those of carbon.

Physical and Chemical Aspects of Weathering

Defined in Section 7.4, *weathering* is discussed here as a geochemical phenomenon. Rocks tend to weather more rapidly when there are pronounced changes over time in physical conditions—alternate freezing and thawing and wet periods alternating with severe drying. Other mechanical aspects are swelling and shrinking of minerals with hydration and dehydration as well as growth of roots through cracks in rocks. The rates of chemical reactions involved in weathering increase with increasing temperature.

As a chemical phenomenon, weathering can be viewed as the result of the tendency of the rock/water/mineral system to attain equilibrium. This occurs through the usual chemical mechanisms of dissolution/precipitation, acid-base reactions, complexation, hydrolysis, and oxidation-reduction.

Weathering is very slow in dry air. Water increases the rate of weathering by many orders of magnitude for several reasons. Water, itself, is a chemically active substance in the weathering process. Furthermore, water holds weathering agents in solution such that they are transported to chemically active sites on rock minerals and contact the mineral surfaces at the molecular and ionic level. Prominent among such weathering agents are CO_2, O_2, organic acids, sulfur acids ($SO_2(aq)$, H_2SO_4), and nitrogen acids (HNO_3, HNO_2). Water provides the source of H^+ ion needed for acid-forming gases to act as acids as shown by the following:

$$CO_2 + H_2O \rightarrow H^+ + HCO_3^- \tag{7.6.1}$$

$$SO_2 + H_2O \rightarrow H^+ + HSO_3^- \tag{7.6.2}$$

Rainwater is essentially free of mineral solutes. It is usually slightly acidic due to the presence of dissolved carbon dioxide or more highly acidic because of acid-rain forming constitutents. As a result of its slight acidity and lack of alkalinity and dissolved calcium salts, rainwater is *chemically aggressive* toward some kinds of mineral matter, which it breaks down by a process called **chemical weathering**.

A typical chemical reaction involved in weathering is the dissolution of calcium carbonate (limestone) by water containing dissolved carbon dioxide.

$$CaCO_3(s) + H_2O + CO_2(aq) \rightarrow Ca^{2+}(aq) + 2HCO_3^-(aq) \tag{7.6.3}$$

Weathering may also involve oxidation reactions, such as occurs when pyrite, FeS_2, dissolves.

$$4FeS_2(s) + 15O_2(g) + (8 + 2x)H_2O \rightarrow 2Fe_2O_3{\cdot}xH_2O + 8SO_4^{2-}(aq) + 16H^+(aq) \tag{7.6.4}$$

Soil Chemistry

A large variety of chemical and biochemical processes occur in soil. In discussing soil chemistry, it is crucial to consider the **soil solution**, which is the aqueous portion of soil that contains dissolved matter from soil chemical and biochemical processes and from exchange with the hydrosphere and biosphere. This medium transports chemical species to and from soil particles and provides intimate contact between the solutes and the soil particles. In addition to providing water for plant growth, it is an essential pathway for the exchange of plant nutrients between roots and solid soil.

Soil acts as a buffer and resists changes in pH. The buffering capacity depends upon the type of soil. The acidity and basicity of soil must be kept within certain ranges to enable the soil to be productive. Usually soil tends to become too acidic through the processes by which plants take up nutrient cations from it (see Reaction 7.6.6 below). Most common plants grow best in soil with a pH near neutrality. If the soil becomes too acidic for optimum plant growth, it may be restored to productivity by liming, ordinarily through the addition of calcium carbonate.

$$\text{Soil}\}(H^+)_2 + CaCO_3 \rightarrow \text{Soil}\}Ca^{2+} + CO_2 + H_2O \quad (7.6.5)$$

Chemical processes involving ions in the soil solution and bound to the soil solids are very important. Cation exchange in soil is the mechanism by which potassium, calcium, magnesium, and essential trace-level metals are made available to plants. When nutrient metal ions are taken up by plant roots, hydrogen ion is exchanged for the metal ions. This process, plus the leaching of calcium, magnesium, and other metal ions from the soil by water containing carbonic acid, tends to make the soil acidic.

$$\text{Soil}\}Ca^{2+} + 2CO_2 + 2H_2O \rightarrow \text{Soil}\}(H^+)_2 + Ca^{2+}(\text{root}) + 2HCO_3^- \quad (7.6.6)$$

A large number of oxidation/reduction processes occur in soil, almost always mediated by microorganisms. The most common of these processes is the biodegradation of biomass, such as that from crop residues, here represented as $\{CH_2O\}$.

$$\{CH_2O\} + O_2 \rightarrow CO_2 + H_2O \quad (7.6.7)$$

This process consumes oxygen and produces CO_2. As a result, the oxygen content of air in soil may be as low as 15%, and the carbon dioxide content may be several percent. Thus, the decay of organic matter in soil increases the equilibrium level of dissolved CO_2 in groundwater. This lowers the pH and contributes to weathering of carbonate minerals, particularly calcium carbonate.

Although the organic fraction of solid soil is usually not greater than 5%, organic matter strongly influences the chemical, physical, and biological properties of the soil. **Soil humus** is by far the most significant organic constituent. Humus, is the residue left when bacteria and fungi biodegrade plant material and results largely from the microbial alteration of lignin, which is the material that binds plant matter together. Humus is composed of soluble humic and fulvic acids and an insoluble fraction called humin. Humus molecules exhibit variable, complex chemical structures. They have acid-base, ion exchanging, and metal chelating properties.

7.7. THE ATMOSPHERE

The **atmosphere** is made up of the thin layer of mixed gases consisting predominantly of nitrogen, oxygen, and water vapor that covers the earth's surface. As shown in Figure 7.7, the atmosphere is divided into several layers on the basis of temperature. Of these, the most significant are the troposphere extending in altitude from the earth's surface to approximately 11 kilometers (km) and the stratosphere from about 11 km to approximately 50 km.

Stratification of the Atmosphere

As shown in Figure 7.7, the atmosphere is stratified on the basis of temperature/density relationships resulting from interrelationships between physical and photochemical (light-induced chemical phenomena) processes in air. The lowest layers of the atmosphere, the troposphere and the stratosphere, are of most significance for human activities.

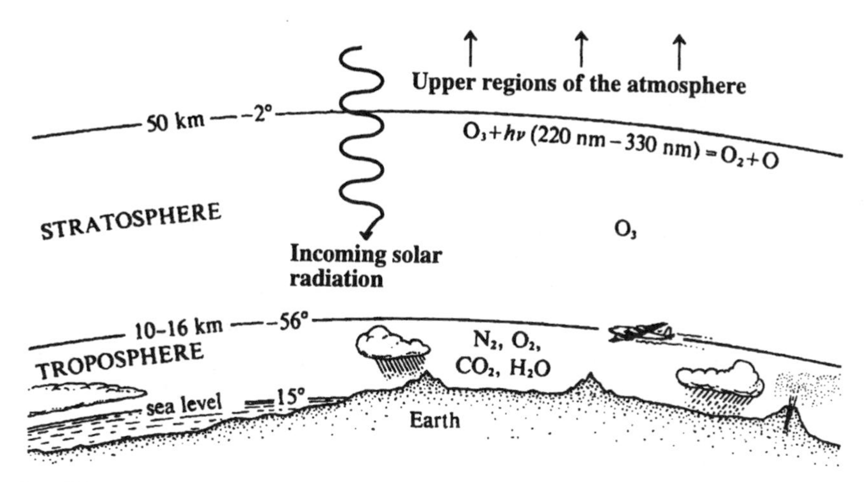

Figure 7.7. The two lower layers of the atmosphere. Above the stratosphere is a region in which the temperature decreases with increasing altitude called the mesosphere and above that is a region characterized by increasing temperature with increasing altitude called the thermosphere. These two regions play a very important role in filtering out high-energy solar electromagnetic radiation before it reaches Earth's surface.

The lowest layer of the atmosphere extending from sea level to an altitude of 10–16 km (depending upon time, temperature, latitude, season, climate conditions, and underlying terrestrial features) is the **troposphere**, characterized by a generally homogeneous composition of major gases other than water and decreasing temperature with increasing altitude from the heat-radiating surface of the earth. The temperature of the troposphere ranges from an average of 15°C at sea level down to an average of -56°C at its upper boundary. The homogeneous composition of the troposphere results from constant mixing by circulating air masses.

However, the water vapor content of the troposphere is extremely variable because of cloud formation, precipitation, and evaporation of water from terrestrial water bodies.

The atmospheric layer directly above the troposphere is the **stratosphere**, the average temperature of which increases from -56˚C at its boundary with the troposphere to -2˚C at its upper boundary. The reason for this increase is absorption of solar ultraviolet energy by ozone (O_3), levels of which may reach around 10 ppm by volume in the mid-range of the stratosphere.

Atmospheric air may contain up to 5% water by volume, although the normal range is 1–3%. The two major constituents of air are nitrogen and oxygen present at levels of 78.08% and 20.95% by volume in dry air. The noble gas argon comprises 0.934% of the volume of dry air. Carbon dioxide makes up about 0.036% by volume of dry air, a figure that fluctuates seasonably because of photosynthesis, and which is increasing steadily as more CO_2 is released to the atmosphere by fossil fuel combustion. There are numerous trace gases in the atmosphere at levels below 0.002%, including neon, helium, krypton, xenon, sulfur dioxide, ozone, CO, N_2O, NO_2, NH_3, CH_4, SO_2, and CCl_2F_2, a persistent chlorofluorocarbon (Freon compound) released from air conditioners and from other sources. Some of the trace gases may have profound effects, such as the role played by chlorofluorcarbons in depletion of the stratosphere's protective ozone layer and that played by methane in global warming.

7.8. ATMOSPHERIC CHEMISTRY

As indicated in Figure 7.8, many kinds of chemical reactions occur in the atmosphere. These reactions take place in the gas phase, on atmospheric particle surfaces, within particulate water droplets and on land and water surfaces in contact with the atmosphere. The most significant feature of atmospheric chemistry is the occurrence of **photochemical reactions** that occur when molecules in the atmosphere absorb energy in the form of light photons, designated $h\nu$. (The energy, E, of a photon of visible or ultraviolet light is given by the equation, $E = h\nu$, where h is Planck's constant and ν is the frequency of light, which is inversely proportional to its wavelength. Ultraviolet radiation has a higher frequency than visible light and is, therefore, more energetic and more likely to break chemical bonds in molecules that absorb it.) One of the most significant photochemical reactions is the one responsible for the presence of ozone in the stratosphere (see above), which is initiated when O_2 absorbs highly energetic ultraviolet radiation in the wavelength ranges of 135–176 nanometers (nm) and 240–260 nm in the stratosphere:

$$O_2 + h\nu \rightarrow O + O \qquad (7.8.1)$$

The oxygen atoms produced by the photochemical dissociation of O_2 react with oxygen molecules to produce ozone, O_3,

$$O + O_2 + M \rightarrow O_3 + M \qquad (7.8.2)$$

where M is a third body, such as a molecule of N_2, which absorbs excess energy from the reaction. The ozone that is formed is very effective in absorbing ultraviolet radiation in the 220–330 nm wavelength range, which causes the temperature increase observed in the stratosphere. The ozone serves as a very valuable filter to remove ultraviolet radiation from the sun's rays. If this radiation reached the earth's surface, it would cause skin cancer and other damage to living organisms.

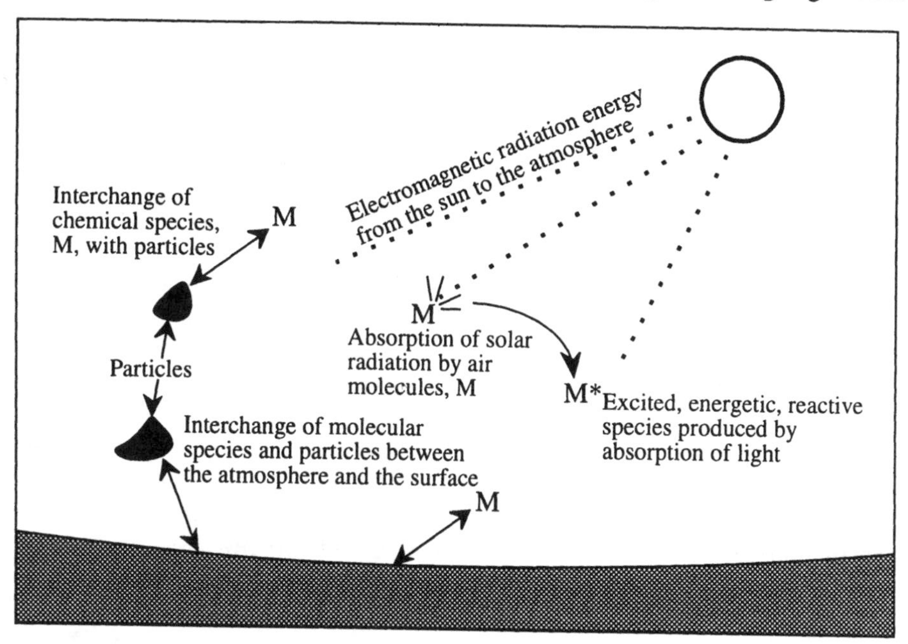

Figure 7.8. Atmospheric chemical processes may occur in the gas phase on or in particles, and on land or water surfaces in contact with the atmosphere. A particular feature of atmospheric chemistry is that of photochemical reactions induced by the absorption of energetic photons of sunlight.

Gaseous Oxides in the Atmosphere

Oxides of carbon, sulfur, and nitrogen are important constituents of the atmosphere and are pollutants at higher levels. Of these, carbon dioxide, CO_2, is the most abundant. It is a natural atmospheric constituent and is required for plant growth. However, the level of carbon dioxide in the atmosphere, now at about 360 parts per million (ppm) by volume, is increasing by about 1 ppm per year. This increase in atmospheric CO_2 may well cause general atmospheric warming — the "greenhouse effect," with potentially very serious consequences for the global atmosphere and for life on earth. One of the central challenges for industrial ecology systems in the future will be to provide energy that they need to keep operating without adding to the global burden of carbon dioxide.

Though not a global threat, carbon monoxide, CO, can be a serious health threat because it prevents blood from transporting oxygen to body tissues (see Section 7.10 and Reaction 7.10.1).

The two most serious nitrogen oxide air pollutants are nitric oxide, NO, and nitrogen dioxide, NO_2, collectively denoted as "NO_x." These tend to enter the atmosphere as NO, and photochemical processes in the atmosphere tend to convert

NO to NO_2. Further reactions can result in the formation of corrosive nitrate salts or nitric acid, HNO_3. Nitrogen dioxide is particularly significant in atmospheric chemistry because of its photochemical dissociation by light with a wavelength less than 430 nm to produce highly reactive O atoms. This is the first step in the formation of photochemical smog (see below). Sulfur dioxide, SO_2, is a reaction product of the combustion of sulfur-containing fuels, such as high-sulfur coal. Part of this sulfur dioxide is converted in the atmosphere to sulfuric acid, H_2SO_4, normally the predominant contributor to acid precipitation.

Hydrocarbons and Photochemical Smog

The most abundant hydrocarbon in the atmosphere is methane, chemical formula CH_4. This gas is released from underground sources as natural gas and produced by the fermentation of organic matter. Methane is one of the least reactive atmospheric hydrocarbons and is produced by diffuse sources, so that its participation in the formation of pollutant photochemical reaction products is minimal.

The most significant atmospheric pollutant hydrocarbons are the reactive ones produced as automobile exhaust emissions. In the presence of NO, under conditions of temperature inversion, which hold masses of air stationary for several days, low humidity, and sunlight, these hydrocarbons produce undesirable **photochemical smog** manifested by the presence of visibility-obscuring particulate matter, oxidants such as ozone, O_3, and noxious organic species, such as aldehydes. The process of smog formation is initiated by the photochemical dissociation of nitrogen dioxide by ultraviolet solar radiation,

$$NO_2 + h\nu \rightarrow NO + O \tag{7.8.3}$$

which produces a reactive oxygen atom. This atom readily reacts with hydrocarbons in the atmosphere,

$$R\text{–}H + O \rightarrow HO\cdot + R\cdot \tag{7.8.4}$$

where R–H represents a generic hydrocarbon molecule and one of the hydrogen atoms bound to it. The dots on the products denote unpaired electrons; species with such electrons are very reactive **free radicals**. The hydroxyl radical, HO•, is particularly important because of the strong role that it plays as a very reactive intermediate in photochemical smog formation and other atmospheric chemical processes. It reacts to initiate series of **chain reactions** in which it is regenerated; the end products of these reactions are the noxious materials characteristic of photochemical smog. The hydrocarbon radical, R•, reacts with molecular O_2,

$$R\cdot + O_2 \rightarrow RO_2\cdot \tag{7.8.5}$$

to produce reactive, strongly oxidizing $RO_2\cdot$ radical. The $RO_2\cdot$ radical in turn reoxidizes NO back to NO_2,

$$NO + RO_2\cdot \rightarrow NO_2 + RO\cdot \tag{7.8.6}$$

which can undergo photodissociation again (Reaction 7.8.3) to further promote the smog-forming process.

Particulate Matter

Particles ranging from aggregates of a few molecules to pieces of dust readily visible to the naked eye are commonly found in the atmosphere. Some of these particles, such as sea salt formed by the evaporation of water from droplets of sea spray, are natural and even beneficial atmospheric constituents. Very small particles called **condensation nuclei** serve as bodies for atmospheric water vapor to condense upon and are essential for the formation of precipitation.

Colloidal-sized particles in the atmosphere are called **aerosols**. Those formed by grinding up bulk matter are known as **dispersion aerosols**, whereas particles formed from chemical reactions of gases are **condensation aerosols**; the latter tend to be smaller. Smaller particles are in general the most harmful because they have a greater tendency to scatter light and are the most respirable (tendency to be inhaled into the lungs).

Much of the mineral particulate matter in a polluted atmosphere is in the form of oxides and other compounds produced during the combustion of high-ash fossil fuel. Smaller particles of **fly ash** enter furnace flues and are efficiently collected in a properly equipped stack system. However, some fly ash escapes through the stack and enters the atmosphere. Unfortunately, the fly ash thus released tends to consist of smaller particles that do the most damage to human health, plants, and visibility.

7.9. THE BIOSPHERE

The **biosphere** is the sphere of the environment occupied by living organisms and in which life processes are carried out. **Biology** is the science of life and of living organisms. A **living organism** is constructed of one or more small units called *cells* and has the following characteristics: (1) It is composed in part of large characteristic macromolecules containing carbon, hydrogen, oxygen, and nitrogen along with phosphorus and sulfur. (2) It is capable of **metabolism**; that is, it mediates chemical processes by which it utilizes energy and synthesizes new materials needed for its structure and function. (3) It regulates itself. (4) It interacts with its environment. (5) It reproduces itself.

The biosphere is important in environmental chemistry for two major reasons. The first of these is that organisms carry out many of the chemical transformations that are part of an environmental chemical system. In aquatic systems (see Figure 7.4) algae carry out photosynthesis that produces biomass from inorganic carbon, bacteria degrade biomass, and bacteria mediate transformations between elemental oxidation states, such as the conversion of NO_3^- to NH_4^+ or the conversion of SO_4^{2-} to H_2S. The second reason that the biosphere is important in environmental chemistry is that the ultimate concern with respect to any pollutant or environmental phenomenon is the effect on living organisms. Such effects are addressed by toxicological chemistry, which is briefly introduced in the following section.

7.10. TOXICOLOGICAL CHEMISTRY

As implied in Figure 7.9, toxicology is the science of poisons, the effects of toxic substances on organisms. **Toxicological chemistry** is the science that deals with the chemical nature and reactions of toxic substances, including their origins, uses, and chemical aspects of exposure, fates, and disposal.[2] Toxicological chemistry addresses the relationships between the chemical properties and molecular structures of molecules and their toxicological effects.

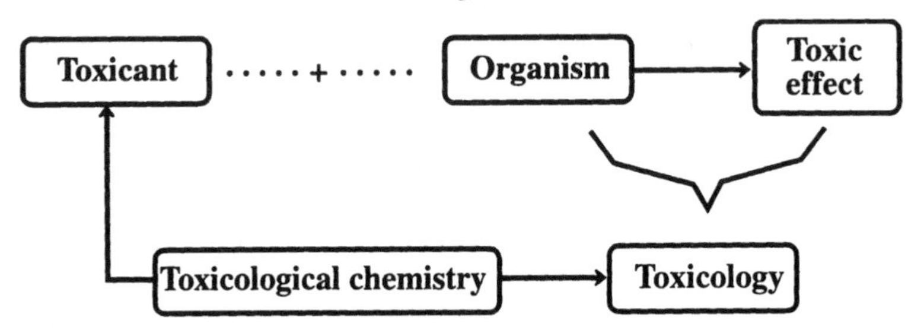

Figure 7.9. Toxicology is the science of poisons. Toxicological chemistry relates toxicology to the chemical nature of toxicants.

In considering the toxicological chemistry of a substance, it is necessary to know the chemical nature of the substance, the organism that is exposed, the routes of exposure, the manner of exposure, and the degree of exposure (dose). Many toxic substances are metabolized in organisms. Although metabolic processes usually tend to reduce toxicities by converting substances to less toxic materials that are more readily eliminated from the organism, in a number of cases the metabolic processes produce more toxic forms. Many substances that produce cancer, for example, do not do so directly, but are first metabolized to species that can initiate cancer, such as by binding to DNA in cells.

The biochemical mechanisms by which toxic substances exert their effects are many and varied. As a simple example, carbon monoxide binds much more strongly to hemoglobin, Hb, in blood than does molecular oxygen,

$$CO + HbO_2 \rightarrow HbCO + O_2 \qquad (7.10.1)$$

and the carboxyhemoglobin product prevents oxygen transport from the lungs through the blood resulting in asphyxiation. Toxic aniline causes the iron(II) in hemoglobin to be oxidized to iron(III). This converts the hemoglobin to methemoglobin, which does not carry oxygen through the bloodstream. Some toxicants react with enzymes causing them to work improperly. Organophosphates in organophosphate insecticides or in much more toxic nerve gases bind with acetylcholinesterase, an enzyme that is required to cause nerve impulses to stop. This can cause paralysis of the breathing process or other lethal effects.

Toxicological chemistry is obviously a very broad and complex subject, and there is not room to discuss it in detail here. Aspects of toxicological chemistry are addressed elsewhere in this book, particularly in Chapter 12, "Toxicological and Biological Hazards."

LITERATURE CITED

1. Manahan, Stanley E., *Environmental Chemistry*, 6th ed., CRC Press/Lewis Publishers, Boca Raton, FL, 1994.
2. Manahan, Stanley E., *Toxicological Chemistry*, 2nd ed., CRC Press/Lewis Publishers, Boca Raton, FL, 1992.

SUPPLEMENTARY REFERENCES

Andrews, J. E., *Environmental Chemistry*, Blackwell Science Publishers, Cambridge, MA, 1996.

Schlesinger, W. H., *Biogeochemistry*, Academic Press, San Diego, CA, 1991.

Spiro, Thomas G. and William M. Stigliani, *Chemistry of the Environment*, Prentice Hall, Upper Saddle River, NJ, 1996.

8 HAZARDOUS SUBSTANCES AND WASTES

8.1. INTRODUCTION AND HISTORY

A **hazardous substance** is a material that may explode, catch fire, be corrosive, exhibit toxicity, or have other dangerous characteristics such that it can endanger the physical environment, living organisms, devices, or materials. If a hazardous substance has been discarded or abandoned, it may be classified as a **hazardous waste**.

One of the more difficult problems in passing legislation and establishing legislation dealing with hazardous wastes has been defining what they really are. Regarding the definition of hazardous wastes, it has been stated that "The discussion on this question is as long as it is fruitless."[1] Three basic approaches to classifying hazardous wastes are (1) a qualitative description by origin, type, and constituents, (2) classification by characteristics largely based upon testing procedures; and (3) by means of concentrations of specific hazardous substances. These aspects are addressed in more detail in this chapter.

Legislation and regulations dealing with wastes often go to great pains to distinguish between hazardous and nonhazardous wastes. In reality, the distinction is blurred. Materials generally regarded as hazardous may be dealt with safely when properly handled and treated. For example, toxic lead from spent storage batteries may be recycled very safely. Generally nonhazardous substances may become dangerous under some conditions. Often, too, nonhazardous materials may interact with hazardous ones in ways that increase their threat to the environment or human welfare.

Hazardous wastes may be individual substances or mixtures. They are dangerous or potentially dangerous to humans or other living organisms. Direct dangers posed by hazardous wastes include fire, explosion, and toxicity. Some hazardous wastes cause cumulative detrimental effects, such as cancer resulting from repeated or prolonged exposure, some may undergo biomagnification in exposed organisms (through food chains), and many hazardous wastes persist in the environment because they do not degrade.

Specific mention should be made of radioactive wastes, which in the U.S. are regulated under the Nuclear Regulatory Commission (NRC) and Department of Energy (DOE). Special problems are posed by **mixed waste** containing both radioactive and chemical wastes.

The management of hazardous wastes and compliance with the regulations pertaining to them occupy an enormous amount of time and effort in the business community. There are several sources of information pertaining to compliance.[2]

History of Hazardous Substances and Wastes

History provides many examples of human exposure to hazardous substances. Wine stored in lead containers is reputed to have caused many cases of lead poisoning in the Roman Empire. Exposure to toxic metals from mining and metal refining operations has poisoned people over many centuries. Coal tar chemicals caused cancer in workers during the 1800s. During the early 1900s, a broad array of metals processing wastes, petroleum refining byproducts, and effluents from inorganic and organic chemicals manufacture caused widespread environmental harm. The accelerating pace of chemicals manufacture during and after World War II increased exposure to chlorinated solvents, pesticide residues, polymers manufacture byproducts, wood preservatives, and other hazardous substances.

The modern environmental movement, which can be dated from the 1962 publication of Rachel Carson's classic book, "The Silent Spring," was at first more concerned with water and air pollution than it was with hazardous substances and wastes. The 1970s in the U.S. saw a variety of new legislation designed to deal with water and air pollution.[3] Several pieces of legislation known as the Clean Water Act originated with the Federal Water Pollution Control Act of 1972 as amended in 1977 and again in 1987. Legislation was generated to control air pollution through the Clean Air Act of 1967, which was amended in 1970, 1977, and 1990. Pesticides were regulated by the Federal Insecticide, Fungicide, and Rodenticide Act of 1972 and toxic substances by the Toxic Substances Control Act of 1976.

With increased environmental awareness, it became apparent that there were a number of locations where improperly discarded chemical substances could pose potential direct hazards to exposed people and the environment. In the U.S. the situation that first focused attention on hazardous wastes occurred in the 1970s with the Love Canal waste disposal site in Niagara Falls, New York. Used since around 1940 for the disposal of about 20,000 metric tons of chemical wastes containing at least 80 different chemicals, this site began to ooze an unpleasant mix of noxious materials and was suspected of causing health problems for nearby residents. Attention was also brought to bear on numerous other places where hazardous wastes were causing problems, including a century-old industrial site used by glue factories, tanneries, and chemical manufacturers in Woburn, Massachusetts; the "Valley of the Drums" in Kentucky; and the Stringfellow Acid Pits near Riverside, California. Many sites in the State of Missouri were contaminated by the illegal disposal of waste oil containing TCDD (dioxin), and the entire community of Times Beach, Missouri, was abandoned because of this contamination.

The landmark U.S. legislation designed to cope with hazardous substances and wastes was the Resource, Conservation, and Recovery Act (RCRA) of 1976. This historical legislation was amended and strengthened by passage of the Hazardous and Solid Wastes Amendments (HSWA) of 1984. These acts gave the U.S. Environmental Protection Agency (EPA) the charge of protecting human health and the environment from the effects of hazardous substances and wastes resulting from their improper management and disposal. This charge included regulatory and enforcement authority pertaining to such materials, requiring "cradle to grave" controls from time of origin to final disposal. The act imposes stringent regulations on treatment, storage, and disposal (TSD) facilities, including requirements for record keeping, tracking, transportation, and treatment.

The 1980 Comprehensive Environmental Response, Compensation, and Liability Act (CERCLA or Superfund) was passed to deal with potential problems from, and release of hazardous materials at uncontrolled or abandoned hazardous waste sites. The legislation was designed to promote identification of waste sites, evaluation of danger from the sites, evaluation of damages to natural resources, monitoring of waste releases, and removal or cleanup. In 1986 CERCLA legislation was extended and strengthened by the Superfund Amendments and Reauthorization Act (SARA). This legislation codified regulations that had become policy under CERCLA, set mandatory schedules and goals for cleanup, and provided new procedures and authorities for enforcement. It increased emphasis upon public health, research, training, and state and citizen involvement. In recognition of the inherent undesirability of land disposal of wastes, SARA legislation provided initiatives that favor permanent solutions reducing volume, mobility, and toxicity of wastes. One of its special features was a program aimed specifically at the problem of leaking underground storage tanks, especially those associated with petroleum products retailing.

As of 1998, there were 1,359 designated Superfund sites, of which 509 sites, 37 percent of the total, had been cleaned up at a total cost to companies of $15 billion.[4] Assuming the same cost for each of the remaining sites, an additional $25 billion in private sector funds would be required to clean them up. The DuPont company, for example, listed an accrued liability of $561 million for waste site cleanup on its 1997 annual report, down from $602 million in 1995.

The problem of hazardous wastes is truly international in scope.[5] As the result of the problem of dumping such wastes in developing countries, the 1989 Basel Convention on the Control of Transboundary Movement of Hazardous Wastes and their Disposal was held in Basel, Switzerland in 1989, and by 1998 had been signed by more than 100 countries. This treaty defines a long List A of hazardous wastes, a List B of nonhazardous wastes, and a List C of as yet unclassified materials. An example of a material on List C is polyvinyl chloride (PVC) coated wire, which is harmless, itself, but may release dioxins or heavy metals when thermally treated.

Relationship to Air and Water Pollution Control

By their nature, air and water pollutants tend to be toxic, corrosive, or otherwise dangerous to health or the environment. Removal of such pollutants from stack gases or water discharges results in concentration of the potentially dangerous substances that must be treated, disposed, or recycled. The paradoxical outcome is that regulations designed to reduce air and water pollution have resulted in the need to deal with more hazardous wastes. These include sludges or concentrated liquors from water treatment processes and solids obtained from baghouses and precipitators used to remove air pollutants.

8.2. KINDS OF HAZARDOUS SUBSTANCES

Hazardous substances come in a wide variety of forms and formulations. These include gases, liquids, and solids. Many hazardous substances are in the form of semisolid sludges and large volumes of hazardous substances are dissolved in water. More often than not, hazardous substances are mixtures and may consist of inorganic or organic materials. An idea of the variety of hazardous substances may be gained by considering various classes of them recognized by the United States Department of Transportation, DOT.

- **Explosives**

  DOT Class A explosives, such as dynamite or black powder, that are sensitive to heat and shock

  DOT Class B explosives, such as rocket propellant powders, in which contaminants may cause explosion

  DOT Class C explosives, such as ammunition, which are subject to thermal or mechanical detonation

- **Gases**

  Compressed gases, such as hydrogen or carbon monoxide

  Special forms of gases, such as acetylene in acetone solution

- **Flammable liquids**, such as gasoline and aluminum alkyls

- **Flammable solids**, such as magnesium metal, sodium hydride, and calcium carbide, that burn readily, are water-reactive, or spontaneously combustible

- **Oxidizing materials**, such as potassium chlorate, that supply oxygen for the combustion of normally nonflammable materials

- **Corrosive materials**, such as oleum, sulfuric acid and caustic soda, which may cause disintegration of metal containers or flesh

- **Poisonous materials**

  Class A poisons, such as hydrocyanic acid, which are toxic by inhalation, ingestion, or absorption through the skin

  Class B poisons, such as aniline; and etiologic agents, including causative agents of anthrax, botulism, or tetanus

- **Radioactive materials** including plutonium, cobalt-60, and uranium hexafluoride.

8.3. REGULATORY CLASSIFICATIONS OF HAZARDOUS WASTES

Definitions promulgated under RCRA attempt to define when a material becomes a "waste" and the circumstances under which a waste is "hazardous."[6] As a result of their complex and overlapping nature, the regulatory definitions of hazardous wastes tend to be confusing and counter-intuitive. This confusion arises in part because all hazardous wastes fall into the category of "solid wastes." However, the definition of solid wastes under RCRA includes not only solids, but semisolids, liquids, and even gases confined in containers! According to this definition, a solid waste is a *discarded material* that is not excluded by certain specified regulations. Even the regulatory definition of "discarded" is not simple. It includes wastes that are *abandoned* by being disposed of, burned or incinerated, or that are being stored with the intent of disposal, burning, or incineration. Also included are wastes that are recycled or that are accumulated, stored, or treated preparatory to recycling. Another category of solid wastes consists of those that are "inherently wastelike," including those with hazardous waste numbers (see *listed wastes*, below) F020, F021, F022, F023, F026, and F028. In general, wastes with these designations are those (except wastewater and spent carbon from hydrogen chloride purification) from the production or manufacturing use (as a reactant, chemical intermediate, or component in a formulating process) of chlorinated benzene or chlorinated phenol compounds as well as residues resulting from the incineration or thermal treatment of soil contaminated with these materials.

Under RCRA regulations, the three main ways of defining hazardous wastes are (1) by **characteristics** of the waste material, (2) **listed wastes** from a compilation by the U.S. Environmental Protection Agency, and (3) **statutory wastes** specified by various statutes, which have generally been written to provide exemptions and exclusions for wastes. To a large extent, wastes are regulated by state agencies in the U.S., which may impose definitions and constraints beyond those of the federal government.

Characteristics

In accordance with the RCRA legislation, the United States Environmental Protection Agency (EPA) defines hazardous wastes according to their behavior, that is, whether they will burn, cause corrosion, are too reactive, or are toxic. In a regulatory sense these categories define **characteristics** of wastes. Presently there are four characteristics that are recognized, although states may have more. The characteristics currently recognized are given below:

- **Ignitability:** Ignitable wastes are liable to cause or accelerate a fire during routine handling. Basically, ignitable substances include those that have a flash point (based on a prescribed test) of less than 60°C; nonliquids that may spontaneously catch fire, by friction, or from exposure to moisture, or that burn vigorously enough when ignited to create a hazard; an ignitable gas as defined by DOT regulations; or an oxidizer as defined by DOT regulations.

- **Corrosivity:** This characteristic is based upon two kinds of behavior (1) tendency to penetrate container walls so that the waste may contact other materials and cause problems or (2) wastes exhibiting pH extremes that may harm living tissue upon contact. Substances that corrode steel at a rate exceeding 6.35 millimeters per year under specified test conditions, or those that have a pH lying outside the range of 2.0 to 12.5 meet the corrosivity characteristic.
- **Reactivity:** This category applies to very unstable wastes which have a trendency to explode or undergo violent, uncontrollable reactions when disturbed by influences such as mechanical shock, heating, or exposure to water. It also includes wastes containing cyanide or sulfide that can generate levels of toxic volatile materials that may endanger human health or the environment when exposed to pH conditions between 2 and 12.5. No single test defines reactivity and EPA has some broadly based descriptions from which this characteristic may be designated.
- **Toxicity:** This characteristic is defined by the Toxicity Characteristic Leaching Procedure (TCLP) designed to simulate conditions under which specified toxic substances may be mobilized so that they have the potential to enter groundwater when exposed to leachant solutions that might be generated in landfills. Leachate from the material generated under specified conditions is analyzed for 25 RCRA-regulated organic compounds, 8 heavy metals, and 6 pesticides.

Listed Wastes

In addition to classification by characteristics, the Environmental Protection Agency designates more than 450 **listed wastes**. The listed wastes consist of either specific substances or classes of substances that are likely to cause problems and are known to be hazardous. In order to designate listed wastes, each such substance is assigned an EPA hazardous waste number in the format of a letter followed by 3 numerals. There are three major categories of listed wastes. Furthermore, the third of these categories is divided into two subcategories. The first category of listed wastes is designated as "F" wastes from nonspecific sources. An example of an "F" waste is plating bath residues from the bottom of plating baths from electroplating operations where cyanides are used in the process (F08); the presence of cyanide obviously poses a toxicity hazard and the threat of release of toxic HCN gas in contact with acid. Wastes designated with a "K" are hazardous wastes from specific sources, such as bottoms from the acetonitrile purification column in the production of acrylonitrile (K014). The third category of listed wastes is based upon commercial chemical products. It includes containers, spill residues, and off-specification species which must be treated as hazardous wastes when discarded. In this category are "P" wastes that are subject to relatively more rigorous management requirements because they are likely to be acutely hazardous. Examples of such wastes include arsenic pentoxide (P011) or strychnine and its salts (P108). The "U" subcategory wastes are regarded as toxic and are predominantly specific compounds such as benzene (U019) or selenium dioxide (U204).

Exempt Wastes

RCRA legislation specifically excludes some materials from the hazardous category. Some of these materials are specifically covered by other legislation, including nuclear material (Atomic Energy Act) and polychlorinated biphenyls (Toxic Substances Control Act). Other excluded materials are household waste, agricultural wastes used as fertilizer, oil and natural gas exploration drilling waste, byproduct brine from petroleum production, waste dust from cement kilns, byproducts (such as ash) of fossil fuel combustion and stack gas cleanup, mining overburden returned to the mine site, waste and sludge from phosphate mining and beneficiation, domestic sewage, return flow from irrigation, wastes created by end users consisting of wood treated with arsenic, and specified wastes containing chromium.

CERCLA defines hazardous substances largely in reference to other environmental statutes.[7] Included are the following:

- Substances designated as hazardous wastes under RCRA, such as characteristic hazardous wastes designated under RCRA Section 3001 (with the exception of those suspended by Congress under the Solid Waste Disposal Act)
- Substances designated by CERCLA Section 102 which “may present substantial danger to public health or welfare or the environment”
- Certain substances or toxic pollutants designated by the Clean Water Act, Sections 307 and 311
- Any hazardous air pollutant listed under Section 112 of the Clean Air Act
- Any imminently hazardous chemical substance or mixture that has been the subject of government action under Section 7 of the Toxic Substances Control Act (TSCA)

8.4. PHYSICAL NATURE AND CLASSIFICATION OF WASTES

Hazardous wastes exhibit a variety of physical forms, a fact that largely determines how they are handled. The following distinct classifications may be recognized:

- **Organic materials**, such as spent chlorinated solvents
- **Aqueous wastes**, such as aqueous solutions of toxic cyanide
- **Sludges**, which are generally hard-to-handle semisolid materials

Management strategies for wastes are determined in part by the **concentration** of wastes. Wastes become diluted when mixed with relatively large quantities of nonhazardous materials, particularly water, thus making it much more difficult to deal with and treat the wastes. If at all possible, wastes should never be diluted or, if dilution is necessary, it should be held to a minimum.

Consideration of waste dilution leads to the concept of **segregation** of wastes, which is very important in treating, storing, and disposing of different kinds of wastes. Segregation of wastes is illustrated in Figure 8.1. It is relatively easy to deal with wastes that are not mixed with other kinds of wastes, that is, those that are highly segregated. For example, spent hydrocarbon solvents can be used as fuel in boilers. However, if these solvents are mixed with spent organochloride solvents, the production of contaminant hydrogen chloride during combustion may prevent fuel use and require disposal in special hazardous waste incinerators. Further mixing with inorganic sludges adds mineral matter and water. These impurities complicate the treatment processes required because they produce mineral ash during incineration or lower the heating value of the material incinerated because of the presence of water. Among the most difficult types of wastes to handle and treat are those with the least segregation, such as "dilute sludge consisting of mixed organic and inorganic wastes," as shown in Figure 8.1.

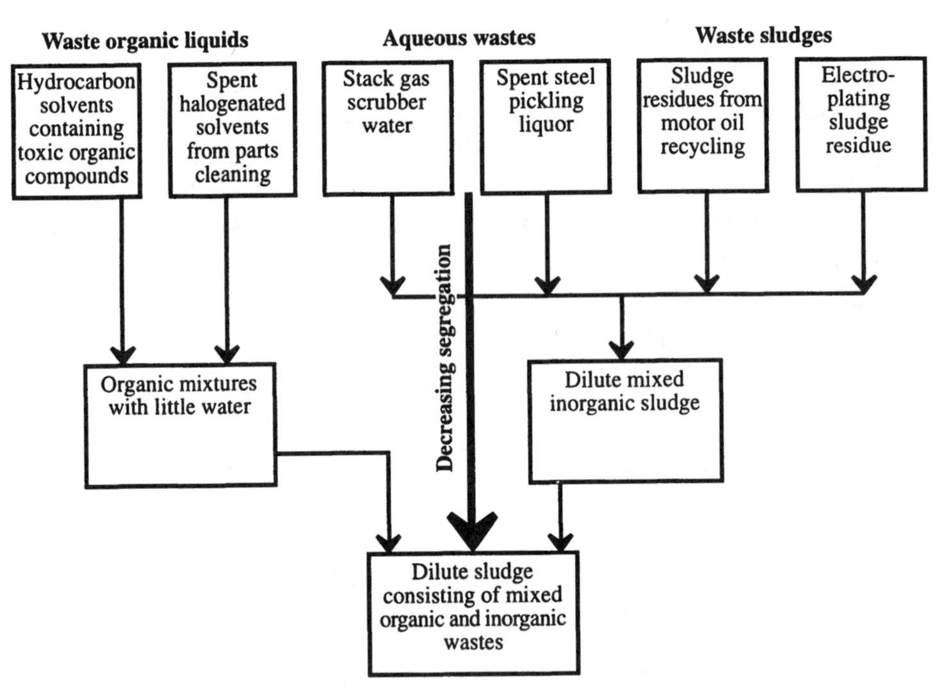

Figure 8.1. Illustration of waste segregation.

8.5. CHEMICAL NATURE AND CLASSIFICATION OF HAZARDOUS SUBSTANCES AND WASTES

Since hazardous substances manifest their hazards largely through their chemical reactions and characteristics, it is convenient to classify them chemically. Although the chemical variability of hazardous substances make such a classification system somewhat inexact, several categories can be defined based upon chemical behavior. These are the following:

- **Combustible** and **flammable** substances, strong reducers that burn readily or violently in the presence of atmospheric oxygen
- **Oxidizers** that provide oxygen for the combustion of reducers
- **Reactive** substances that are likely to undergo rapid, violent reactions, often in an unpredictable manner
- **Corrosive** substances that are generally sources of H^+ ion or OH^- ion and that tend to react destructively with materials, particularly metals

Some hazardous substances fall into more than one of these groups, which increases the dangers that they pose. Each of these categories is addressed in this section.

Combustion Hazards

A **flammable substance** burns more readily than one classified as a **combustible substance**. The designation of **flammable solid**, which excludes explosives, is given to a solid substance that (1) may ignite from heat left over from its manufacture, (2) may ignite from friction, or (3) may cause a serious hazard if ignited. A **flammable compressed gas** is designated according to the minimum percentage of the gas that will burn in air, the range of percentage in air over which it burns, and specified criteria for flame projection.

Most hazardous combustible substances are liquids that produce **vapors**. Usually more dense than air, such vapors may settle and ignite when exposed to a suitable ignition source. This has lead to standard tests in which a liquid is heated in proximity to a flame under specified conditions, and the temperature of the liquid at which the vapor ignites is designated as the **flash point**. A flash point below 37.8°C (100°F) defines a **flammable liquid** and a flash point above 37.8°C but below 93.3°C defines a **combustible liquid**.

There are upper and lower limits of the percentage of vapor in air between which the vapor will support combustion. These are known as **flammability limits**, the difference between them defines the **flammability range**, and along with flash point values, they express the flammability hazards of liquids. Another characteristic that can be cited is the optimal percentage of vapor in air at which it burns best (is most explosive). Table 8.1 gives lower and upper flammability limits and flash points of some typical liquids. Examination of the table reveals that notoriously flammable diethyl ether combines a very low flash point with a very low lower flammability limit and a wide flammability range.

Finely divided particles of both liquids and solids in air may burn or explode. A spray of liquid particles behaves much like the vapor of the liquid and may ignite at a temperature much below the flash point of the bulk liquid. Very finely divided solid particles of coal, grain, metals, carbon, polymers (such as cellulose acetate, polyethylene, and polystyrene), or any other solid capable of burning can catch fire. This can occur explosively resulting in **dust explosions**.

Some substances may catch fire without an ignition source, a process called **spontaneous ignition**. Such **pyrophoric substances** include white phosphorus; the alkali metals (sodium, potassium); powdered forms of some metals, including magnesium, calcium, cobalt, or aluminum; organometallic compounds, such as lithium ethyl (LiC_2H_4); and metal hydrides, such as lithium hydride, LiH, which gets heated to ignition by reaction with moisture in air. **Hypergolic mixtures** of oxidizers and oxidizable chemicals, such as a mixture of nitric acid and phenol, spontaneously catch fire.

Toxic combustion products often kill victims who are relatively unharmed by the fire, itself. A wide variety of substances with varying toxicities are produced in a fire. The most common such product is carbon monoxide, CO, which binds to hemoglobin and prevents it from carrying oxygen, thus depriving body tissue of oxygen. Toxic and corrosive oxides, including SO_2 and P_4O_{10}, are often dangerous combustion products, as is corrosive HCl from the combustion of organochloride compounds.

Table 8.1. Flash Points and Flammability Limits of Some Organic Liquids

		Volume percent in air	
Liquid	Flash point (°C)[1]	LFL[2]	UFL[2]
Acetone	-20	2.6	13
Diethyl ether	-43	1.9	36
Gasoline (2,2,4-trimethylpentane)	—	1.4	7.6
Methanol	12	6.0	37
Pentane	-40	1.5	7.8
Toluene	4	1.27	7.1
Naphthalene	157	0.9	5.9

[1] Closed-cup flash point test

[2] LFL, lower flammability limit; UFL, upper flammability limit at 25°C.

Oxidizers

Oxidizers are electron acceptor chemical species that react with combustible substances to produce heat, as shown by the reaction,

$$4HNO_3(\textit{hot concentrated solution}) + Cu(s) \rightarrow Cu(NO_3)_2(aq) + 2H_2O(l) + 2NO_2(g) \quad (8.5.1)$$

in which oxidant nitric acid reacts with copper metal to release toxic NO_2 gas. In this reaction, the copper metal is oxidized, losing 2 electrons per atom, and the nitrogen in the nitric acid is the oxidizing agent that is reduced. Air is the source of the most common oxidizer, diatomic molecular oxygen, O_2; many other oxidizers contain oxygen. Oxidizers can be divided into several chemical categories as follows:

- Elemental oxidizers

  O_2(gas) O_3(gas) F_2(gas) Cl_2(gas stored as liquid) Br_2(liquid)

- Oxidizer compounds containing available oxygen

  NH_4NO_3(solid) NH_4ClO_4(solid) $KMnO_4$(solid) $Na_2Cr_2O_7$(solid)

  H_2O_2(solution) HNO_3(concentrated solution)

  $HClO_4$(concentrated solution) N_2O(gas stored as liquid)

- Oxidizer compounds without oxygen

  ClF_3 (gas) BrF_5(liquid) BrCl(highly unstable liquid and gas)

The action of an oxidizer depends upon the reducer with which it is in contact. For example, carbon dioxide acts as an oxidizer that will support the combustion of elemental magnesium.

$$2Mg + CO_2 \rightarrow MgO + C \qquad (8.5.2)$$

However, the oxygen in CO_2 is so tightly bound that carbon dioxide can be used as an extinguisher for most common fires in which, for example, hydrocarbons are being oxidized.

Solid oxygen-containing oxidizers provide concentrated, readily available sources of O_2. This happened, for example, in the tragic 1996 crash of a Valujet DC-9 in the Florida Everglades when oxidant $KClO_3$ being transported in the airplane's baggage compartment released O_2 that caused tires in the same compartment to burn very vigorously. An intimate mixture of an oxidant and reductant can be used to produce an explosive; a lethal mixture of oxidant NH_4NO_3 with diesel fuel was used as the explosive in the dastardly 1996 bombing of the Oklahoma City Federal Building.

Reactive Substances

Substances that contain both an oxidant and reductant in the same compound may be **reactive substances** that are prone to undergo sudden, violent reactions. Nitroglycerin, chemical formula $C_3H_5(ONO_2)_3$, is a very sensitive explosive that spontaneously forms large quantities of CO_2, H_2O, O_2, and N_2 gases with a rapid release of a very high amount of energy. Another kind of reactive substance consists of those, such as potassium metal, that react violently with water.

$$2K + 2H_2O \rightarrow 2KOH + H_2 + \text{heat} \qquad (8.5.3)$$

Reactions of reactive substances evolve heat, which in turn increases the reaction rate, often leading to an uncontrollable event. The physical form of a reactant may play a strong role in reactivity. For example, a metal powder tends to be much more reactive than a bulk metal. Other factors include pressure, degree of mixing of reactants, and presence or absence of a catalyst.

Some chemical structures are associated with high reactivity. Among inorganic materials, these include the presence in the same compound of any two of the elements nitrogen, oxygen, or halogen. Examples of such reactive compounds are nitrous oxide (N_2O), which contains both oxygen and nitrogen; nitrogen halides (NCl_3 and NI_3) and halogen azides (ClN_3), in which both halogens and nitrogen are present; interhalogen compounds (BrCl), consisting of two different halogens; and halogen oxides (ClO_2) and compounds with oxyanions of the halogens (NaClO, $KClO_3$), which contain both a halogen and oxygen. Highly reactive ammonium perchlorate, NH_4ClO_4, which is used in solid rocket fuel formulations, contains nitrogen, chlorine, and oxygen all in the same compound. Compounds with metal-nitrogen bonds are also often highly reactive.

High reactivity in organic compounds is associated with multiple bonds to carbon (especially on adjacent carbon atoms or doubly bonded carbon atoms separated by only one other carbon atom), the presence of nitrogen, double bonds to nitrogen, the presence of both nitrogen and oxygen in the same functional group, and the presence of oxygen in peroxide groups. In the category of multiply bonded carbon are allenes (C=C=C) and dienes (C=C-C=C). Azo compounds with the C=N-N=C group contain multiple bonds to both carbon and oxygen. Peroxide may be present in the hydroperoxide group (R–OOH, where R represents a hydrocarbon moiety) or in R–OO–R' groups, where the peroxide is located between two carbon atoms. Oxiranes, such as ethylene oxide, may be very reactive.

```
        O
       / \
   H—C———C—H    Ethylene oxide
     |   |
     H   H
```

A special case of reactive compounds consists of explosives, which contain both oxidizing and reducing groups in the same compound in approximately the right stoichiometric ratios to convert all carbon, hydrogen, and nitrogen present to CO_2, H_2O, and N_2, respectively. Nitroglycerin actually contains an excess of oxygen so that the reaction for its explosion is the following:

$$4C_3H_5N_3O_9 \rightarrow 12CO_2 + 10H_2O + 6N_2 + O_2 \quad (8.5.4)$$

Corrosive Substances

Narrowly defined, **corrosive substances** dissolve metals or form oxide surfaces on metals. The most straightforward example of a corrosive substance is an acid, which can provide H^+ ion to react with iron in steel.

$$2H^+ + Fe \rightarrow Fe^{2+} + H_2 \quad (8.5.5)$$

In some cases strong bases also dissolve metals, such as is the case when hydroxide ion attacks aluminum.

$$6OH^- + 2Al \rightarrow 2AlO_3^{3-} + 3H_2 \quad (8.5.6)$$

A broader definition of corrosive substances is that they are substances that attack materials in general, including flesh.

In addition to strong acids and bases, corrosive substances include oxidants and dehydrating agents. A notably corrosive substance is hot, concentrated sulfuric acid, H_2SO_4, which, in addition to acting as a strong acid, is also an oxidizing agent and a potent dehydrating agent. Concentrated nitric acid, HNO_3, is a strong oxidizer, as well as an acid, so that it attacks metals, such as copper, that non-oxidizing acids do not attack. In addition, it reacts with protein in tissue to form yellow xanthoproteic acid, producing lesions that are slow to heal. Hydrofluoric acid, HF, dissolves even glass, and causes particularly bad flesh burns. The strong metal hydroxides, such as NaOH and KOH, will react with aluminum, lead, and zinc metals and are severe caustic poisons to flesh. Hydrogen peroxide, H_2O_2, is a potent oxidizing agent that severely burns flesh at concentrations greater than a few percent in solution. The elemental halogens (F_2, Cl_2, Br_2), interhalogen compounds (ClF, BrF_3), and halogen oxides (OF_2, Cl_2O, Cl_2O_7) are powerful corrosive irritants that acidify, oxidize, and dehydrate tissue.

8.6. TOXIC WASTES

The U.S. Environmental Protection Agency specifies a test called the **Toxicity Characteristic Leaching Procedure (TCLP)** designed to determine the toxicity hazard of wastes. The test was designed to estimate the availability to organisms of both inorganic and organic species in hazardous materials present as liquids, solids, or multiple phase mixtures.[8] The analyte in which these constituents is measured is called the TCLP extract. If the waste is a liquid containing less than 0.5% solids, it is filtered through a 0.6–0.8 µm glass fiber filter to give a liquid designated as the TCLP extract. At solids levels exceeding 0.5%, any liquid present is filtered off for separate analysis and the solid is extracted to provide a TCLP extract (after size reduction, if the particles exceed certain size limitations). The choice of the extraction fluid is determined by the pH of the aqueous solution produced from shaking a mixture of 5 g of solids and 96.5 mL of water. If the pH is less than 5.0, a pH 4.93 acetic acid/sodium acetate buffer is used for extraction; otherwise, the extraction fluid used is a pH of 2.88±0.05 solution of dilute acetic acid. Extractions are carried out in a sealed container rotated end-over-end for 18 hours. The liquid portion is then separated and analyzed for the specific substances given in Table 8.2. If values exceed the regulatory limits, the waste is designated as "toxic."

8.7. WASTE MANAGEMENT

As shown in Table 8.3, there are many kinds of hazardous wastes from a number of different sources. Hazardous waste **management** occurs through a carefully organized system in which wastes go through appropriate pathways from **generation**, through **treatment**, and ultimately to proper **disposal**. Throughout these steps the primary objective is protection of human health and the environment. Hazardous wastes are managed through **treatment, storage, and disposal facilities** (TSDF) in which the following definitions apply:[9]

Table 8.2. Contaminants Determined in TCLP Procedure

EPA hazardous waste number	Contaminant	Regulatory level, mg/L
Heavy metals (metalloids)		
D004	Arsenic	5.0
D005	Barium	100.0
D006	Cadmium	1.0
D007	Chromium	5.0
D008	Lead	5.0
D009	Mercury	0.2
D010	Selenium	1.0
D011	Silver	5.0
Organics		
D018	Benzene	0.5
D019	Carbon tetrachloride	0.5
D021	Chlorobenzene	100.0
D022	Chloroform	6.0
D023	*o*-Cresol	200.0[1]
D024	*m*-Cresol	200.0[1]
D025	*p*-Cresol	200.0[1]
D026	Cresol	200.0[1]
D027	1,4-Dichlorobenzene	7.5
D028	1,2-Dichloroethane	0.5
D029	1,1-Dichloroethylene	0.7
D030	2,4-Dinitrotoluene	0.13[2]
D032	Hexachlorobenzene	0.13[2]
D033	Hexachlorobutadiene	0.5
D034	Hexachloroethane	3.0
D035	Methylethyl ketone	200.0
D036	Nitrobenzene	2.0[2]
D037	Pentachlorophenol	100.0
D038	Pyridine	5.0[2]
D039	Tetrachloroethylene	0.7
D040	Trichloroethylene	0.5
D041	2,4,5-Trichlorophenol	400.0
D042	2,4,6-Trichlorophenol	2.0
D043	Vinyl chloride	0.2
Pesticides		
D012	Endrin	0.02
D013	Lindane	0.4
D014	Methoxychlor	10.0
D015	Toxaphene	0.5
D016	2,4-D	10.0
D017	2,4,5-TP (Silvex)	1.0
D020	Chlordane	0.03
D031	Heptachlor (and its epoxide)	0.008

[1] If *o*-, *m*-, and *p*-Cresol concentrations cannot be differentiated, the total cresol (D026) concentration is used. The regulatory level of total cresol is 200 mg/L.

[2] Quantitation limit is greater than the calculated regulatory level. The quantitation limit therefore becomes the regulatory level.

- Treatment: "Any method, technique, or process, including neutralization, designed to change the physical, chemical, or biological character or composition of any hazardous waste so as to neutralize such waste, or so as to recover energy or material resources from the waste, or so as to render such waste nonhazardous, or less hazardous; safer to transport, store, or dispose of; or amenable for recovery, amenable for storage, or reduced in volume."
- Storage: "The holding of hazardous wastes for a temporary period, at the end of which the hazardous waste is treated, disposed of, or stored elsewhere."
- Disposal: "The discharge, deposit, injection, dumping, spilling, leaking or placing of any solid waste or hazardous waste into or on any land or water so that such solid waste or hazardous waste or any constituent thereof may enter the environment or be emitted into the air or discharged into any waters, including ground waters."

Table 8.3. Sources and Nature of Wastes

Source	Examples of wastes[1]
Chemicals and allied products	Acids, alkalies, salts, heavy metals, organometallics, organics
Petroleum production	Brines, organics, metals (largely excluded from the "hazardous" category by legislation)
Petroleum refining and products	Organics, metals, spent catalysts that may contain toxic metals
Metals refining	Heavy metals, salts (including toxic cyanide)
Fabricated metals and metal plating	Heavy metals, organics (such as cutting oils), cyanide, chelating agents (heavy metal carriers), wastewater treatment sludges
Electronics	Heavy metals, cyanides, organic solvents
Electrical equipment manufacture	Heavy metals, sludges
Municipal waste treatment	Heavy metals and organics in sludges (largely excluded from the "hazardous" category by legislation)
Paint and pigment manufacture	Heavy metals, organic sludges, wastewater treatment sludge from manufacture of pigments such as molybdate orange pigments
Wood processing	Heavy metals, organic sludges, sludge from treatment of wastewater from wood treatment

[1] In the form of solids, liquids, sludges, contained gases

Permits are required for treatment, storage, and disposal facilities. The permitting system is designed to make sure that the facility operates safely, protects personnel, and protects the environment. A facility has to comply with both general requirements for accepted waste management as well as requirements specific to the facility. Permits require rather stringent reporting and inspection procedures and must remain in effect through closure of the facility. Post-closure permits may be required unless removal or decontamination can be demonstrated.

The **effectiveness** of a hazardous waste management system is a measure of how well it reduces the quantities and hazards of wastes.[10] As shown in Figure 8.2, the best management option consists of measures that prevent generation of wastes. Next in order of desirability is recovery and recycle of waste constituents. Next is destruction and treatment with conversion to nonhazardous waste forms. The least desirable option is disposal of hazardous materials in storage or landfill.

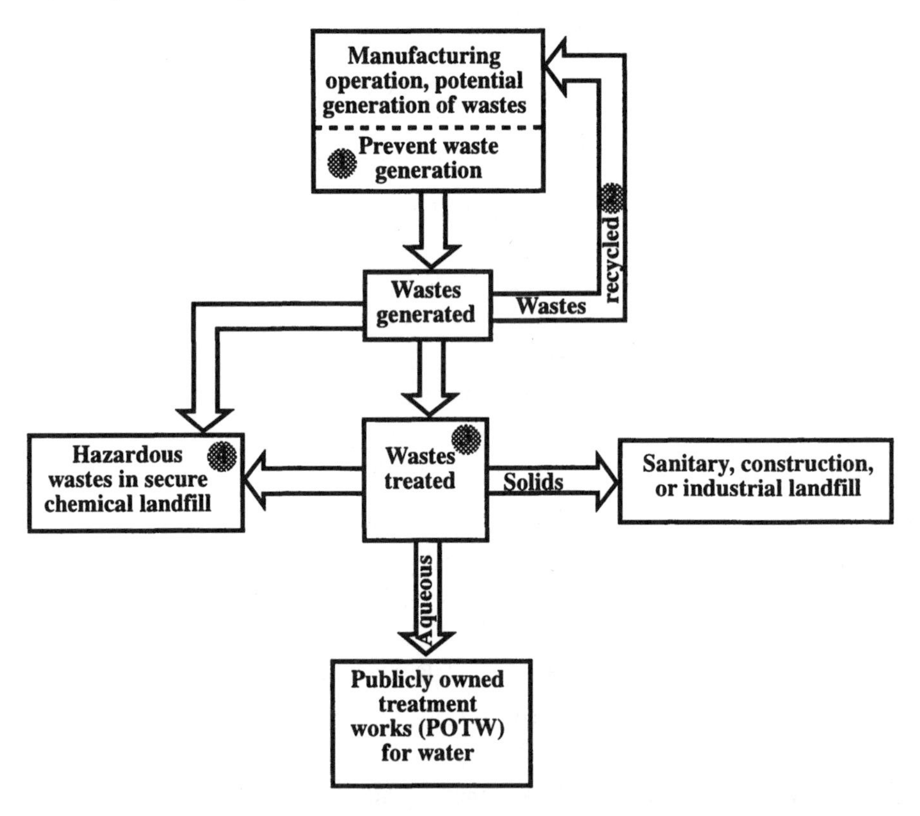

Figure 8.2. Order of effectiveness of waste treatment management options. The darkened circles indicate the degree of effectiveness from the most desirable (1) to the least (4).

8.8. SOURCES AND GENERATORS OF WASTES

Hazardous wastes come from many sources including manufacturing industries, agriculture, laboratories, and households. More than 95% of the quantities of hazardous wastes come from only about 10% of the generators. Around 75% of the quantities of hazardous wastes are generated by the petroleum and chemical industries; about 3/4 of the remainder comes from metal-related industries and only about 1/4 from all other industries. Table 8.3 lists some major sources of hazardous wastes and the kinds and forms of wastes that they typically generate.

Numerous kinds of industrial processes generate hazardous wastes. Distillation leaves a residue of still bottoms, such as those from the distillation of phthalic anhydride from naphthalene (K024). Hazardous residues may remain from extraction processes. Filter cakes, such as filter cake from the filtration of diethylphosphorodithioic acid in the production of phorate (K039) may be hazardous. Hazardous products of air pollution control include wastes such as baghouse dusts from the production of carbamates (K158).

LITERATURE CITED

1. Wolbeck, Bernd, "Political Dimensions and Implications of Hazardous Waste Disposal," in *Hazardous Waste Disposal*, Lehman, John P., Ed., Plenum Press, New York, 1982, pp. 7–18.
2. Karnofsky, Brian, Ed., *Hazardous Waste Management Compliance Handbook*, 2nd ed., Van Nostrand Reinhold, New York, NY, 1996.
3. Sullivan, Thomas F. P., Ed., *Environmental Law Handbook*, 14th ed., Government Institutes, Rockville, MD, 1997.
4. Petersen, Melody, "Cleaning Up in the Dark," New York Times, May 14, 1998, p. C1.
5. Hileman, Bette, "Treaty Grows Less Contentious," *Chemical and Engineering News*, April 6, 1998, pp. 29–30.
6. Cheremisinoff, Nicholas P. and Paul N. Cheremisinoff, *Hazardous Materials and Waste Management*, Noyes Publications, Park Ridge, New Jersey, 1995.
7. Lee, Robert T., "Comprehensive Environmental Response, Compensation and Liability Act," Chapter 8 in *Environmental Law Handbook*, 14th ed., Thomas F. P. Sullivan, Ed., Government Institutes, Rockville, MD, 1997, pp. 225–277.
8. "Toxicity Characteristic Leaching Procedure," Test Method 1311 in *Test Methods for Evaluating Solid Waste, Physical/Chemical Methods*, EPA Publication SW-846, 3rd ed., (November, 1986), as amended by Updates I, II, IIA, U.S. Government Printing Office, Washington, D.C.
9. "Hazardous Waste Management System: General," *Code of Federal Regulations*, **40**, July 1, 1986, Part 260, U.S. Government Printing Office, Washington, D.C., pp. 339–358.
10. Andrews, Richard N. L. and Francis M. Lynn, "Siting of Hazardous Waste Facilities," Section 3.1 in *Standard Handbook of Hazardous Waste Treatment and Disposal*, Harry M. Freeman, Ed., McGraw-Hill, New York, 1989, pp. 3.3–3.16.

9 ENVIRONMENTAL CHEMISTRY OF HAZARDOUS MATERIALS

9.1. INTRODUCTION

The properties of hazardous materials, their production, and what makes a hazardous substance a hazardous waste were discussed in Chapter 8. Hazardous materials normally cause problems when they enter the environment. Therefore, the present chapter deals with the environmental chemistry of hazardous materials. In discussing the environmental chemistry of hazardous materials, it is convenient to consider the following five aspects based upon the definition of environmental chemistry:

- Origins • Transport • Reactions • Effects • Fates

It is also useful to consider the five environmental spheres as defined and outlined in Chapter 1:

- Anthrosphere • Geosphere • Hydrosphere • Atmosphere • Biosphere

Hazardous materials almost always originate in the anthrosphere, are often discarded to the geosphere, and are frequently transported through the hydrosphere or the atmosphere. The greatest concern for their effects is usually on the biosphere, particularly human beings. Figure 9.1 summarizes these relationships.

As discussed in Chapter 8, there are numerous producers of hazardous wastes. There are also a variety of ways that they get into the environment. Although now much more controlled by pollution prevention laws, hazardous substances have been deliberately added to the environment by humans. Wastewater containing a variety of toxic substances has been discharged in large quantities into waterways. Hazardous gases and particulate matter have been discharged to the atmosphere through stacks from power plants, incinerators, and a variety of industrial operations. Hazardous wastes have been deliberately spread on soil or placed in landfills in the geosphere. Evaporation and wind erosion may move hazardous materials from wastes dumps into the atmosphere, or they may be leached from waste dumps into groundwater or surface waters. Underground storage tanks or pipelines have leaked a variety of materials into soil. Accidents, fires, and explosions may distribute dangerous materials into the environment. Another source of such materials consists of improperly operated waste treatment or storage facilities.

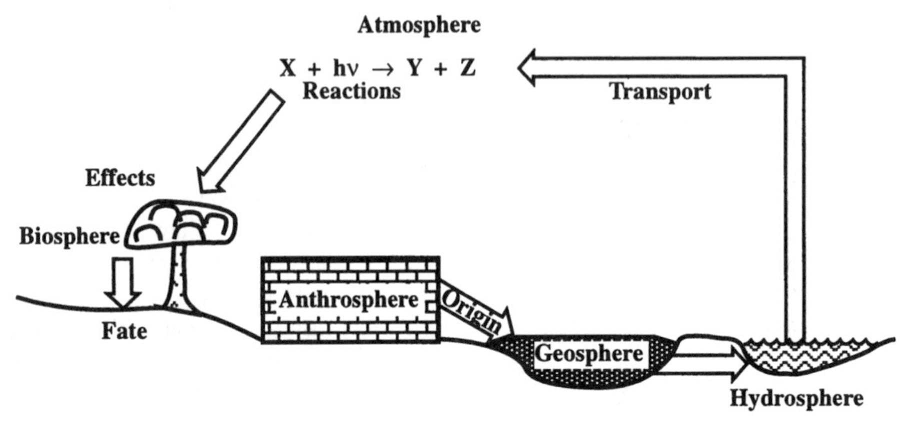

Figure 9.1. Scheme of interactions of hazardous wastes in the environment.

9.2. HAZARDOUS WASTES AND THE ANTHROSPHERE

As the part of the environment where humans process substances, the anthrosphere is the source of most hazardous wastes. These materials may come from manufacturing, transportation activities, agriculture, and any one of a number of activities in the anthrosphere. Hazardous wastes may be in any physical form and may include liquids, such as spent halogenated solvents used in degreasing parts; semisolid sludges, such as those generated from the gravitation separation of oil/water/solids mixtures in petroleum refining; and solids, such as baghouse dusts from the production of pesticides.

Releases of hazardous wastes from the anthrosphere commonly occur through incidents such as spills of liquids, accidental discharge of gases or vapors, fires and explosions.[1] Resource Conservation and Recovery Act (RCRA) regulations designed to minimize such accidental releases from the anthrosphere and to deal with them when they occur are contained in 40 CFR 265.31 (Title 40 of the Code of Federal Regulations, Part 265.31). Under these regulations, hazardous waste generators are required to have specified equipment, trained personnel, and procedures that protect human health in the event of a release and that facilitate remediation if a release occurs. An effective means of communication for summoning help and giving emergency instruction must be available. Also required are firefighting capabilities including fire extinguishers and adequate water. To deal with spills, a facility is required to have on hand absorbents, such as granular vermiculite clay, or absorbents in the form of pillows or pads. Neutralizing agents for corrosive substances that may be used should be available as well.

9.3. PHYSICAL AND CHEMICAL PROPERTIES OF HAZARDOUS WASTES

Having considered the generation of hazardous wastes from the anthrosphere, the next thing to consider is their properties, which determine movement and other kinds of behaviors. These properties can be generally divided into physical and chemical properties.

The behavior of waste substances in the atmosphere is largely determined by their volatilities. In addition, their solubilities in water determine the degree to which they are likely to be removed with precipitation. Water solubility is the most important physical property in the hydrosphere. The movement of substances through the action of water in the geosphere is largely determined by the degree of sorption to soil, mineral strata, and sediments.

Volatility is a function of the vapor pressure of a compound. Vapor pressures at a particular temperature can vary over many orders of magnitude. Of common organic liquids, diethyl ether has one of the highest vapor pressures, whereas those of polychlorinated biphenyls (PCBs) are very low. When a volatile liquid is present in soil or in water, its water solubility also determines how well it evaporates. For example, although methanol boils at a lower temperature than benzene, the much lower solubility of benzene in water means that it has the greater tendency to go from the hydrosphere or geosphere into the atmosphere.

The environmental movement, effects, and fates of hazardous waste compounds are strongly related to their chemical properties. For example, a toxic heavy metal cationic species, such as Pb^{2+} ion, may be strongly held by negatively charged soil solids. If the lead is chelated by the chelating EDTA anion, represented Y^{4-}, it becomes much more mobile in the anionic PbY^{2-} form. Oxidation state can be very important in the movement of hazardous substances. The reduced states of iron and manganese, Fe^{2+} and Mn^{2+}, respectively, are water soluble and relatively mobile in the hydrosphere and geosphere. However, in their common oxidized states, Fe(III) and Mn(IV), these elements are present as insoluble $Fe_2O_3{\cdot}xH_2O$ and MnO_2, which have virtually no tendency to move. Furthermore, these iron and manganese oxides will sequester heavy metal ions, such as Pb^{2+} and Cd^{2+}, preventing their movement in the soluble form.

The major properties of hazardous substances and their surroundings that determine the environmental transport of such substances are the following:

- Physical properties of the substances, including vapor pressure and solubility.
- Physical properties of the surrounding matrix.
- Physical conditions to which wastes are subjected. Higher temperatures and erosive wind conditions enable volatile substances to move more readily.
- Chemical and biochemical properties of wastes. Substances that are less chemically reactive and less biodegradable will tend to move farther before breaking down.

9.4. TRANSPORT, EFFECTS, AND FATES OF HAZARDOUS WASTE SUBSTANCES IN THE ANTHROSPHERE

As noted above, hazardous wastes originate in the anthrosphere. However, to a large extent they move, have effects, and end up in the anthrosphere as well. Large quantities of hazardous substances are moved by truck, rail, ship, and pipeline. Spills and releases from such movement, ranging from minor leaks from small containers to catastrophic releases of petroleum from wrecked tanker ships are a common occurrence. Much effort in the area of environmental protection can be profitably devoted to minimizing and increasing the safety of the transport of hazardous substances through the anthrosphere.

In the United States the transportation of hazardous substances is regulated through the U.S. Department of Transportation (DOT). One of the ways in which this is done is through the **manifest** system of documentation designed to accomplish the following goals:

- Acts as a tracking device to establish responsibility for the generation, movement, treatment, and disposal of the waste.
- By requiring the manifest to accompany the waste, such as during truck transport, it provides information regarding appropriate actions to take during emergencies, such as collisions, spills, fires, or explosions.
- Acts as the basic documentation for record keeping and reporting

Many of the adverse effects of hazardous substances occur in the anthrosphere. One of the main examples of such effects occurs as corrosion of materials that are strongly acidic or basic or that otherwise attack materials. Fire and explosion of hazardous materials can cause severe damage to anthrospheric infrastructure.

The fate of hazardous materials is often in the anthrosphere. One of the main examples of a material dispersed in the anthrosphere consists of lead-based anti-corrosive paints that are spread on steel structural members.

9.5. TRANSPORT, EFFECTS, AND FATES OF HAZARDOUS WASTE SUBSTANCES IN THE GEOSPHERE

The sources, transport, interactions, and fates of contaminant hazardous wastes in the geosphere involve a complex scheme, some aspects of which are illustrated in Figure 9.2. As illustrated in the figure, there are numerous vectors by which hazardous wastes can get into groundwater. Leachate from a landfill can move as a waste plume carried along by groundwater, in severe cases draining into a stream or into an aquifer, where it may contaminate well water. Sewers and pipelines may leak hazardous substances into the geosphere. Such substances seep from waste lagoons into geological strata, eventually contaminating groundwater. Wastes leaching from sites where they have been spread on land for disposal or as a means of treatment can contaminate the geosphere and groundwater. In some cases wastes are pumped into deep wells as a means of disposal.

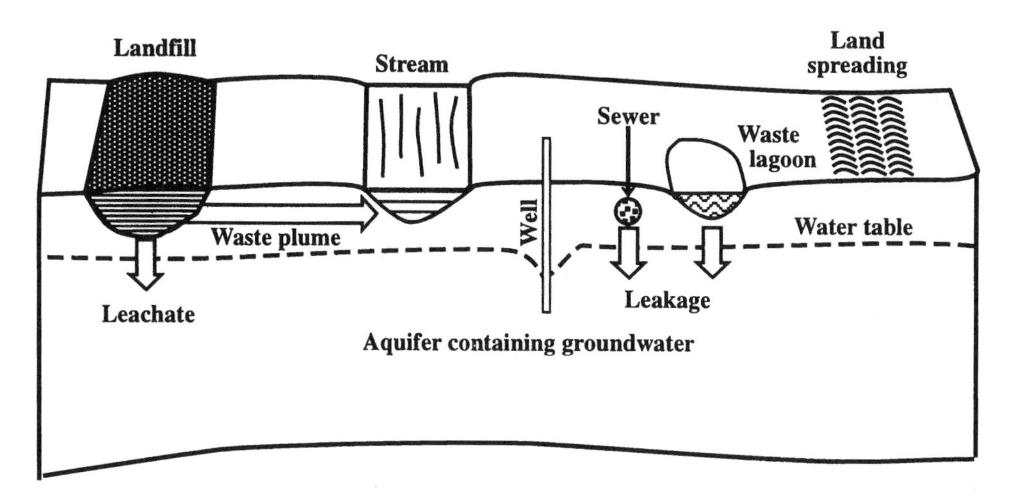

Figure 9.2. Sources and movement of hazardous wastes in the geosphere.

The movement of hazardous waste constituents in the geosphere is largely by the action of flowing water in a waste plume as shown in Figure 9.2. The speed and degree of waste flow depend upon numerous factors. Hydrologic factors, such as water gradient and permeability of the solid formations through which the waste plume moves are important. The rate of flow is usually rather slow, typically several centimeters per day. An important aspect of the movement of wastes through the geosphere is **attenuation** by the mineral strata. This occurs because waste compounds are sorbed to solids by various mechanisms. A measure of the attenuation can be expressed by a **distribution coefficient**, K_d,

$$K_d = \frac{C_S}{C_W} \tag{9.5.1}$$

where C_S and C_W are the equilibrium concentrations of the constituent on solids and in solution, respectively. This relationship assumes relatively ideal behavior of the hazardous substance that is partitioned between water and solids (the sorbate). A more empirical expression is based on the Freundlich equation,

$$C_S = K_F C_{eq}^{\ 1/n} \tag{9.5.2}$$

where and K_F and $1/n$ are empirical constants.

Several important properties of the solid determine the degree of sorption. One obvious factor is surface area. The chemical nature of the surface is also important. Among the important chemical factors are presence of sorptive clays, hydrous metal oxides, and humus (particularly important for the sorption of organic substances).

In general, sorption of hazardous waste solutes is higher above the water table in the unsaturated zone of soil. This region tends to have a higher surface area and to favor aerobic biodegradation processes.

The chemical nature of the leachate is important in sorptive processes of hazardous substances in the geosphere. Organic solvents or detergents in leachates will solubilize organic materials, preventing their retention by solids. Acidic leachates tend to dissolve metal oxides,

$$M(OH)_2(s) + 2H^+ \rightarrow M^{2+} + 2H_2O \qquad (9.5.3)$$

thus preventing sorption of metals in insoluble forms. This is a reason that leachates from municipal landfills, which contain weak organic acids, are particularly prone to transport metals. Solubilization by acids is particularly important in the movement of heavy-metal ions.

Heavy metals are among the most dangerous hazardous waste constituents that are transported through the geosphere. Many factors affect their movement and attenuation. The temperature, pH, reducing nature (as expressed by the negative log of the electron activity, pE) of the solvent medium are important. The nature of the solids, especially the inorganic and organic chemical functional groups on the surface, the cation-exchange capacity, and the surface area of the solids largely determine the attenuation of heavy metal ions. In addition to being sorbed and undergoing ion exchange with geospheric solids, heavy metals may undergo oxidation-reduction processes, precipitate as slightly soluble solids (especially sulfides), and in some cases, such as occurs with mercury, undergo microbial methylation reactions that produce mobile organometallic species.

The importance of chelating agents interacting with metals and increasing their mobilities has been illustrated by the effects of chelating ethylenediaminetetraacetic acid (EDTA) on the mobility of radioactive heavy metals, especially ^{60}Co.[2] The EDTA and other chelating agents, such as diethylenetriaminepentaacetic acid (DTPA) and nitrilotriacetic acid (NTA), were used to dissolve metals in the decontamination of radioactive facilities and were codisposed with radioactive materials at Oak Ridge National Laboratory (Tennessee) during the period 1951–1965. Unexpectedly high rates of radioactive metal mobility were observed, which was attributed to the formation of anionic species, such as $^{60}CoT^-$ (where T^{3-} is the chelating NTA anion). Whereas unchelated cationic metal species are strongly retained on soil by precipitation reactions and cation exchange processes,

$$Co^{2+} + 2OH^- \rightarrow Co(OH)_2(s) \qquad (9.5.4)$$

$$2Soil\}^-H^+ + Co^{2+} \rightarrow (Soil\}^-)_2Co^{2+} + 2H^+ \qquad (9.5.5)$$

anion bonding processes are very weak, so that the chelated anionic metal species are not strongly bound. Naturally occurring humic acid chelating agents may also be involved in the subsurface movement of radioactive metals. Other species that may also facilitate mobility include citrate, fluoride, oxalate, and gluconate salts. Although biodegradation of the chelating agents does occur,[3] it is a very slow process under subsurface conditions.

Soil can be severely damaged by hazardous waste substances. Such materials may alter the physical and chemical properties of soil and thus its ability to support plants. Some of the more catastrophic incidents in which soil has been damaged by exposure to hazardous materials have arisen from soil contamination from SO_2

emitted from copper or lead smelters or from brines from petroleum production. Both of these contaminants stop the growth of plants and, without the binding effects of viable plant root systems, topsoil is rapidly lost by erosion.

9.6. TRANSPORT, EFFECTS, AND FATES OF HAZARDOUS WASTE SUBSTANCES IN THE HYDROSPHERE

As mentioned in the preceding section, one of the more common ways for hazardous waste substances to enter the hydrosphere is as leachate from waste landfills, drainage from waste ponds, seepage from sewer lines, or runoff from soil. Deliberate release to waterways also occurs, and is a particular problem in countries with lax environmental enforcement. There are, therefore, numerous ways by which hazardous materials may enter the hydrosphere.

For the most part, the hydrosphere is a dynamic, moving system, so that it provides perhaps the most important variety of pathways for moving hazardous waste species in the environment. Once in the hydrosphere, hazardous waste species can undergo a number of processes by which they are degraded, retained, and transformed. These include the common chemical processes of precipitation-dissolution, acid-base reactions, hydrolysis, and oxidation-reduction reactions. Also included are a wide variety of biochemical processes which, in most cases, reduce hazards, but in some cases, such as the biomethylation of mercury, greatly increase the risks posed by hazardous wastes.

Physically, chemically, and biologically, water is a unique substance, a fact that has a strong influence on the environmental chemistry of hazardous wastes in the hydrosphere. Aquatic systems are subject to constant change. Water moves with groundwater flow, stream flow, and convection currents. Bodies of water become stratified so that low-oxygen reducing conditions may prevail in the bottom regions of a body of water, and there is a constant interaction of the hydrosphere with the other environmental spheres. There is a continuing exchange of materials between water and the other environmental spheres. Organisms in water may have a strong influence on even poorly biodegradable hazardous waste species through bioaccumulation mechanisms.

Figure 9.3 shows some of the pertinent aspects of hazardous waste materials in bodies of water, with emphasis upon the strong role played by sediments. An interesting kind of hazardous waste material that may accumulate in sediments consists of dense, water-immiscible liquids. These may simply sink to the bottom of bodies of water or aquifers and accumulate there as "blobs" of liquid. Hundreds of tons of PCB wastes have accumulated in sediments in the Hudson River in New York State and are the subject of a heated debate regarding how to remediate the problem.

Hazardous waste species undergo a number of physical, chemical, and biochemical processes in the hydrosphere which strongly influence their effects and fates. The major ones of these are listed below:

- **Hydrolysis reactions** are those in which a molecule is cleaved with addition of a molecule of H_2O. An example of a hydrolysis reaction is the hydrolysis of dibutyl phthalate, Hazardous Waste Number U069:

$$C_6H_4(COOC_4H_9)_2 + 2H_2O \rightarrow C_6H_4(COOH)_2 + 2HOC_4H_9$$

Another example is the hydrolysis of bis(chloromethyl)ether to produce HCl and formaldehyde:

$$Cl{-}CH_2{-}O{-}CH_2{-}Cl + H_2O \rightarrow 2H{-}C(=O){-}H + 2HCl$$

Compounds that hydrolyze are normally those, such as esters and acid anhydrides, originally formed by joining two other molecules with the loss of H_2O.

- **Precipitation reactions**, such as the formation of insoluble lead sulfide from soluble lead(II) ion in the anaerobic regions of a body of water:

$$Pb^{2+} + HS^- \rightarrow PbS(s) + H^+$$

An important part of the precipitation process is normally **aggregation** of the colloidal particles first formed to produce a cohesive mass. Precipitates are often relatively complicated species, such as the basic salt of lead carbonate, $2PbCO_3{\cdot}Pb(OH)_2$). Heavy metals, a common ingredient of hazardous waste species precipitated in the hydrosphere, tend to form hydroxides, carbonates, and sulfates with the OH^-, HCO_3^-, and SO_4^{2-} ions

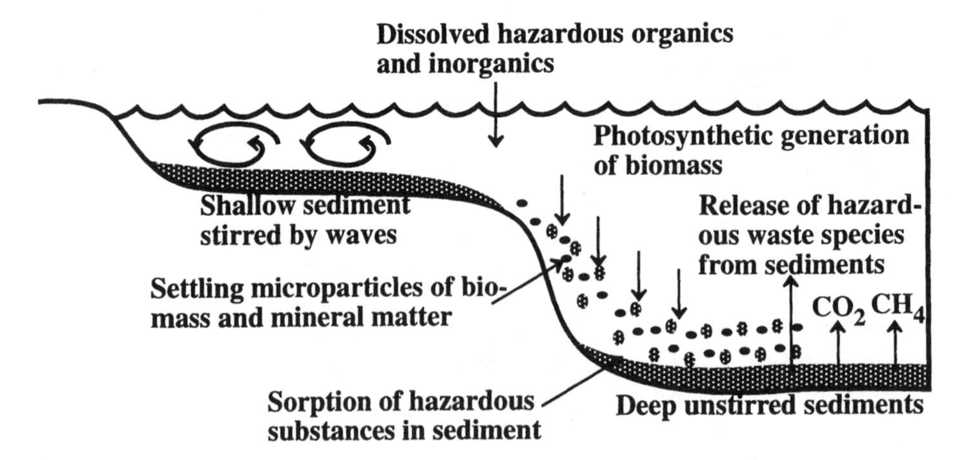

Figure 9.3. Aspects of hazardous wastes in surface water in the hydrosphere. The deep unstirred sediments are anaerobic and the site of hydrolysis reactions and reductive processes that may act on hazardous waste constituents sorbed to the sediment.

that commonly are present in water, and sulfides are likely to be formed in bottom regions of bodies of water where sulfide is generated by anaerobic bacteria. Heavy metals are often coprecipitated as a minor constituent of some other compound, or are sorbed by the surface of another solid.

- **Oxidation-reduction reactions** commonly occur with hazardous waste materials in the hydrosphere, generally mediated by microorganisms. An example of such a process is the oxidation of ammonia to toxic nitrite ion mediated by *Nitrosomonas* bacteria:

$$NH_3 + {}^3/_2O_2 \rightarrow H^+ + NO_2^-(s) + H_2O$$

- **Biochemical processes**, which often involve hydrolysis and oxidation-reduction reactions. Organic acids and chelating agents, such as citrate, produced by bacterial action may solubilize heavy metal ions. Bacteria also produce methylated forms of metals, particularly mercury and arsenic.

- **Photolysis reactions** and miscellaneous chemical phenomena. Photolysis of hazardous waste compounds in the hydrosphere commonly occurs on surface films exposed to sunlight on the top of water.

Hazardous waste compounds have a number of effects on the hydrosphere. Perhaps the most serious of these is the contamination of groundwater, which in some cases can be almost irreversible. Waste compounds accumulate in sediments, such as river or estuary sediments. Hazardous waste compounds dissolved in, suspended in, or floating as surface films on the surface of water can render it unfit for use and for sustenance of aquatic organisms.

Many factors determine the fate of a hazardous waste substance in water. Among these are the substance's solubility, density, biodegradability, and chemical reactivity. As discussed above and in Section 9.8, biodegradation largely determines the fates of hazardous waste substances in the hydrosphere. In addition to biodegradation, some substances are concentrated in organisms by bioaccumulation processes and may become deposited in sediments as a result. Organophilic materials may be sorbed by organic matter in sediments. Cation-exchanging sediments have the ability to bind cationic species, including organics that form cations.

9.7. TRANSPORT, EFFECTS, AND FATES OF HAZARDOUS WASTE SUBSTANCES IN THE ATMOSPHERE

Hazardous waste chemicals can enter the atmosphere by evaporation from hazardous waste sites, by wind erosion, or by direct release. Hazardous waste chemicals usually are not evolved in large enough quantities to produce secondary air pollutants. (Secondary air pollutants are formed by chemical processes in the atmosphere. Examples are sulfuric acid formed from emissions of sulfur oxides and oxidizing photochemical smog formed under sunny conditions from nitrogen oxides and hydrocarbons.) Therefore, species from hazardous waste sources are usually of most concern in the atmosphere as primary pollutants emitted in localized areas at a hazardous waste site. Plausible examples of primary air pollutant hazardous waste chemicals include corrosive acid gases, particularly HCl; toxic organic vapors, such as vinyl chloride (U043); corrosive acid gases (HCl); and toxic inorganic gases, such as HCN potentially released by the accidental mixing of waste cyanides:

$$H_2SO_4 + 2NaCN \rightarrow Na_2SO_4 + 2HCN(g) \qquad (9.7.1)$$

Primary air pollutants such as these are almost always of concern only adjacent to the site or to workers involved in site remediation. One such substance that has been responsible for fatal poisonings at hazardous waste sites, usually tanks that are undergoing cleanup or demolition, it highly toxic hydrogen sulfide gas, H_2S.

An important characteristic of a hazardous waste material that enters the atmosphere is its **pollution potential**. This refers to the degree of environmental threat posed by the substance acting as a primary pollutant, or to its potential to cause harm from secondary pollutants.

Another characteristic of a hazardous waste material that determines its threat to the atmosphere is its **residence time**, which can be expressed by an estimated atmospheric half-life, $\tau_{1/2}$. Among the factors that go into estimating atmospheric half lives are water solubilities, rainfall levels, and atmospheric mixing rates.

Hazardous waste compounds in the atmosphere that have significant water solubilities are commonly removed from the atmosphere by **dissolution** in water. The water may be in the form of very small cloud or fog particles or as rain droplets.

Some hazardous waste species in the atmosphere are removed by **adsorption onto aerosol particles**. Typically, the adsorption process is rapid so that the lifetime of the species is that of the aerosol particles (typically a few days). Adsorption onto solid particles is the most common removal mechanism for highly nonvolatile constituents such as benzo[a]pyrene.

Dry deposition is the name given to the process by which hazardous waste species are removed from the atmosphere by impingement onto soil, water, or plants on the Earth's surface. These rates are dependent upon the type of substance, the nature of the surface that they contact, and weather conditions.

A significant number of hazardous waste substances leave the atmosphere much more rapidly than predicted by dissolution, adsorption onto particles, and dry deposition, meaning that chemical processes must be involved. The most important of these are photochemical reactions, commonly involving hydroxyl radical, HO•. Other reactive atmospheric species that may act to remove hazardous waste compounds are ozone (O_3), atomic oxygen (O), peroxyl radicals (HOO•), alkylperoxyl radicals (ROO•), and NO_3. Although its concentration in the troposphere is relatively low, HO• is so reactive that it tends to predominate in the chemical processes that remove hazardous waste species from air. Hydroxyl radical undergoes *abstraction reactions* that remove H atoms from organic compounds,

$$R\text{-}H + HO\bullet \rightarrow R\bullet + H_2O \qquad (9.7.2)$$

and may react with those containing unsaturated bonds by addition as illustrated in Reaction 9.7.3.

$$\begin{matrix} R & & H \\ & C{=}C & \\ H & & H \end{matrix} + HO\cdot \longrightarrow H{-}\overset{R}{\underset{\bullet}{C}}{-}\overset{H}{\underset{H}{C}}{-}OH \tag{9.7.3}$$

The free radical products are very reactive. They react further to form oxygenated species, such as aldehydes, ketones, and dehalogenated organics, eventually leading to the formation of particles or water-soluble materials that are readily scavenged from the atmosphere.

Direct photodissociation of hazardous waste compounds in the atmosphere may occur by the action of the shorter wavelength light that reaches to the troposphere and is absorbed by a molecule with a light-absorbing group called a **chromophore**:

$$R{-}X + h\nu \rightarrow R\bullet + X\bullet \tag{9.7.4}$$

Among the factors involved in assessing the effectiveness of direct absorption of light to remove species from the atmosphere are light intensity, quantum yields (chemical reactions per quantum absorbed), and atmospheric mixing. The requirement of a suitable chromophore limits direct photolysis as a removal mechanism for most compounds other than conjugated alkenes, carbonyl compounds, some halides, and some nitrogen compounds, particularly nitro compounds, all of which commonly occur in hazardous wastes.

9.8. TRANSPORT, EFFECTS, AND FATES OF HAZARDOUS WASTE SUBSTANCES IN THE BIOSPHERE

Microorganisms, bacteria, fungi, and to a certain extent protozoa, may act metabolically on hazardous wastes substances in the environment. Most of these subtances are *anthropogenic* (made by human activities), and most are classified as *xenobiotic* molecules that are foreign to living systems. Although by their nature xenobiotic compounds are degradation resistant, almost all classes of them—nonhalogenated alkanes, halogenated alkanes (trichloroethane, dichloromethane), nonhalogenated aryl compounds (benzene, naphthalene, benzo[a]pyrene), halogenated aryl compounds (hexachlorobenzene, pentachlorophenol), phenols (phenol, cresols), polychlorinated biphenyls, phthalate esters, and pesticides (chlordane, parathion)—can be at least partially degraded by various microorganisms.

Bioaccumulation occurs in which wastes are concentrated in the tissue of organisms. It is an important mechanism by which wastes enter food chains. **Biodegradation** occurs when wastes are converted by biological processes to generally simpler molecules; the complete conversion to simple inorganic species, such as CO_2, NH_3, SO_4^{2-}, and $H_2PO_4^-/HPO_4^-$, is called **mineralization.** The production of a less toxic product by biochemical processes is called **detoxification**. An example is the bioconversion of highly toxic organophosphate paraoxon to *p*-nitrophenol, which is only about 1/200 as toxic:

$$H_5C_2\text{-O-}\overset{\overset{\text{O}}{\|}}{\underset{\underset{H_5C_2\text{-O}}{|}}{\text{P}}}\text{-O-}C_6H_4\text{-}NO_2 \xrightarrow[\text{Enzymes}]{H_2O,\{O\}} HO\text{-}C_6H_4\text{-}NO_2 + \text{other products} \quad (9.8.1)$$

Microbial Metabolism in Waste Degradation

The following terms and concepts apply to the metabolic processes by which microorganisms biodegrade hazardous wastes substances:

- **Biotransformation** is the enzymatic alteration of a substance by microorganisms.
- **Metabolism** is the biochemical process by which biotransformation is carried out.
- **Catabolism** is an enzymatic process by which more complex molecules are broken down into less complex ones.
- **Anabolism** is an enzymatic process by which simple molecules are assembled into more complex biomolecules.

Two major divisions of biochemical metabolism that operate on hazardous waste species are **aerobic processes** that use molecular O_2 as an oxygen source and **anaerobic processes**, which make use of another oxidant. For example, when sulfate ion acts as an oxidant (electron receptor) the transformation $SO_4^{2-} \rightarrow H_2S$ occurs. (This has the benefit of providing sulfide, which precipitates insoluble metal sulfides in the presence of hazardous waste heavy metals.) Because molecular oxygen does not penetrate to such depths, anaerobic processes predominate in the deep sediments as shown in Figure 9.3.

For the most part, anthropogenic compounds resist biodegradation much more strongly than do naturally occurring compounds. Given the nature of xenobiotic substances, there are very few enzyme systems in microorganisms that act directly on these substances, especially in making an intial attack on the molecule. Therefore, most xenobiotic compounds are acted upon by a process called **cometabolism**, which occurs concurrently with normal metabolic processes. An interesting example of cometabolism is provided by the white rot fungus, *Phanerochaete chrysosporium*, which has been promoted for the treatment of hazardous organochlorides, such as PCBs, DDT, and chlorodioxins. This fungus uses dead wood as a carbon source and has an enzyme system that breaks down wood lignin, a degradation-resistant biopolymer that binds the cellulose in wood. Under appropriate conditions, this enzyme system attacks organochloride compounds and enables their mineralization.

The susceptibility of a xenobiotic hazardous waste compound to biodegradation depends upon its physical and chemical characteristics. Important physical characteristics include water solubility, hydrophobicity (aversion to water), volatility, and lipophilicity (affinity for lipids). In organic compounds, certain structural groups — branched carbon chains, ether linkages, meta-substituted benzene rings, chlorine, amines, methoxy groups, sulfonates, and nitro groups — impart particular resistance to biodegradation.

Microorganisms vary in their ability to degrade hazardous waste compounds; virtually never does a single microorganism have the ability to completely mineralize a waste compound. Abundant aerobic bacteria of the ***Pseudomonas*** family are particularly adept at degrading synthetic compounds, such as biphenyl, naphthalene, DDT, and many other compounds. ***Actinomycetes***, microorganisms that are morphologically similar to both bacteria and fungi, degrade a variety of organic compounds, including degradation-resistant alkanes, and lignocellulose, as well as pyridines, phenols, nonchlorinated aryls, and chlorinated aryls.

Because of their requirement for oxygen-free (anoxic) conditions, anaerobic bacteria are fastidious and difficult to study. However, they can play an important role in degrading biomass, particularly through hydrolytic processes in which molecules are cleaved with addition of H_2O. Anaerobic bacteria reduce oxygenated organic functional groups. As examples, they convert nitro compounds to amines, degrade nitrosamines, promote reductive dechlorination, reduce epoxide groups to alkenes, and break down aryl structures. Partial dechlorination of PCBs has been reported by bacteria growing anaerobically in PCB-contaminated river sediments.[4]

Fungi are particularly noted for their ability to attack long-chain and complex hydrocarbons and are more successful than bacteria in the initial attack on PCB compounds. The potential of the white rot fungus, *Phanerochaete chrysosporium*, to degrade biodegradation-resistant compounds, especially organochloride species, was previously noted.

Phototrophic microorganisms, algae, photosynthetic bacteria, and cyanobacteria that perform photosynthesis, have lipid bodies that accumulate lipophilic compounds. There is some evidence to suggest that these organisms can induce photochemical degradation of the stored compounds.

Biologically, the greatest concern with wastes has to do with their toxic effects on animals, plants, and microbes. Virtually all hazardous waste substances are poisonous to a degree, some extremely so. Toxicities vary markedly with the physical and chemical nature of the waste, the matrix in which it is contained, the type and condition of the species exposed, and the manner, degree, and time of exposure.

LITERATURE CITED

1. "Spills, Fires, Explosions, and Other Releases," Chapter 9 in *Hazardous Waste Management Compliance Handbook*, 2nd ed., Van Nostrand Reinhold Publishing Company, New York, NY, 1997, pp. 107–116.

2. Means, J. L., D. A. Crear, and J. O. Duguid, "Migration of Radioactive Wastes: Radionuclide Mobilization by Complexing Agents," *Science* **200**, 1477–81 (1978).

3. Bolton, H., Jr., S. W. Li, D. J. Workman, and D. C. Girvin, "Biodegradation of Synthetic Chelates in Subsurface Sediments from the Southeast Coastal Plane," *Journal of Environmental Quality*, **22**, 125–132 (1993).

4. Rhee, G.-Yull, Roger C. Sokol, Brian Bush, and Charlotte M. Bethoney, "Long-Term Study of the Anaerobic Dechlorination of Arochlor 1254 With and Without Biphenyl Enrichment," *Environmental Science and Technology*, **27**, 714–719 (1993).

SUPPLEMENTARY REFERENCES

Environmental Restoration and Waste Management. Five-Year Plan Fiscal Years 1992–1996, U.S. Department of Energy, Washington, D.C., 1990.

Freeman, Harry M., Ed., *Standard Handbook of Hazardous Waste Treatment and Disposal*, 2nd ed., McGraw-Hill, New York, NY, 1998.

Grosse, Douglas, *Managing Hazardous Wastes Containing Heavy Metals*, SciTech Publishers, Matawan, NJ, 1990.

Higgins, Thomas E., *Hazardous Waste Minimization Handbook*, Lewis Publishers, Chelsea, MI, 1989.

Howard, Philip H., *Handbook of Environmental Fate and Exposure Data for Organic Chemicals. Vol. 1, Large Production and Priority Pollutants*, Lewis Publishers, Chelsea, MI, 1989.

Racke, Kenneth D. and Joel R. Coats, Eds., *Enhanced Biodegradation of Pesticides in the Environment*, American Chemical Society, Washington, D.C., 1990.

Suffet, I. H. and Patrick MacCarthy, Eds., *Aquatic Humic Substances: Influence on Fate and Treatment of Pollutants*, Advances in Chemistry Series **219**, American Chemical Society, Washington, D.C., 1989.

10 HAZARDOUS INORGANIC AND ORGANOMETALLIC MATERIALS

10.1. INTRODUCTION

This chapter addresses hazardous inorganic materials along with hazardous organometallic compounds. Organic materials are covered in Chapter 11. Chapter 10 is further divided into sections addressing (1) hazardous nonmetallic elements, (2) hazardous metals and metalloids, (3) hazardous inorganic compounds, and (4) hazardous organometallic compounds. These divisions are, of course, somewhat artificial. In addition, there is significant overlap among them. It should also be pointed out that there are interactions among all classes of hazardous chemical substances that affect the hazards posed by each. For example, organic chelating agents bind with heavy metals greatly affecting their mobilities and toxicities and the presence of humic substances in hazardous waste leachate may affect the transport of hazardous organic constituents in the leachate.

The first of two criteria for hazardous elements is that some elemental forms are particularly dangerous. For example, elemental chlorine is a very toxic, corrosive gas, although in the combined form of chloride ion, it is a constituent of sodium chloride, ordinary table salt. Elemental mercury vapor, white phosphorus, and ozone, a form of elemental oxygen, are all quite toxic. Other elements are notable for their toxicities in chemically bound forms. Most of these are heavy metals or metalloids, such as lead, cadmium, or arsenic.

Inorganic compounds present a wide variety of toxicities. For the most part these compounds are composed of elements that are not by themselves notably toxic. For example the elements in deadly hydrogen cyanide, HCN, Hazardous Waste Number P063, are all essential ingredients of proteins. Inorganic compounds often pose more than one hazard. For example, in addition to its toxicity, HCN is also flammable.

Organometallic compounds include a variety of species. They are usually hazardous because of their toxicities; dimethyl mercury and tetraethyl lead (P110) are examples.

10.2. HAZARDOUS NONMETALLIC ELEMENTS

Elemental **hydrogen, H_2**, is a gas that is sometimes handled and transported as a liquid at very low temperatures. It has many industrial uses, including synthesis of ammonia, NH_3, organic synthesis, and petroleum upgrading. It is used to hydrogenate unsaturated oils in the food industry. Since its combustion product is water, it is a nonpolluting fuel and can be used in fuel cells to directly generate electricity. It is used along with oxygen as a propellant in the NASA Space Shuttle program. The same flammability property that make it such a good fuel also makes it a hazardous flammable substance, and it forms extremely explosive mixtures with air. Residual absorbed hydrogen can make discarded hydrogenation catalyst hazardous.

Carbon can be hazardous in the form of **carbon black**, which is produced by the partial combustion of hydrocarbons, such as methane:

$$CH_4 + O_2 \rightarrow C\ (\textit{finely divided, amorphous}) + 2H_2O \qquad (10.2.1)$$

Carbon black is widely used in the manufacture of pigments, inks, toners, plastics, and carbon electrodes, and as an additive in rubber tire manufacture. It can form explosive dusts and explosive mixtures when mixed with oxidants, such as NH_4NO_3. It conducts electricity and can result in fires by causing shorts in electrical equipment.

The most common form of elemental oxygen is the diatomic gas, O_2. Essentially pure O_2 may be extracted from air by distillation of liquid air or by membrane processes operating on gaseous air. It is one of the more common chemical commodities, usually used as the gas, but commonly stored and transported as the cryogenic liquid. Exposure to pure oxygen may cause substances to burn violently, and extremely explosive mixtures are formed when combustible vapors or combustible solid dusts are present in an oxygen atmosphere.

A much more rare form of oxygen is ozone, O_3, produced when oxygen atoms combine with molecular oxygen.

$$O + O_2 \rightarrow O_3 \qquad (10.2.2)$$

Oxygen atoms required to generate ozone may be produced by photochemical dissociation of molecular oxygen, but are generated in the commercial production of ozone by an electrical discharge through dry oxygen gas or air:

$$O_2 \xrightarrow[\text{discharge}]{\text{Electrical}} O + O \qquad (10.2.3)$$

Ozone generated by an electrical discharge can be is used as a water disinfectant in place of chlorine and as a water and air deodorant. Some specialized chemical synthesis reactions (ozonation) use ozone as an oxidizing agent, and it is an effective bleaching agent.

Ozone is a potent oxidant that can react explosively with oxidizable organic substances. Ozone is very toxic, causing severe irritation to the eyes and respiratory tract at levels of a part per million or less in air. Severe exposures may cause fatal pulmonary edema. The toxicological action of ozone is due in part to its ability to form reactive free radicals (species such as HO•, with an unpaired electron) in tissue.

The three lighter halogens — **fluorine** (F_2), **chlorine** (Cl_2), and **bromine** (Br_2) are widely produced and used in the elemental form for a variety of purposes, including chemical synthesis. Fluorine is generated by the electrolysis of molten KHF_2 containing some HF. Chlorine is generated by the electrolysis of sodium chloride, NaCl. Bromine can also be generated by electrolysis or by displacement from its salts with elemental Cl_2. Under normal conditions, elemental fluorine and chlorine are gases and bromine is a volatile liquid. Fluorine and chlorine are commonly stored and transported as liquids.

Fluorine is the most chemically reactive element and elemental oxidizing agent; it combines violently with inorganic and organic reducing agents. It is a highly toxic material and reacts with water to produce extremely toxic hydrogen fluoride, HF.

Ingestion of excessive amounts of fluoride ion, F^-, causes bone abnormalities and the formation of soft, mottled teeth. This condition is called **fluorosis**. The possibility of fluorosis is the reason that deliberate fluoridation of water to prevent tooth decay is so strongly opposed by some groups and individuals. Some water supplies are naturally too high in fluoride. Industrial fallout of fluoride on grazing land has led to conditions in which grazing animals become lame or die.

Elemental chlorine is a widely used industrial chemical with annual U.S. production of about 10 million tons per year. Chlorine is used as a water disinfectant, as a bleach, and to make organochloride compounds (chlorinated hydrocarbons). It is a reactive, potent oxidizing agent, though less so than elemental fluorine. It is toxic and was the first poisonous gas used in warfare.

Volatile brown liquid bromine is used to synthesize pharmaceuticals, gasoline additives, agricultural chemicals, and other chemicals. When spilled as the liquid form on a strong reducing agent, such as finely divided aluminum, it can act violently as an oxidizing agent. It is corrosive to skin as either liquid or vapor.

There are several elemental forms of **phosphorus** of which the most common by far is **white phosphorus**, widely used to make phosphate fertilizer, food-grade phosphoric acids and phosphates, organophosphate insecticides, and a variety of phosphorus-containing organic compounds. White phosphorus is very dangerous because of its reactivity and toxicity. It ignites spontaneously in air to form a dense white fog of P_4O_{10}, a highly deliquescent and corrosive material that reacts with water or water vapor in the air to give orthophosphoric acid, H_3PO_4. Elemental white phosphorus is a systemic poison that can be inhaled or get into the body through the skin or orally. Exposure to white phosphorus causes a number of maladies, of which deterioration and fracture of the jaw bone, a condition called **phossy jaw**, is the most characteristic of phosphorus poisoning. White phosphorus also causes bone brittleness, eye damage, anemia, and gastrointestinal system dysfunction.

Elemental sulfur, S, is normally encountered as a yellow solid. It burns to produce toxic sulfur dioxide gas, SO_2, and finely divided sulfur can be explosive in air. One of the worst industrial accidents of all time involved the reaction of strongly reducing elemental sulfur with ammonium nitrate, NH_4NO_3. This catastrophe occurred in Texas City, Texas, in 1947 when sulfur in a ship caught on fire and melted onto ammonium nitrate, which was also in the ship's hold, causing an explosion that killed almost 600 people. Elemental sulfur is not particularly toxic, although inhalation of sulfur dust can irritate mucous membranes.

10.3. HAZARDOUS METALS AND METALLOIDS

Metals compose most of the elements in the periodic table. Several elements, the **metalloids**, are on the border between metals and nonmetals in the periodic table and are discussed here along with metals. All the metals and metalloids have forms that are hazardous. Most notably, a number of the higher atomic mass elements, the heavy metals, stand out for the toxicities of their compounds and one of them, mercury, is uniquely hazardous as the elemental metal vapor. Very reactive metals are hazardous in their elemental forms because they are extremely strong reducing agents. The hazards of metals and metalloids are summarized in Table 10.1, and several that are notably hazardous are discussed separately in this section.

Arsenic (As) is a toxic metalloid contained in wastes from arsenic compound processing and wastes from other sources, such as As_2S_3 removed from food-grade phosphoric acid. Arsenic compounds are present in several listed wastes, including arsenic acid, H_3AsO_4 (P010); arsenic(III) oxide, As_2O_3 (P012); and arsenic(V), As_2O_5 oxide (P010). It is also present in some organometallic compounds, for example, diethylarsine (P038) and dimethyl arsinic acid (U136). The manufacture of arsenic-containing veterinary pharmaceuticals produces hazardous wastewater treatment sludges (K084), distillation tar residues (K101), and activated carbon residues (K102).

Chemically combined arsenic can exist in the +5 and +3 oxidation states, of which the latter is generally more toxic. A major biochemical effect of arsenic is that it inhibits the production of adenosine triphosphate (ATP), a key intermediate in energy-yielding metabolic processes. Arsenic also complexes with coenzymes, preventing them from carrying out their functions, and it coagulates proteins.

Although **beryllium** (Be) is in Group 2A of the periodic table, its chemistry is atypical of that group, such as is manifested by its tendency to form covalent compounds, for example, nonionic $BeCl_2$. Beryllium has some important applications, especially in the formulation of alloys and ceramics. Beryllium powder is designated as a hazardous waste (P015). Beryllium is a very toxic metal, the most serious effect of which is berylliosis, an often fatal respiratory condition manifested by lung fibrosis and pneumonitis. This insidious disease may have latency periods of up to 20 years. Skin exposed to beryllium may develop granulomas and become ulcerated. Beryllium may also cause hypersensitivity in exposed individuals.

Cadmium is widely used for corrosion-resistant metal plating, in pigments, in catalysts, in plastic stabilizers, and with nickel in rechargeable batteries. It is also contained in some wastes from steel production. Cadmium is one of the more toxic of the common heavy metals. Exposure can damage the kidneys and cause a painful bone disease called osteomalacia. Finely divided cadmium oxide particles in air are

Table 10.1. Hazardous Metals and Metalloids

Metal or class of metals or metalloids	Hazard
Group 1A[1] alkali metals, lithium (Li), sodium (Na), potassium (K)	Highly reactive reductants that react strongly with water to produce their strongly basic hydroxides and generate so much heat that they may catch fire. Fires are very difficult to extinguish. Corrosive to skin.
Group 2a metals, magnesium (Mg), calcium (Ca)	Strong reducing agents that burn very vigorously. Magnesium is commonly used as a structural metal, and its turnings are dangerously flammable.
Antimony	Constituent of hazardous waste No. K021, aqueous spent antimony catalyst waste from fluoromethanes production.
Barium (Ba)[2]	Toxic
Beryllium (Be)	Toxic light metal contained in beryllium dust (waste No. P015).
Chromium (Cr)[2] (K090)	Heavy metal toxic in the Cr(VI) form, which is the hazardous constituent responsible for listing a number of hazardous wastes (K002-K006, K048-K051).
Nickel (Ni)	Toxic heavy metal contained in some metal plating wastes and as the particularly toxic nickel carbonyl, $Ni(CO)_4$ (P073).
Osmium	Constituent of osmium tetroxide (P087)
Selenium (Se)[2]	Toxic metalloid present in selenium dioxide (U204) and selenium sulfide (U205).
Silver (Ag)[2]	Toxic heavy metal
Thallium (Tl)	Toxic heavy metal present in thallium oxide, Tl_2O_3 (P113); thallium(I) selenite (P114); and thallium(I) sulfate (P115).
Vanadium	Contained in vanadium pentoxide (P120)

[1] Refers to groups in the periodic table.
[2] Measured as part of toxicity characteristics leaching procedure (TCLP).

particularly dangerous. Inhalation of such dusts or fumes can cause pulmonary epithelium necrosis and edema (cadmium pneumonitis).

Lead is the most widely used of the toxic heavy metals. It is an ingredient of listed wastes, including those from lead smelting, pigment production, and processing other metals. Several lead compounds including lead acetate (U144), lead phosphate (U145), and lead subacetate (U146), are listed hazardous wastes. Other lead-containing hazardous wastes are surface impoundment solids contained in and dredged from surface impoundments at primary lead smelting facilities (K065), emission control dust/sludge from secondary lead smelting (K069), and waste leaching solution from acid leaching of emission control dust/sludge from secondary lead lead smelting (K100).

The hazard from lead is due to its toxicity, aggravated by its widespread distribution as the metal and in inorganic and organometallic compounds. Lead poisoning damages the central and peripheral nervous systems. It can cause damage to the kidneys and inhibit the synthesis of hemoglobin.

Elemental **mercury** metal (U151) is used as the liquid in various instruments, thermometers, vacuum apparatus, switches and seals, and as the vapor in mercury vapor lamps and ultraviolet radiation sources. Mercury-containing hazardous wastes include brine purification muds from the mercury cell process in chlorine production, where separately pre-purified brine is not used (K071) and wastewater treatment sludge from the mercury cell process in chlorine production (K106).

By far the greatest hazard posed by mercury is that of its toxicity. Almost unique among metals, mercury metal has a significant vapor pressure. When inhaled, mercury vapor can be carried by the blood to the brain where it readily penetrates the blood-brain barrier, disrupting metabolic processes in the brain. In addition to tremor, exposure of the brain to mercury causes a variety of psychopathological symptoms, including depression, shyness, insomnia, and irritability. Inorganic mercury compounds may be quite toxic. Both inorganic and organometallic mercury compounds (most notoriously, dimethyl mercury) are hazardous. Mercury(II) chloride, $HgCl_2$, known as "corrosive sublimate," is a deadly poison, and Hg^{2+} ion damages the kidney.

Although one of the less toxic metals, **zinc** has caused some problems when dispersed to the environment. One place where this has occurred is Palmerton, PA, where a zinc smelter operation left deposits of zinc and other metals deposited on the soil on a mountain overlooking the town.[1] As a result, soil on parts of the mountain is sterile and does not support plant life, or even the microorganisms that would normally degrade dead tree matter, which remains where it fell years after the trees were killed. Some contaminated yards in Palmerton have had to be resodded to permit growth of grass and to reduce metal exposure to children.

10.4. HAZARDOUS INORGANIC COMPOUNDS

There is a wide variety of hazardous inorganic compounds. Some of them are reactive or strong oxidizers, particularly those exhibiting extremes of pH may be corrosive; most are toxic to some degree, many powerfully so. Some inorganic compounds are hazardous because of more than one characteristic, one of which is usually toxicity.

Prominent among the toxic inorganic species are **arsenic salts**, of which arsenic acid (P010), arsenic(V) oxide (P011), and arsenic(III) oxide (P012) were previously mentioned. Aluminum phosphide, AlP, used as an insecticide and fumigant is hazardous; it releases toxic phosphine (PH_3) in contact with water. Boron trichloride, BCl_3, is reactive and toxic, hydrolyzing to HCl in contact with water.

Phosgene (P095), $COCl_2$, boils at 8.3°C and may be encountered as a highly volatile liquid or as a gas. Despite the fact that it is so toxic that it has been used as a military poison, it is required in some processes for the manufacture of polyurethane, polycarbonates, and herbicides. In addition to its toxicity, it is hazardous because of its high reactivity with a number of subtances, such as aluminum.

Carbon monoxide, CO, is a flammable gas. It is used in chemical synthesis and is one of the two major ingredients of synthesis gas, a mixture with H_2 which is produced by the reaction of steam with hot carbon. Carbon monoxide is a common byproduct of the partial oxidation of carbon. It is hazardous because of its flammability, ability to form explosive mixtures, and toxicity. The toxic effects of carbon monoxide are manifested by impairment of judgement and visual perception at levels of around 10 parts per million (ppm) in air; fatigue, headache, and dizziness at 100 ppm; unconsciousness and eventual death at 250 ppm; and rapid death at 1,000 pm. Carbon monoxide acts by depriving the body of oxygen by binding strongly to hemoglobin (Hb) in blood:

$$CO + O_2Hb \rightarrow O_2 + COHb \qquad (10.4.1)$$

Although the reaction is reversible, COHb (carboxyhemoglobin) is much more stable than O_2Hb (oxyhemoglobin), and, when a substantial fraction of the hemoglobin is converted to carboxyhemoglobin, oxygen deprivation occurs.

Interhalogen Compounds

Interhalogen compounds are those in which two halogens form a compound. There are numerous examples of interhalogen compounds. The most common contain fluorine and include ClF, a colorless gas, mp –154°C, bp –101°C; ClF_5, a colorless gas, mp –83°C, bp 12°C; BrF, a pale brown gas, bp 20°C; BrF_5, a colorless liquid, mp –61.3°C, bp 40°C; and IF_7, a colorless sublimable solid, mp 5.5°C. Other examples of interhalogen compounds include BrCl, a red/yellow highly unstable liquid and gas; IBr, a gray sublimable solid, mp 42°C; ICl, a red-brown solid, mp 27°C, bp 9°C; and ICl_3, an orange-yellow solid subliming at 64°C.

Interahalogen compounds are powerful, reactive oxidizing agents. They react with water to produce acidic HX solutions (where X represents a halogen) and strongly oxidizing nascent oxygen, {O}. Toxicologically, they are corrosive poisons that acidify, oxidize, and dehydrate tissue at the point of contact. They are especially damaging to the eyes and to the mucous membranes of the mouth, throat, and pulmonary systems. The fluorides produce highly toxic HF in contact with tissue.

Halogen Oxides

Examples of halogen oxides include oxygen difluoride, OF_2, a colorless gas, mp –224°C, bp –145°C; Cl_2O_7, a colorless oil, mp –91.5°C, bp 82°C; Br_2O, a brown solid which decomposes at –18°C; and I_2O_5, a colorless oil, which decomposes at 325°C. The most commonly used halogen oxide is chlorine dioxide, ClO_2. This compound is used in place of elemental chlorine for bleaching (of wood pulp), oxidation, odor control, and disinfection. Its advantage over chlorine is that it forms fewer undesirable byproducts, particularly trihalomethanes, such as chloroform, $HCCl_3$. Because of its extreme chemical reactivity, chlorine dioxide is generated on site and is not shipped.

The hazards presented by the halogen oxides are largely due to their extreme reactivity and potent oxidizing abilities, in which respect they resemble the elemental halogens and interhalogen compounds. They react with moisture in tissue to produce the hydrohalic acids (HF, HCl) and nascent oxygen, {O}. Therefore, they are strong irritants to eye, skin, and mucous membrane tissue.

Cyanide Compounds

Because of the extreme toxicity of cyanide as HCN (P063) or cyanide ion, CN^-, **cyanide compounds** are among the most prominent hazardous wastes. These include barium cyanide (P013), calcium cyanide (P021), potassium cyanide (P098), sodium cyanide (P106) and zinc cyanide (P121). Cyanide may be encountered in any one of several forms, including hydrogen cyanide gas, HCN (P063); hydrocyanic acid (dissolved unionized HCN(*aq*)); dissolved cyanide ion, CN^-; solid cyanide salts, such as NaCN or KCN; and complexed cyanide in ions such as $Fe(CN)_6^{4-}$ or $Ni(CN)_4^{2-}$. A variety of different wastes are considered to be hazardous because of their cyanide contents. These include spent cyanide plating bath solutions from electroplating operations (F007), plating bath residues from the bottom of plating baths from electroplating operations where cyanides are used in the process (F008), spent stripping and cleaning bath solutions from electroplating operations where cyanides are used in the process (F009), quenching bath residues from oil baths from metal heat treating operations where cyanides are used in the process (F010), and spent cyanide solutions from salt bath pot cleaning from metal heat treating operations (F011).

Cyanide is toxic both as HCN and as the cyanide anion. Ingestion of as little as 60–90 mg of cyanide is sufficient to kill a human. It acts by binding to iron(III) in the ferricytochrome oxidase enzyme. This keeps the iron in the enzyme from being reduced to iron(II) during the metabolic utilization of O_2 by oxidative phosphorylation, thus preventing utilization of oxygen. Death by cyanide poisoning is rapid.

Nitrogen Compounds

In addition to cyanide discussed above, there are several other simple nitrogen compounds that are prominent among hazardous waste substances. Nitrous oxide provides oxygen for combustion, thus posing a flammability hazard. Used as an anesthetic, it is a central nervous system depressant and can cause asphyxiation. As a reactive oxidant, it can form explosive mixtures with reductants such as aluminum, hydrazine, and phosphine. Nitric oxide, NO, is toxic, though not as much so as the dioxide, NO_2, to which it may be oxidized, thus increasing its hazard. Nitric oxide reacts strongly with reductants, such as aluminum metal, carbon disulfide, or phosphine. Nitrogen dioxide, NO_2, which dimerizes reversibly to N_2O_4 is more toxic than either N_2O or NO. It can react explosively with some substances including F_2 and formaldehyde.

Of the nitrogen oxides, nitrogen dioxide is the most toxic. It is a lung irritant, causing pulmonary edema. Fatal poisoning can occur from relatively brief inhalation of 200–700 ppm of NO_2. A fatal condition called bronchiolitis fibrosa obliterans develops about three weeks after exposure. Nitrogen dioxide disrupts lactic dehydrogenase enzyme and some other enzymes.

Two hydrides of nitrogen are important hazardous waste substances. Large quantities of ammonia, NH_3, are manufactured and transported for use as a chemical raw material, fertilizer, or refrigerant fluid. It is toxic and reactive with some substances. Its salt ammonium nitrate, NH_4NO_3, is a hazardous oxidizer that has been involved in a number of damaging explosions. The other significant nitrogen hydride is hydrazine, N_2H_4, a colorless, fuming liquid. It has been used in rocket fuels. It is used as a reductant in some water treatment processes, including the Lancy process for treating copper plating rinsewaters. Its major advantage as a reductant is that it is oxidized to nitrogen gas and water, and does not produce waste byproducts that require treatment or disposal. Pure hydrazine is a fuming liquid that may undergo autoignition in contact with iron rust at temperatures as low as 23°C. It is a strong reductant that explodes in contact with some oxidizers, such as liquid oxygen, elemental fluorine or chlorine, or chromium(VI) salts. Because of the danger in handling pure hydrazine, it is usually obtained as the hydrate or as a salt. Hydrazine is a systemic poison that is transported through the body and can cause a toxic response in a location remote from the exposure site. Skin exposure may cause hypersensitivity. It is a strong reductant and its autoignition is catalyzed by iron rust at temperatures as low as 23°C. Hydrazine explodes in contact with a number of substances such as liquid oxygen, chromium(VI) salts, sodium, and elemental fluorine and chlorine.

The common **cyanogen** compounds are gaseous cyanogen (NCCN), volatile liquid cyanogen bromide (BrCN), and highly volatile cyanogen chloride (ClCN, b.p. 13.1°C). They are reactive compounds, and cyanogen is readily oxidized. They react with water to produce toxic HCN. Toxicologically they act as respiratory tract irritants.

The inorganic **azides** contain N_3^- ion. This group is bound to hydrogen in hydrogen azide or azoic acid, HN_3 or to metal cations in compounds such as sodium azide, NaN_3 (P105). Azides are unstable and may be explosive.

The **chloramines** are chlorinated derivatives of ammonia. They are formed by reaction of dissolved ammonium ion with hypochlorite or HOCl

$$NH_4^+ + HOCl \rightarrow NH_2Cl + H^+ + H_2O \quad (10.4.2)$$

$$NH_2Cl + HOCl \rightarrow NHCl_2 + H^+ + H_2O \quad (10.4.3)$$

Monochloramine and dichloramine are formed deliberately in drinking water to retain disinfection properties of the water in the distribution system.

Sulfur Compounds

Hydrogen sulfide, H_2S (U135), is a toxic, flammable, hazardous gas with a foul rotten-egg odor. It occurs in huge quantities in "sour" natural gas and is widely generated in petroleum refining and coal coking. It produces a number of toxic effects, which, at lower doses, include headache, dizziness, excitement resulting from damage to the central nervous system, and general debility. It kills even more rapidly than hydrogen cyanide in relatively high doses of around 1000 ppm in air. Some sulfide salts are regarded as hazardous. The hazards of sulfide salts result in part from their ability to release hydrogen sulfide gas in contact with acids:

$$FeS(s) + 2HCl(aq) \rightarrow H_2S(g) + Fe^{2+}(aq) + 2Cl^-(aq) \qquad (10.4.5)$$

Phosphorus sulfide (U189) and selenium sulfide, Se_2S (U205), are listed specifically as hazardous wastes.

Carbon disulfide (CS_2) is a volatile colorless liquid that has some industrial uses, including use as a solvent in viscose rayon and cellophane manufacture. It is also used as a solvent in some chemical analyses. Formerly, it was applied as an insecticide and fumigant. Carbon disulfide is highly flammable. It has an extremely wide explosive range of 1.3–50% in air. Its reaction with a number of substances, including aluminum, chlorine, fluorine, potassium, and zinc can occur with explosive speed. Toxicologically, it is a narcotic and central nervous system anesthetic. Carbon oxysulfide, COS, is a volatile liquid that is a common byproduct of natural gas production and petroleum refining. It is a toxic narcotic.

Sulfur dioxide, SO_2, is the most common oxidation product of elemental sulfur and sulfur compounds. It is an irritant to mucous membrane tissue, eyes, and the respiratory tract. Because of its water solubility, it is largely removed from inhaled air in the upper respiratory tract, reacting with water in tissue to produce acid:

$$SO_2 + H_2O(aq) \rightarrow H^+ + HSO_3^- \qquad (10.4.6)$$

Sulfur trioxide, SO_3, is the solid anhydride of sulfuric acid. Its toxicological properties are generally those of concentrated sulfuric acid, which it forms in contact with moist tissue.

There are several halides and oxyhalides of sulfur that are hazardous. Both the monofluoride and the tetrafluoride, S_2F_2, and SF_4, are strong reactive irritants. However, the gaseous hexafluoride, SF_6, is so stable that it is used as a tracer gas; its toxicity is very low, although it may be contaminated with hazardous S_2F_2 or SF_4. Sulfuryl chloride (SO_2Cl_2) and thionyl chloride ($SOCl_2$) are toxic liquids that act as corrosive irritants.

Acids

Because they exhibit the characteristic of *corrosivity*, characterized by low pH and the ability to corrode steel, strong mineral acids are classified as hazardous wastes. The most abundant of these acids are sulfuric acid (H_2SO_4), hydrochloric acid (HCl), phosphoric acid (H_3PO_4), and nitric acid (HNO_3). There are numerous sources of strong acid wastes, one of the most common of which is steel pickling liquor composed largely of HCl or other strong acids. Strong mineral acids are also used in etching and anodizing metals. In addition to being hazardous in their own right, strong mineral acid wastes are often contaminated with metals, including heavy metals, because of their applications to metal processing. The H^+ ion in strong mineral acid solution causes toxic heavy metals to dissolve from the metal salts, or even from the metal. Furthermore, H^+ reacts with cyanide and sulfide salts evolving toxic HCN or H_2S gases, respectively. In addition to corroding metals, strong acids injure exposed skin and eye tissue and are damaging to the respiratory tract when inhaled as aerosol droplets.

Silicon Compounds

Numerous silicon compounds are used industrially for a variety of purposes. Under some circumstances some of them may be hazardous. Their use in the semiconductor industry means that silicon compounds are rather widely dispersed, though not necessarily in very large quantities.

The most commonly used silicon compound is the dioxide, **silica**, SiO_2. As the indgredient of sand, it is used in sandblasting and is an ingredient of diatomaceous earth, which has a number of applications including filtration. The greatest hazard from silica arises from inhalation which can result in the formation of lung nodules and a condition known as pulmonary fibrosis. Silicosis is one of the most common occupational diseases.

Much more dangerous than silica is **asbestos**, a group of silicon-containing fibrous (serpentine) minerals with a formula of approximately $Mg_3P(Si_2O_5)(OH)_4$. Because of its superb insulating and heat-resistant properties, asbestos was once widely used as insulation, in brake linings, to manufacture pipe, and in other applications. Unfortunately, it causes injury to the pulmonary system including asbestos-induced pneumonia (asbestosis), cancer in the air passages of the lungs (brochogenic carcinoma), and mesothelioma, a condition characterized by tumorous growth in the mesothelial tissue that lines the chest cavity in which the lungs are contained. Uses of asbestos have been sharply curtailed, and the greatest danger now comes from cleanup of asbestos in buildings and other structures.

Several hydrides and chlorine-substituted hydrides are among commonly used chemicals, especially in the semiconductor industry. The two common hydrides are silane, SiH_4, and disilane, H_3SiSiH_3. Substitution of hydrocarbon groups for H on these compounds gives several organosilicon derivatives. The chlorine-substituted derivative of silane, silicon tetrachloride, $SiCl_4$, is used along with trichlorosilane, $SiHCl_3$, and dichlorosilane, SiH_2Cl_2, in the synthesis of other silicon compounds and for the preparation of high-purity silicon for semiconductor manufacture. Silicon tetrachloride is a fuming liquid that reacts with moisture in air to produce HCl vapor. This compound and trichlorosilane are both strong irritants to exposed tissue, especially in the eyes and respiratory tract.

Phosphorus Compounds

Phosphine, PH_3, is used to make organophosphorus compounds and is sometimes generated as an unintentional byproduct of chemical processes. This simple compound is a colorless gas that autoignites to produce voluminous choking fumes of P_4O_{10} and H_3PO_4. Phosphine is very toxic causing pulmonary tract irritation, central nervous system depression, and related symptoms. The P_4O_{10} product of the combustion of phosphine, elemental phosphorus, and other phosphorus species is a corrosive skin, eye, and mucous membrane irritant that reacts strongly with water vapor to produce H_3PO_4. Phosphorous pentachloride, PCl_5, is used to make other phosphorus compounds and as a chlorinating agent. It is a corrosive skin, eye, and mucous membrane irritant that reacts with water to produce H_3PO_4 and HCl acids. Another commonly used industrial phosphorus compound with similar properties is phosphorus oxychloride, $POCl_3$, which reacts with water to produce hydrochloric acid and phosphoric acid:

$$POCl_3 + 3H_2O \rightarrow 3HCl + H_3PO_4 \quad (10.4.6)$$

Alkaline Wastes

Substances that produce a solution with a pH higher than 12.5 may be classified as corrosive alkaline wastes. Highly basic materials produce a high concentration of hydroxide ion, OH^-. They are corrosive because they attack materials, causing cellulose in wood to hydrolyze and disintegrate and even dissolving some metals, such as aluminum. Strong bases are corrosive to flesh. Toxic ammonia gas, NH_3, can be liberated when alkaline wastes react with ammonium salts:

$$NH_4NO_3 + NaOH \rightarrow NH_3 + NaNO_3 + H_2O \quad (10.4.7)$$

Alkaline wastes are generated by a number of industrial processes, including those used to remove acidic gases, such as H_2S, from waste gas streams. There are a number of wastes that are acids, so alkaline wastes often find uses in treating acidic wastes.

10.5. HAZARDOUS ORGANOMETALLIC COMPOUNDS

In this section, the term **organometallic compounds** will be used to designate compounds in which a metal, such as lead, or a metalloid, such as arsenic, is covalently bound to carbon on a hydrocarbon group. Organometallic compounds are very important in consideration of hazardous wastes because of their widespread use and their hazardous properties, particularly their toxicities. In addition, compared to the metals from which they are formed, organometallic compounds are often very mobile in the environment and in biological systems.

There are several kinds of organometallic compounds, depending upon the nature of the hydrocarbon group bound to the metal. One of these classes consists of **alkyl compounds**, such as tetraethyllead,

```
    H H        H H
    | |        | |
  H-C-C        C-C-H
    | |\      /| |
    H H \    / H H
         Pb           Tetrethyllead
    H H /    \ H H
    | |/      \| |
  H-C-C        C-C-H
    | |        | |
    H H        H H
```

in which an alkyl group, such as methyl, $-CH_3$, or ethyl, $-C_2H_5$, is bonded to a metal by a sigma bond, as compared to those in which the organic group is a π electron donor (see below). In cases where the metal is very electropositive, the metal-carbon bond may be ionic:

$$Na^{+-}C_3H_7 \text{ Propylsodium}$$

A second major class of organometallic compounds consists of those in which the organic group is a **π electron donor** as shown below for the dibenzene compound of chromium:

Carbonyl compounds are those that contain CO bonded to the metal. Organometallics in which the metal is bonded both to several CO molecules and π-bonded to a π electron donor are common. One of the more notably hazardous carbonyls is nickel carbonyl, $Ni(CO)_4$ (P073). This volatile compound is rather widely generated and is very hazardous because of its high toxicity, flammability, and tendency to react violently with oxidizing agents, including O_2.

Many significant compounds exist in which part of the molecule is organometallic, with at least one C-metal bond, and the rest of the molecule consists of a nonorganic group covalently or ionically bonded to the metal. Methylmercury chloride, $H_3CHg^{+-}Cl$ is a compound with one C-metal bond and an ionic bond between the organometallic cation and the inorganic anion. Phenyldichloroarsine, $C_6H_5AsCl_2$, has a C-metalloid bond between the phenyl group and arsenic, as well as two covalent As-Cl bonds. Another such compound is phenylmercurydimethyldithiocarbamate,

$$C_6H_5-Hg-S-\overset{\overset{S}{\|}}{C}-N(CH_3)_2 \quad \text{Phenylmercurydimethyldithiocarbamate}$$

formerly used as a mold retardant for paper and slimicide for wood. In this case a phenyl group (benzene minus a hydrogen) is sigma-bonded to mercury and the rest of the molecule is attached to the mercury by a covalent bond. Because of the C-metal bond and because of their general chemical, environmental, and biological similarities to organometallic compounds, it is convenient to consider these compounds as hazardous wastes along with organometallic compounds.

Grignard reagents constitute a special class of organometallic compounds in which a hydrocarbon group is bound to an active metal, usually magnesium as shown for methylmagnesium iodide below:

$$H-\underset{H}{\overset{H}{\overset{|}{\underset{|}{C}}}}-MgI \qquad \text{Methylmagnesium iodide}$$

These compounds are used to tranfer hydrocarbon groups, such as the $-CH_3$ group in methylmagnesium iodide, in organic synthesis. Grignard reagents, which are reactive with water or air, are kept in ethylether solutions. Ether solutions of methylmagnesium bromide (CH_3MgBr) catch fire spontaneously in contact with water, igniting the ether and causing bad fires. Grignard reagents are corrosive to skin, and inhalation can damage pulmonary tissue.

Isopropyl titanate, $Ti(OC_3H_7)_4$, also called (titanium isopropylate),

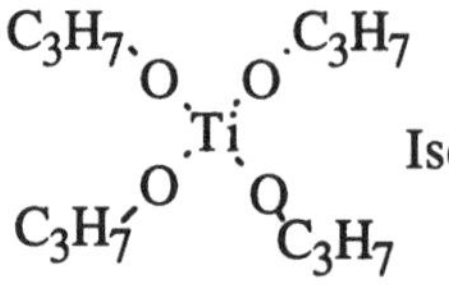

C_3H_7–O, O–C_3H_7, Ti, O–C_3H_7, O–C_3H_7 Isopropyl titanate

is an example of a class of compounds containing both metals and organic groups that have organometallic character. These compounds have no C-metal bonds, but possess bulky organic groups that cause the compound to behave like an organometallic species. Whereas an inorganic compound of titanium most likely would be a high-melting-point solid, this compound is a colorless liquid that melts at only 14.8°C and boils at 104°C, values that are typical of many organometallic species.

The **alkoxides**, such as potassium methoxide, $K^{+-}OCH_3$ are produced by reacting alkali metals (Na, K) with alcohols. These compounds react strongly with water, as shown below for sodium ethoxide:

$$Na^{+}\,{}^{-}OC_2H_5 + H_2O \rightarrow Na^{+} + OH^{-} + C_2H_5OH \tag{10.5.1}$$

Because of their formation of strongly basic solutions with water, the alkoxides are corrosive alkaline substances. Because of the presence of hydrocarbon groups in their formulas, the alkoxides behave similarly to organometallic compounds and can be discussed with them as hazardous substances.

Organometallic compounds were widely synthesized beginning around 1900, and the waste problems posed by them have evolved since then. Some of the earlier organometallic compounds manufactured were organoarsenic compounds made as pharmaceuticals in the early 1900s. The first of these was atoxyl (the sodium salt of 4-aminophenylarsinic acid), prescribed to victims of sleeping sickness. In 1907, Dr. Paul Erlich produced Salvarsan, which was effective for the treatment of syphilis. Indeed, the use of Salvarsan marks the beginning of modern **chemotherapy** (chemical treatment of disease).

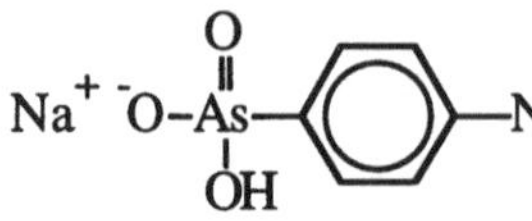

$Na^{+}\,{}^{-}O$–As(=O)(OH)–C_6H_4–NH_2 Atoxyl

$Cl^{-}\,{}^{+}H_3N$, HO–C_6H_3–As=As–C_6H_3–OH, $NH_3^{+}Cl^{-}$ Salvarsan

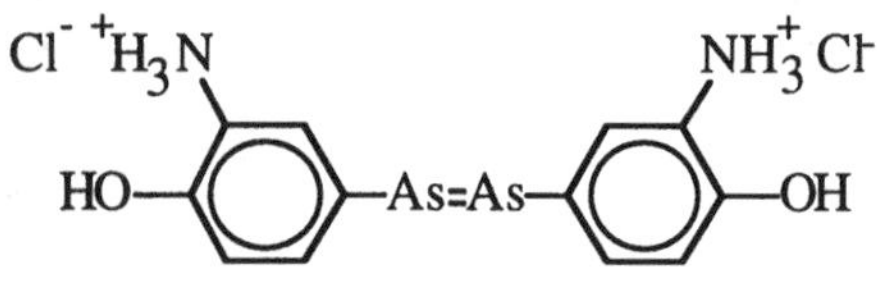

Although the use of organoarsenic compounds to treat human diseases has essentially ceased because of the development of better compounds with fewer

adverse side effects, organoarsenic compounds are used on animals. Four organoarsenic compounds added to animal feed are shown in Figure 10.1. Of these, arsanilic acid and Roxarsone are used to control swine dysentery and increase the rate of gain relative to the amount of feed in swine and chickens. Carbarsone and nitarsone (4-nitrophenylarsanilic acid) act as antihistomonads in chickens.

The manufacture of organoarsenic compounds used as veterinary pharmaceuticals may produce hazardous wastes. The hazardous wastes from specific sources involving veterinary pharmaceuticals manufacture are the following:

- K084 Wastewater treatment sludges generated during the production of veterinary pharmaceuticals from arsenic or organo-arsenic compounds.
- K101 Distillation tar residues from the distillation of aniline-based compounds in the production of veterinary pharmaceuticals from arsenic or organo-arsenic compounds.
- K102 Residue from the use of activated carbon for decolorization in the production of veterinary pharmaceuticals from arsenic or organo-arsenic compounds.

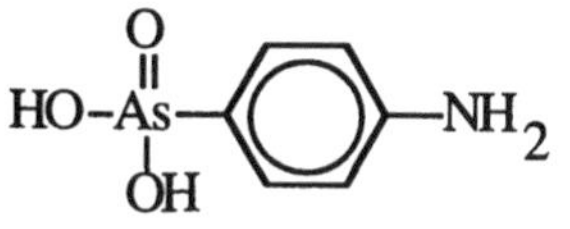

Arsanilic acid

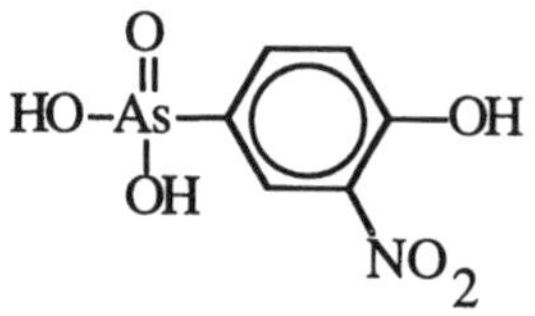

3-Nitro-4-hydroxyphenylarsinic acid (Roxarsone)

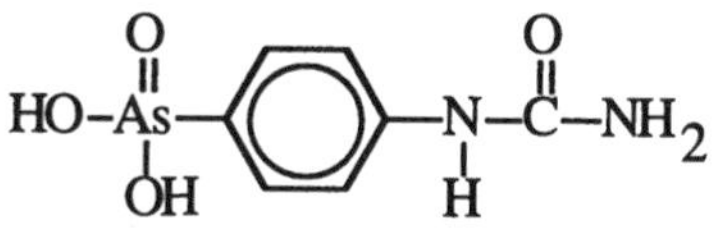

N-carbamoylarsinic acid (Carbarsone)

Figure 10.1. Organoarsenic compounds added to animal feeds.

Several organomercury compounds were developed and used to kill fungi and treat slimes and molds. Phenylmercurydimethyldithiocarbamate was mentioned above. Ethylmercury chloride, $C_2H_5Hg^{+-}Cl$, was once widely used as a seed fungicide and posed a danger to people who unknowingly consumed the treated grain for food.

During the mid-1900s, tetraethyllead (P110), the formula of which is shown earlier in the chapter, was widely manufactured and used as a gasoline octane booster. As a result, large quantities of lead were spewed from automobile exhausts and distributed into the environment. However, lead poisons automobile exhaust catalysts, so tetraethyllead has been largely phased out. Tetraethyllead is one of the most toxic compounds that was ever widely used in a commercial product, and many workers were exposed to it during manufacture and handling. It can enter the body through the skin, by inhalation, and by ingestion. Its toxic action is much different from that of inorganic lead compounds, in part because of its high affinity for lipids. Symptoms of tetraethyllead poisoning include those resulting from damage to the central nervous system, including fatigue, weakness, restlessness, ataxia, psychosis, and convulsions. Fatalities have occurred within one or two days of fatal exposure.

More recently, one of the major problems with organometallic compounds has come from organotin compounds. Organotins have been used widely used as industrial biocides in fungicides, acaricides, disinfectants, and antifouling paints. They have also been used as catalysts, in the formation of SnO_2 films on glass, and as stabilizers to reduce the adverse effects of light and heat on polyvinyl chloride (PVC) plastics.

The greatest environmental problems from organotin compounds have come from the environmental release of biocidal organotin compounds, including tin naphthenate, bis(tributyltin) oxide, tris(tributylstannyl) phosphate, tributyltin hydroxide, and tributyltin chloride:

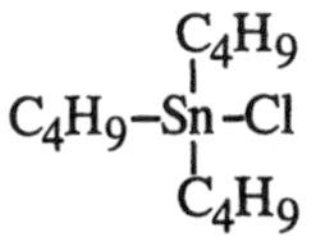

$$\begin{array}{c} C_4H_9 \\ | \\ C_4H_9\text{-}Sn\text{-}Cl \\ | \\ C_4H_9 \end{array} \quad \text{Tributyltin chloride}$$

Among the many biocidal applications of tributyltin compounds have been their uses in boat and ship hull coatings to prevent the growth of fouling organisms; as antifungal slimicides in cooling tower water; and as preservatives for wood, leather, paper, and textiles.

There are several organosilicon compounds. One of the more widely produced is trimethylchlorosilane, $(CH_3)_3SiCl$, used to make high purity silicon.

Reactivity and Combustion Hazards of Organometallic Compounds

Some organometallic compounds are extremely reactive. Regarding an organometallic compound as a species with a general formula C_cH_hM, the products of combustion with O_2 are highly stable CO_2, H_2O, and metal oxide, all of which have high heats of formation. Therefore, when an organometallic compound, such as diethylmagnesium burns,

$$Mg(C_2H_5)_2 + 7O_2 \rightarrow MgO + 5H_2O + 4CO_2 \qquad (10.5.2)$$

huge amounts of heat are released. Indeed, both this organomagnesium compound and the dimethyl analog are pyrophoric, undergoing self-ignition in air and reacting violently with water. They will even burn in a carbon dioxide atmosphere, extracting O from CO_2 to produce MgO.

Another example of an extremely reactive organometallic compound is liquid trimethylaluminum, $Al(CH_3)_3$, which has an almost explosive reaction with water releasing large quantities of heat and forming $Al(OH)_3$ as well as potentially noxious organic products.

Secondary Organometallic Species

There are several significant secondary organometallic species that are produced from discarded metal wastes by biological action. Specifically, anaerobic bacteria acting in sediments of bodies of water can attach the methyl group to metal atoms. The most important example is mercury upon which methylating bacteria can act to produce water-soluble CH_3Hg^+ ion or soluble and volatile dimethylmercury ($(CH_3)_2Hg$). This phenomenon, which was discovered around 1970, provides a means for mercury, which is largely insoluble in its environmentally stable inorganic forms, to get into natural waters. The ability of the methyl mercuries to dissolve and bioaccumulate in lipid tissues results in fish tissue concentrations as much as three orders of magnitude higher than the mercury concentrations in the surrounding water.

The other element that is noted for biological generation of methylated organometallic species is arsenic. In the reducing environment that exists in sediments in bodies of water, arsenic(V) can be reduced to arsenic(III):

$$H_3AsO_4 + 2H^+ + 2e \rightarrow H_3AsO_3 + H_2O \tag{10.5.3}$$

Methylcobalamin in anaerobic bacteria can methylate arsenic(III), first to methylarsinic acid,

$$H_3AsO_4 \rightarrow \begin{array}{c} \text{H}\;\;\text{O} \\ |\;\;\;\; \| \\ \text{H—C–As–OH} \\ |\;\;\;\; | \\ \text{H}\;\;\text{OH} \end{array} \tag{10.5.4}$$

then to dimethylarsinic acid,

$$\begin{array}{c} \text{H}\;\;\text{O} \\ |\;\;\;\; \| \\ \text{H—C–As–OH} \\ |\;\;\;\; | \\ \text{H}\;\;\text{OH} \end{array} \rightarrow \begin{array}{c} \text{H}\;\;\text{O}\;\;\;\;\text{H} \\ |\;\;\;\; \|\;\;\;\;\;\; | \\ \text{H—C–As—C–H} \\ |\;\;\;\; |\;\;\;\;\;\; | \\ \text{H}\;\;\text{OH}\;\;\text{H} \end{array} \tag{10.5.5}$$

Although methyl- and dimethylarsinic acids are the most common organoarsenic compounds found in the environment, dimethylarsine can also be formed by biological reduction:

$$\begin{array}{c} \text{H}\;\;\text{O}\;\;\;\;\text{H} \\ |\;\;\;\; \|\;\;\;\;\;\; | \\ \text{H—C–As—C–H} \\ |\;\;\;\; |\;\;\;\;\;\; | \\ \text{H}\;\;\text{OH}\;\;\text{H} \end{array} + 4H^+ + 4e^- \rightarrow \begin{array}{c} \text{H}\;\;\text{H}\;\;\text{H} \\ |\;\;\;\; |\;\;\;\; | \\ \text{H—C–As–C–H} \\ |\;\;\;\;\;\;\;\;\;\; | \\ \text{H}\;\;\;\;\;\;\;\;\;\text{H} \end{array} + 2H_2O \tag{10.5.6}$$

Methylarsinic acid and dimethyarsinic acid are the two organoarsenic compounds that are most likely to be encountered in the environment.

Arsenic used to be widely used in pigments contained in plaster and wallpaper. The formation of volatile methylated forms of arsenic in plaster and wallpaper under warm, humid conditions caused a significant number of cases of arsenic poisoning in Europe during the 1800s. The strong garlic odor of methylated arsenic compounds was reported in rooms so afflicted.

LITERATURE CITED

1. Petersen, Melody, "Cleaning Up in the Dark," *New York Times*, May 14, 1998, p. C 1.

11 HAZARDOUS ORGANIC MATERIALS

11.1. ORGANIC HAZARDOUS WASTE FORMS

Organic materials are encountered in hazardous wastes as organic solvents, solutions in organic solvents, solutions of organic compounds in water, emulsions, sludges, still bottoms, residuals, oils, greases, paint, solids, and organic pesticide wastes. There are millions of known organic compounds, most of which can be hazardous in some way and to some degree. Some of these substances are discussed in this chapter. A three-digit number following a letter (D, F, K, P, U) in parentheses after the name of a compound or waste denotes an EPA Hazardous Waste Number.

Organic/Inorganic Interactions

Organic wastes often strongly interact with inorganic consituents of hazardous wastes. Because of these interactions, organic compounds codisposed with inorganic substances often significantly affect the properties of the inorganic wastes. For example, organic surfactants, complexing agents, and chelating agents (nitrilotriacetate (NTA) ion, ethylenedinitrilotetraacetate (EDTA) anion, quadrol, citrates, gluconates) contained in metal electroplating and finishing wastes may increase the mobility and solubility of heavy metals and make the metals more difficult to remove in waste treatment processes.

Chemical Classification of Organic Wastes

Chemically, most organic compounds can be divided among hydrocarbons, oxygen-containing compounds, nitrogen-containing compounds, organohalides, sulfur-containing compounds, phosphorus-containing compounds, or combinations thereof. Each of these classes of organic compounds is discussed briefly here.

11.2. HYDROCARBONS

Hydrocarbons are organic compounds of carbon and hydrogen. The major types of hydrocarbons are alkanes, alkenes, alkynes, and aryl compounds. Structural formulas of examples of each are shown in Figure 11.1.

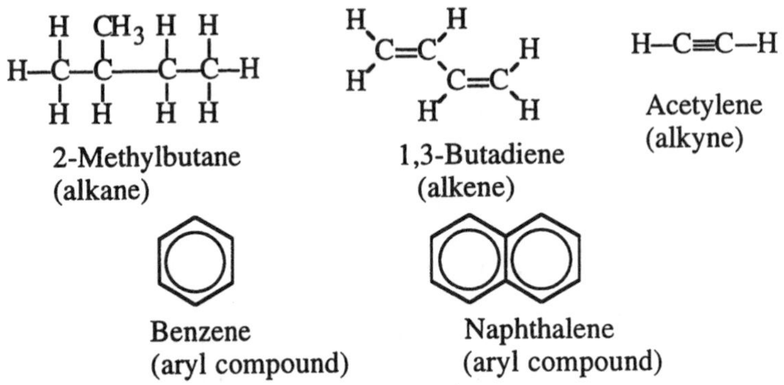

2-Methylbutane (alkane)

1,3-Butadiene (alkene)

H–C≡C–H

Acetylene (alkyne)

Benzene (aryl compound)

Naphthalene (aryl compound)

Figure 11.1. Examples of major types of hydrocarbons.

Alkanes

Alkanes, also called **paraffins** or **aliphatic hydrocarbons**, are hydrocarbons in which the C atoms are joined by single covalent bonds (sigma bonds) consisting of two shared electrons. The carbon atoms in hydrocarbons may form straight chains or branched chains. As shown in Figure 11.1, a typical branched chain alkane is 2-methylbutane, a volatile, highly flammable liquid. It is a component of gasoline and is commonly found as a pollutant in urban ambient air. Alkanes may also have cyclic structures, as in cyclohexane (C_6H_{12}). Each of the 6 carbon atoms in a cyclohexane molecule has 2 H atoms bonded to it. The general molecular formula for straight- and branched-chain alkanes is C_nH_{2n+2}, and that of cyclic alkanes is C_nH_{2n}.

Reactions of Alkanes

One of the more significant chemical reactions of alkanes is **oxidation** with molecular oxygen in air as shown for the following combustion reaction of propane:

$$C_3H_8 + 5O_2 \rightarrow 3CO_2 + 4H_2O + \text{heat} \qquad (11.2.1)$$

Common alkanes are highly flammable and the more volatile lower molecular mass alkanes form explosive mixtures with air. Furthermore, combustion of alkanes in an oxygen-deficient atmosphere or in an automobile engine produces significant quantities of carbon monoxide, CO, the toxic properties of which are discussed in Section 7.3.

Alkanes also undergo **substitution reactions** in which one or more H atoms on an alkane are replaced by atoms of another element. The most common such reaction is the replacement of H by chlorine, to yield **organochloride** compounds. For example, methane reacts with chlorine to give carbon tetrachloride, an **organochloride** compound:

$$\begin{array}{c}H\\|\\H-C-H\\|\\H\end{array} + 4Cl_2 \rightarrow \begin{array}{c}Cl\\|\\Cl-C-Cl\\|\\Cl\end{array} + 4HCl \qquad (11.2.2)$$

Substitution for H of other halogens can give fluorine, bromine, and iodine halocarbons as well.

Alkenes and Alkynes

Alkenes or olefins are hydrocarbons that have double bonds consisting of 4 shared electrons. The simplest and most widely manufactured alkene is ethylene,

$$\begin{array}{ccc}H & & H\\ & C=C & \\H & & H\end{array} \quad \text{Ethylene (ethene)}$$

widely used for the production of polyethylene polymer. Another example of an important alkene is 1,3-butadiene (Figure 11.1), widely used in the manufacture of polymers, particularly synthetic rubber. The lighter alkenes, including ethylene and 1,3-butadiene, are highly flammable and form explosive mixtures with air. These compounds have been involved in several bad industrial explosions and fires.

Acetylene (Figure 11.1) is an **alkyne**, a class of hydrocarbons characterized by carbon-carbon triple bonds consisting of 6 shared electrons. Highly flammable acetylene is used in large quantities as a chemical raw material and fuel for oxyacetylene torches. It forms dangerously explosive mixtures with air.

Addition Reactions

Alkenes and alkynes both undergo **addition reactions** in which pairs of atoms are added across unsaturated bonds as shown in the reaction of ethylene with hydrogen to give ethane,

$$\begin{array}{ccc}H & & H\\ & C=C & \\H & & H\end{array} + H-H \longrightarrow \begin{array}{c}H\ \ H\\|\ \ \ |\\H-C-C-H\\|\ \ \ |\\H\ \ H\end{array} \qquad (11.2.3)$$

or that of HCl gas with acetylene to give vinyl chloride,

$$H-C\equiv C-H + H-H \longrightarrow \begin{array}{ccc}H & & Cl\\ & C=C & \\H & & H\end{array} \qquad (11.2.4)$$

This kind of reaction, which is not possible with alkanes, adds to the chemical and metabolic versatility of compounds containing unsaturated bonds and is a factor contributing to their generally higher toxicities.

Aryl Hydrocarbons

Benzene and naphthalene shown in Figure 11.1 are **aryl** or **aromatic** hydrocarbons. Aryl compounds have special characteristics of **aromaticity**, which include a low hydrogen:carbon atomic ratio; C–C bonds that are quite strong and of intermediate length between such bonds in alkanes and those in alkenes; tendency to undergo substitution reactions rather than the addition reactions characteristic of alkenes; and delocalization of π electrons over several carbon atoms resulting in resonance stabilization of the molecule.

Benzene (U019) is a volatile, colorless, highly flammable liquid that is consumed as a raw material for the manufacture of phenolic and polyester resins, polystyrene plastics, alkylbenzene surfactants, chlorobenzenes, insecticides, and dyes. It is hazardous both for its ignitability and toxicity (exposure to benzene causes blood abnormalities that may develop into leukemia). **Naphthalene** (U165) is the simplest member of a large number of multicyclic aryl hydrocarbons having two or more fused rings. It is a volatile white crystalline solid with a characteristic odor and has been used to make mothballs. The most important of the many chemical derivatives made from naphthalene is phthalic anhydride, from which phthalate ester plasticizers are synthesized.

Polycyclic Aromatic Hydrocarbons

Benzo(a)pyrene,

is the most studied of the polycyclic aromatic hydrocarbons (PAHs), which are characterized by condensed ring systems ("chicken wire" structures). These compounds are formed by the incomplete combustion of other hydrocarbons, a process that consumes hydrogen in preference to carbon. The carbon residue is left in the thermodynamically favored condensed aryl ring system of the PAHs.

Because there are so many partial combustion and pyrolysis processes that favor production of PAHs, these compounds are encountered abundantly in the atmosphere, soil, and elsewhere in the environment from sources that include engine exhausts, wood stove smoke, cigarette smoke, and char-broiled food. Coal tars and petroleum residues, such as road and roofing asphalt, have high levels of PAHs. Some PAH compounds, including benzo(a)pyrene, are of toxicological concern because they are precursors to cancer-causing metabolites.

11.3. ORGANOOXYGEN COMPOUNDS

As shown in Figure 11.2, numerous hazardous organic compounds contain oxygen in various fuctional groups. The major types of these compounds are epoxides, alcohols, phenols, ethers, aldehydes, ketones, and carboxylic acids. The functional groups characteristic of these compounds are illustrated by the examples of oxygen-containing compounds shown in Figure 11.2.

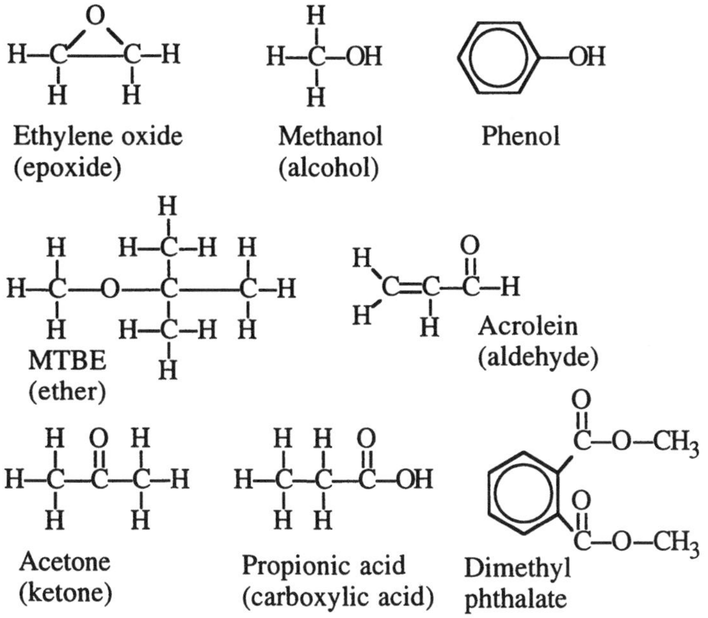

Ethylene oxide (epoxide)

Methanol (alcohol)

Phenol

MTBE (ether)

Acrolein (aldehyde)

Acetone (ketone)

Propionic acid (carboxylic acid)

Dimethyl phthalate

Figure 11.2. Examples of oxygen-containing hazardous waste compounds.

Ethylene oxide (U115) is a moderately to highly toxic sweet-smelling, colorless, flammable, explosive gas used as a chemical intermediate, sterilant, and fumigant. It is a mutagen and a carcinogen to experimental animals. It is classified as hazardous for both its toxicity and ignitability. **Methanol** (U154) is a clear, volatile, flammable liquid alcohol used for chemical synthesis, as a solvent, and as a fuel. It has been strongly advocated in some quarters as an alternative to gasoline that could result in significantly less photochemical smog formation and CO production than currently used gasoline formulations. Ingestion of methanol can be fatal and blindness can result from sublethal doses. **Phenol** is a dangerously toxic aryl alcohol widely used for chemical synthesis and polymer manufacture. **Methyltertiarybutyl ether**, MTBE, is an ether that has become the octane booster of choice to replace tetraethyllead in gasoline. **Acrolein**, (P003) is an alkenic aldehyde and a volatile, flammable, highly reactive chemical. It forms explosive peroxides upon prolonged contact with O_2. An extreme lachrimator and strong irritant, acrolein is quite toxic by all routes of exposure. **Acetone** is the lightest of the ketones. **Propionic acid** is a typical organic carboxylic acid.

11.4. ORGANONITROGEN COMPOUNDS

Figure 11.3 shows examples of three major classes of the many kinds of compounds that contain N (amines, nitrosamines, and nitro compounds). Nitrogen occurs in many functional groups in organic compounds, some of which contain nitrogen in ring structures, or along with oxygen.

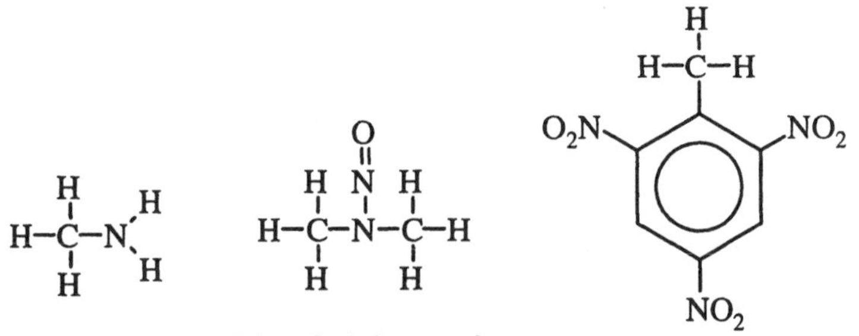

Methylamine | Dimethylnitrosamine (N-nitrosodimethylamine) | Trinitrotoluene (TNT)

Figure 11.3. Examples of hazardous waste organic compounds containing nitrogen.

Methylamine is a colorless, highly flammable gas with a strong odor. It is a severe irritant affecting eyes, skin, and mucous membranes. Methylamine is the simplest of the **amine** compounds, which have the general formula,

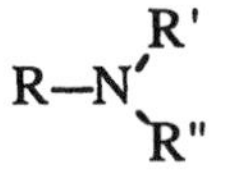

$$R{-}N(R')(R'')$$

where at least one of the Rs is a hydrocarbon group.

Dimethylnitrosamine is an N-nitroso compound; all N-nitroso compounds contain the N–N=O functional group. Dimethylnitrosamine was once widely used as an industrial solvent, but was observed to cause liver damage and jaundice in exposed workers. Subsequently numerous other N-nitroso compounds, many produced as byproducts of industrial operations and food and alcoholic beverage processing, were found to be carcinogenic.

Solid **trinitrotoluene** (TNT) has been widely used as a military explosive. TNT is moderately to very toxic and has caused toxic hepatitis or aplastic anemia in exposed individuals, a few of whom died from its toxic effects. It belongs to the general class of nitro compounds characterized by the presence of $-NO_2$ groups bonded to a hydrocarbon structure.

Some organonitrogen compounds are chelating agents that strongly bind to metal ions and play a role in the solubilization and transport of heavy metal wastes. Prominent among these are salts of the aminocarboxylic acids which, in the acid form, have $-CH_2CO_2H$ groups bonded to nitrogen atoms. A prominent example of such a compound is the monohydrate of trisodium nitrilotriacetate (NTA), the structural formula of which is shown below:

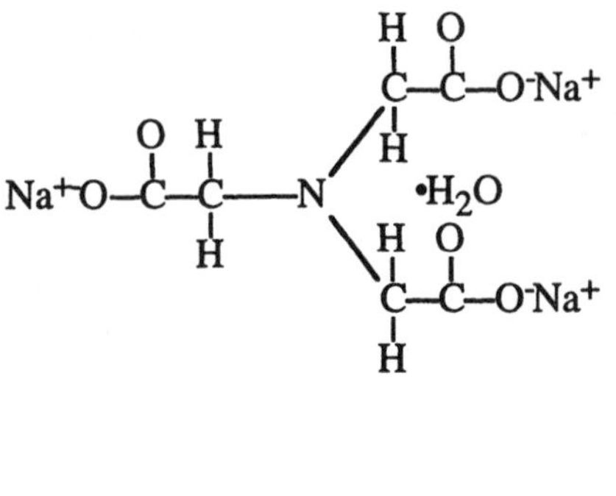

$$N(CH_2COO^-Na^+)_3 \cdot H_2O$$

Trisodium nitrilotriacetate

This compound has been widely used in Canada and Sweden as a substitute for detergent phosphates to bind to calcium ion and make the detergent solution chemically basic. NTA is used in metal plating formulations. It is highly water soluble and quickly eliminated with urine when ingested. It has a low acute toxicity and no chronic effects have been shown for plausible doses. However, concern does exist over its interaction with heavy metals in waste treatment processes and in the environment.

11.5. ORGANOHALIDE COMPOUNDS

Organohalides exhibit a wide range of physical and chemical properties. These compounds consist of halogen-substituted hydrocarbon molecules, each of which contains at least one atom of F, Cl, Br, or I. They may be saturated (**alkyl halides**), unsaturated (**alkenyl halides**), or aromatic (**aryl halides**). The most widely manufactured organohalide compounds are chlorinated hydrocarbons, many of which are listed as hazardous substances and hazardous wastes. Organohalide compounds are commonly found as contaminants in waste hydrocarbon oils.[1]

Alkyl Halides

Substitution of halogen atoms for one or more hydrogen atoms on alkanes gives **alkyl halides**, for which example structural formulas are given in Figure 11.4. Most of the commercially important alkyl halides are derivatives of alkanes of low molecular mass. A brief discussion of the uses of the compounds listed in Figure 11.4 is given here to provide an idea of the versatility of the alkyl halides. Volatile **chloromethane** (methyl chloride) is consumed in the manufacture of silicones. **Dichloromethane** is a volatile liquid with excellent solvent properties for nonpolar organic solutes. It has been used as a solvent for the decaffeination of

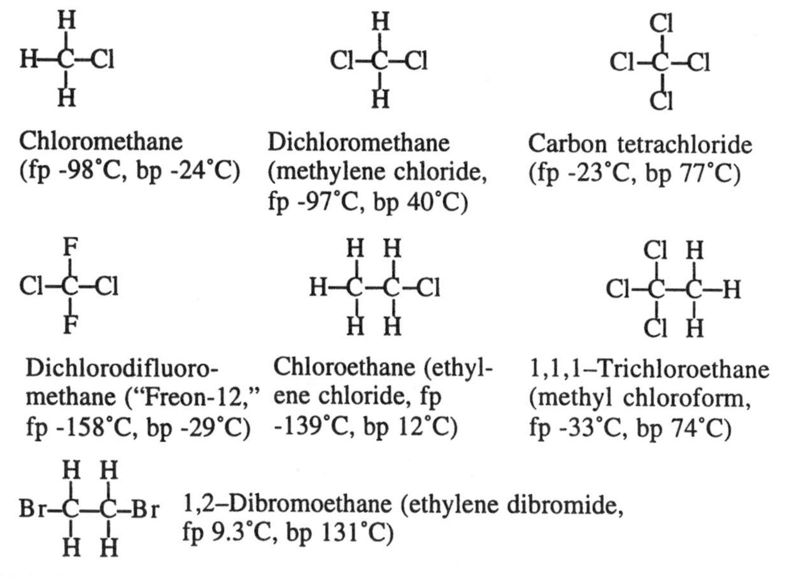

Figure 11.4. Some typical low-molecular-mass alkyl halides.

coffee, in paint strippers, as a blowing agent in urethane polymer manufacture, and to depress vapor pressure in aerosol formulations. Once commonly sold as a solvent and stain remover, highly toxic **carbon tetrachloride** is now largely restricted to use as a chemical intermediate under controlled conditions, primarily to manufacture chlorofluorocarbon refrigerant fluid compounds, which are also discussed in this section. **Chloroethane** is an intermediate in the manufacture of tetraethyllead and is an ethylating agent in chemical synthesis. One of the more common industrial chlorinated solvents is **1,1,1-trichloroethane**. Insecticidal **1,2-dibromoethane** has been consumed in large quantities as a lead scavenger in leaded gasoline and to fumigate soil, grain, and fruit (fumigation with this compound has been discontinued because of toxicological concerns). An effective solvent for resins, gums, and waxes, it serves as a chemical intermediate in the syntheses of some pharmaceutical compounds and dyes.

Alkenyl Halides

Viewed as hydrocarbon-substituted derivatives of alkenes, the **alkenyl** or **olefinic organohalides** contain at least one halogen atom and at least one carbon-carbon double bond. The most significant of these are the lighter chlorinated compounds, such as those illustrated in Figure 11.5.

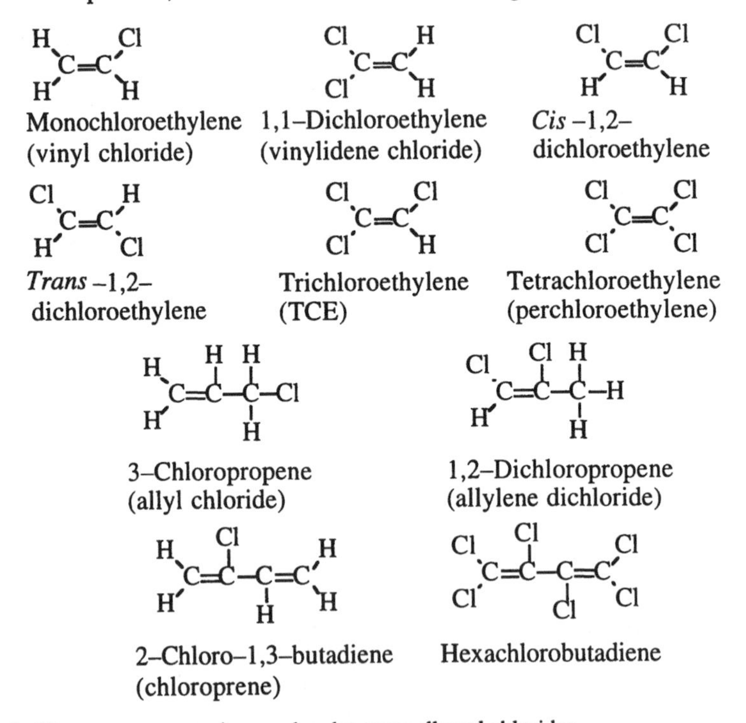

Figure 11.5. The more common low-molecular-mass alkenyl chlorides.

Vinyl chloride is used in large quantities as a raw material to manufacture pipe, hose, wrapping, and other products fabricated from polyvinylchloride plastic. This highly flammable, volatile, sweet-smelling gas is a known human carcinogen.

As shown in Figure 11.5, there are three possible dichloroethylene compounds, all clear, colorless liquids. Vinylidene chloride is used to produce a copolymer with vinyl chloride that finds applications in some kinds of coating materials. The geometrically isomeric 1,2-dichloroethylenes are used as organic synthesis intermediates and as solvents.

Trichloroethylene is a clear, colorless, nonflammable, volatile liquid. It is an excellent degreasing and drycleaning solvent and has been used as a household solvent and for food extraction (for example, in decaffeination of coffee). Colorless, nonflammable liquid **tetrachloroethylene** has properties and uses similar to those of trichloroethylene.

The two chlorinated propene compounds shown are colorless liquids with pungent, irritating odors. **Allyl chloride** is an intermediate in the manufacture of allyl alcohol and other allyl compounds, including pharmaceuticals, insecticides, and thermosetting varnish and plastic resins. **Dichloropropene** compounds, of which one isomer is shown, can be used as soil fumigants, as well as solvents for oil, fat, drycleaning, and metal degreasing.

Produced in large quantities for the manufacture of neoprene rubber, **chloroprene** is a colorless liquid with an ethereal odor. **Hexachlorobutadiene**, a colorless liquid with an odor somewhat like that of turpentine, is used as a solvent for higher hydrocarbons and elastomers, as a hydraulic fluid, in transformers, and for heat transfer.

Although the alkyl halides have generally low reactivities, they may pyrolyze in flames to produce HCl gas and other hazardous products. Alkenyl halides are more reactive. As shown by the example below, they may burn to produce highly toxic phosgene.

$$\mathrm{(H)(Cl)C{=}C(Cl)_2} + O_2 \longrightarrow HCl + \mathrm{Cl{-}\overset{\overset{\displaystyle O}{\|}}{C}{-}Cl} + CO \qquad (11.5.1)$$

Aryl Halides

Aryl halide derivatives of benzene and toluene have many uses, which have resulted in substantial human exposure and environmental contamination. Some of these compounds, their properties, and major applications are summarized in Table 11.1.

Two major classes of halogenated aryl compounds containing two benzene rings are synthesized by the chlorination of naphthalene and biphenyl and have been sold as mixtures with varying degrees of chlorine content. Examples of chlorinated naphthalenes, and polychlorinated biphenyls (PCBs discussed further in Section 11.8), are shown in Figure 11.6. The less highly chlorinated of these compounds are liquids and those with higher chlorine contents are solids. Because of their physical and chemical stabilities and other desirable qualities, these compounds have had many uses, including heat transfer fluids, hydraulic fluids, and dielectrics. Polybrominated biphenyls (PBBs) have served as flame retardants. However, because chlorinated naphthalenes, PCBs, and PBBs are environmentally extremely persistent, their uses have been severely curtailed.

Table 11.1. Examples of Single-Ring Aryl Halides

Structural formula	Name	Properties
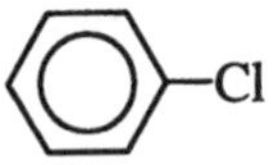 (benzene ring)–Cl	Monochlorobenzene	Flammable liquid (fp –45°C, bp 132°C), solvent, heat transfer fluid, synthetic reagent
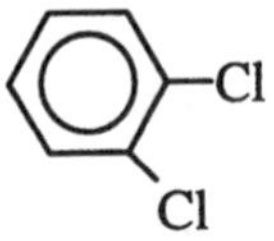 (benzene ring)–Cl, Cl	1,2-Dichlorobenzene	Solvent for degreasing hides and wool, synthetic reagent for dye manufacture
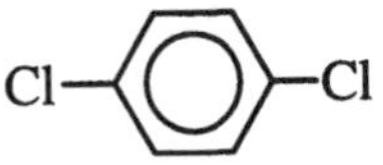 Cl–(benzene ring)–Cl	1,4-Dichlorobenzene	White sublimable solid, dye manufacture, germicide, moth repellant
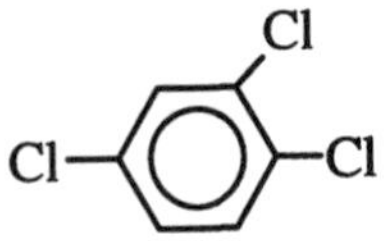 Cl, Cl–(benzene ring)–Cl	1,2,4-Trichloro-benzene	Liquid (fp 17°C, bp 213°C), solvent, lubricant, dielectric fluid, formerly used as a termiticide
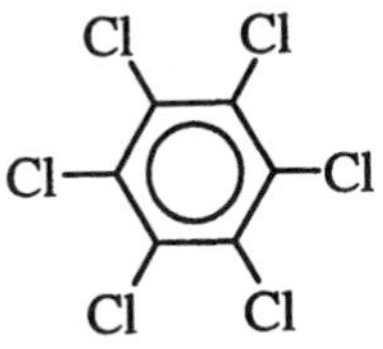 Cl, Cl, Cl–(benzene ring)–Cl, Cl, Cl	Hexachlorobenzene	High-melting-point solid, seed fungicide, wood preservative, intermediate for organic synthesis
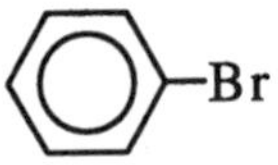 (benzene ring)–Br	Bromobenzene	Liquid (fp –31°C, bp 156°C), solvent, motor oil additive, intermediate for organic synthesis
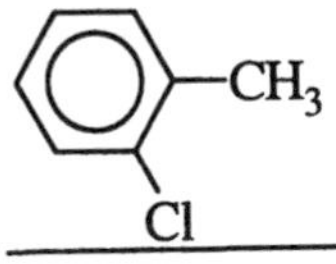 (benzene ring)–CH_3, Cl	1-Chloro-2-methyl-benzene	Intermediate for the synthesis of 1-chlorobenzotrifluoride

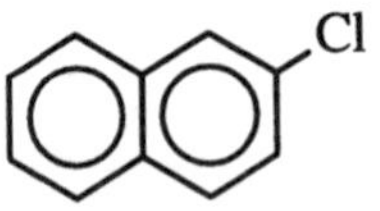

Cl

2-Chloronaphthalene

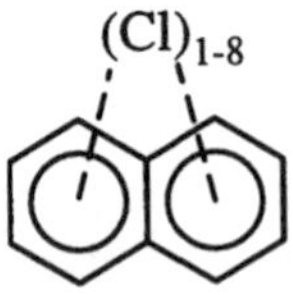

$(Cl)_{1-8}$

Polychlorinated naphthalenes

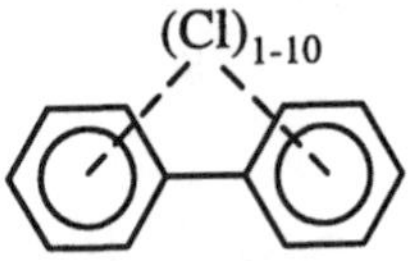

$(Cl)_{1-10}$

Polychlorinated biphenyls (PCBs)

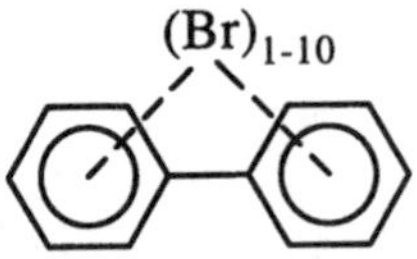

$(Br)_{1-10}$

Polybrominated biphenyls (PBBs)

Figure 11.6. Halogenated naphthalenes and biphenyls.

Chlorofluorocarbons, Halons, and Hydrochlorofluorocarbons

Chlorofluorocarbons (CFCs) are volatile 1- and 2-carbon compounds that contain Cl and F bonded to carbon. These compounds are notably stable and non-toxic. They have been widely used in recent decades in the fabrication of flexible and rigid foams and as fluids for refrigeration and air conditioning. The most widely manufactured of these compounds are CCl_3F (CFC-11), CCl_2F_2 (CFC-12), $C_2Cl_3F_3$ (CFC-113), $C_2Cl_2F_4$ (CFC-114), and C_2ClF_5 (CFC-115).

Halons are related compounds that contain bromine and are used in fire extinguisher systems. The major commercial halons are $CBrClF_2$ (Halon-1211), $CBrF_3$ (Halon-1301), and $C_2Br_2F_4$ (Halon-2402), where the sequence of numbers denotes the number of carbon, fluorine, chlorine, and bromine atoms, respectively, per molecule. Halons are particularly effective fire extinguishing agents because of the way in which they stop combustion. Some fire supressants, such as carbon dioxide, act by depriving the flame of oxygen by a smothering effect, whereas water cools a burning substance to a temperature below which combustion is supported. Halons act by chain reactions that destroy hydrogen atoms that sustain combustion. The basic sequence of reactions by which halons quench combustion is outlined below:

$$CBrClF_2 + H\cdot \rightarrow CClF_2\cdot + HBr \quad (11.5.2)$$

$$HBr + H\cdot \longrightarrow Br\cdot + H_2 \quad (11.5.3)$$

$$Br\cdot + H\cdot \longrightarrow HBr \quad (11.5.4)$$

Halons are used in automatic fire extinguishing systems, such as those located in flammable solvent storage areas, and in specialty fire extinguishers, such as those on aircraft. It has proven difficult to find suitable substitutes for halons, especially in their critical role in aircraft.

All of the chlorofluorocarbons and halons discussed above have been implicated in the halogen-atom-catalyzed destruction of atmospheric ozone. As a result of U.S. Environmental Protection Agency regulations imposed in accordance with the 1986 Montreal Protocol on Substances that Deplete the Ozone Layer, production of CFCs and halocarbons in the U.S. was curtailed starting in 1989. The substitutes for these halocarbons are hydrogen-containing chlorofluorocarbons (HCFCs) and hydrogen-containing fluorocarbons (HFCs). These include CH_2FCF_3 (HFC-134a, a substitute for CFC-12 in automobile air conditioners and refrigeration equipment), $CHCl_2CF_3$ (HCFC-123, substitute for CFC-11 in plastic foam-blowing), CH_3CCl_2F (HCFC-141b, substitute for CFC-11 in plastic foam-blowing), and $CHClF_2$ (HCFC-22, air conditioners and manufacture of plastic foam food containers). Because of the more readily broken H–C bonds that they contain, these compounds are more easily destroyed by atmospheric chemical reactions (particularly with hydroxyl radical, see Section 7.8) before they reach the stratosphere. Relative to a value of 1.0 for CFC-11, the ozone-depletion potentials of these substitutes are HFC-134a, 0; HCFC-123, 0.016; HCFC-141b, 0.081; and HCFC-22, 0.053. Concern has been expressed over toxicities and fire hazards of HCFCs, which are more reactive both chemically and biochemically than the CFCs and halons that they are designed to replace, although few, if any, adverse effects have been documented.

Chlorinated Phenols

The chlorinated phenols, particularly **pentachlorophenol**,

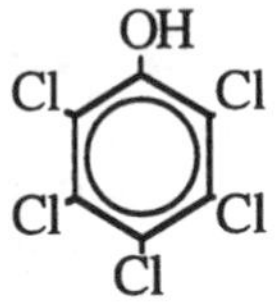

Pentachlorophenol

and the trichlorophenol isomers are significant hazardous wastes. These compounds are biocides that are used to treat wood to prevent rot by fungi and termite infestation. They are toxic, causing liver malfunction and dermatitis; contaminant polychlorinated dibenzodioxins may be responsible for some of the observed effects.

Wood preservative chemicals such as pentachlorophenol may be encountered at hazardous waste sites in wastewaters and sludges. EPA listed waste F027, discarded unused formulations containing tri-, tetra-, or pentachlorophenol or discarded unused formulations containing compounds derived from these chlorophenols, may contain pentachlorophenol and related compounds. (Wood preservatives from wood treatment sites can be very troublesome organic hazardous wastes. One of the materials widely used for wood treatment is cresote. A fluid derived from coal coking, creosote is a mixture of almost 300 organic compounds, including a number of polycyclic aromatic hydrocarbons.)

11.6. ORGANOSULFUR COMPOUNDS

Sulfur is chemically similar to, but more diverse than oxygen. Whereas, with the exception of peroxides, most chemically combined organic oxygen is in the -2 oxidation state, sulfur occurs in the -2, +4, and +6 oxidation states. Many organosulfur compounds are noted for their foul "rotten egg" or garlic odors, and some produce intolerable odors at remarkably low levels in the atmosphere.

Thiols and Thioethers

Substitution of alkyl or aryl hydrocarbon groups such as phenyl and methyl for H on hydrogen sulfide, H_2S, leads to a number of different organosulfur **thiols** (mercaptans, R–SH) and **sulfides**, also called thioethers (R–S–R). Structural formulas of examples of these compounds are shown in Figure 11.7. Of these compounds, benzenethiol (thiophenol) is EPA listed waste No. P014.

Methanethiol and other lighter alkyl thiols are fairly common air pollutants that have "ultragarlic" odors; both 1- and 2-butanethiol are associated with skunk odor. Gaseous methanethiol and volatile liquid ethanethiol are used as odorant leak-detecting additives for natural gas, propane, and butane; they are also employed as intermediates in pesticide synthesis. Although information about their toxicities to humans is lacking, methanethiol, ethanethiol, and 1-propanethiol should be considered dangerously toxic, especially by inhalation.

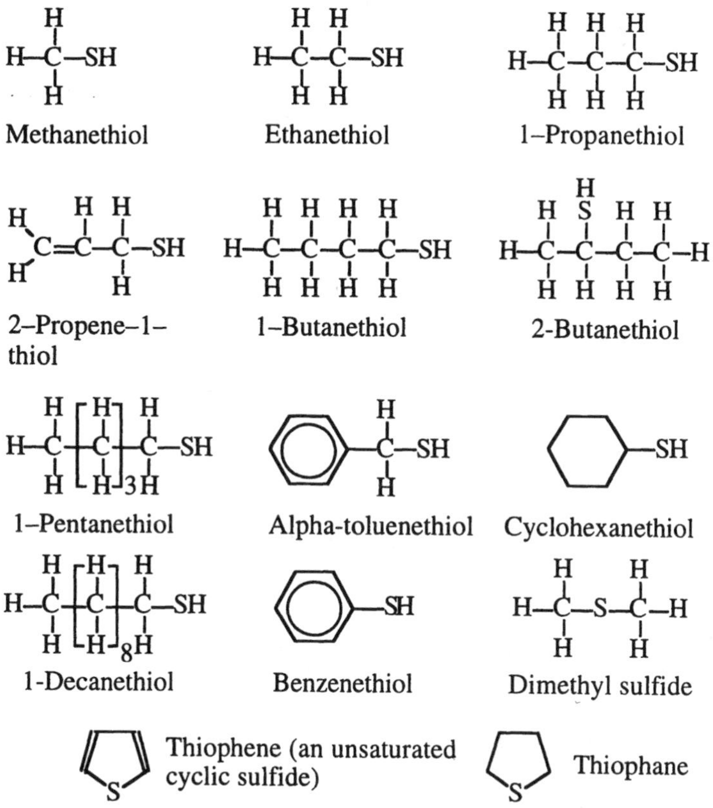

Methanethiol

Ethanethiol

1–Propanethiol

2–Propene–1–thiol

1–Butanethiol

2-Butanethiol

1–Pentanethiol

Alpha-toluenethiol

Cyclohexanethiol

1-Decanethiol

Benzenethiol

Dimethyl sulfide

Thiophene (an unsaturated cyclic sulfide)

Thiophane

Figure 11.7. Common low-molecular-mass thiols and sulfides. All are liquids at room temperature, except for methanethiol, which boils at 5.9°C.

A toxic, irritating volatile liquid with a strong garlic odor, 2-propene-1-thiol (allyl mercaptan) is a typical alkenyl mercaptan. Alpha-toluenethiol (benzyl mercaptan, bp 195°C) is very toxic and is an experimental carcinogen. Benzenethiol (phenyl mercaptan) is the simplest of the aryl thiols. It is a toxic liquid with a severely "repulsive" odor.

Alkyl sulfides (thioethers) contain the C–S–C functional group. The lightest of these compounds is moderately toxic dimethyl sulfide, a volatile liquid (bp 38°C). Cyclic sulfides contain the C–S–C group in a ring structure. The most common of these compounds is thiophene, a heat-stable liquid (bp 84°C) with a solvent action much like that of benzene that is used in the manufacture of pharmaceuticals, dyes, and resins. Its saturated analog is tetrahydrothiophene or thiophane.

Nitrogen-Containing Organosulfur Compounds

Many important organosulfur compounds also contain nitrogen. One such compound is **thiourea**, the sulfur analog of urea. Its structural formula is shown in Figure 11.8 along with other thiourea compounds. Thiourea and **phenylthiourea** (EPA listed waste number P093) have been used as rodenticides. Commonly called ANTU, **1-naphthylthiourea** (EPA listed waste number P072) is an excellent rodenticide that is virtually tasteless and has a very high rodent:human toxicity ratio.

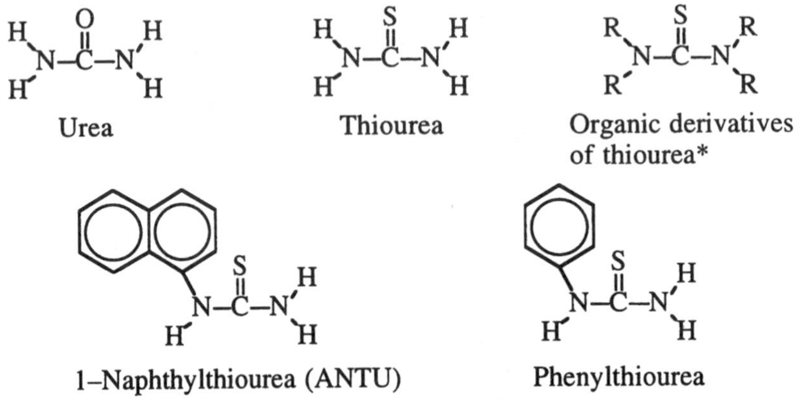

Urea: $H_2N{-}C({=}O){-}NH_2$

Thiourea: $H_2N{-}C({=}S){-}NH_2$

Organic derivatives of thiourea*: $R_2N{-}C({=}S){-}NR_2$

1–Naphthylthiourea (ANTU): $C_{10}H_7{-}NH{-}C({=}S){-}NH_2$

Phenylthiourea: $C_6H_5{-}NH{-}C({=}S){-}NH_2$

Figure 11.8. Structural formulas of urea, thiourea, and organic derivatives of thiourea.

* At least one R group is an alkyl, alkenyl, or aryl substituent.

Substitution of hydrocarbon groups such as the methyl group for H on thiocyanic acid (HSCN) yields organic **thiocyanates**. First used for insect control during the 1930s, some thiocyanates are regarded as the first synthetic organic insecticides. Volatile methyl, ethyl, and isopropyl thiocyanates kill insects upon contact and are effective fumigants for insect control.

Methylisothiocyanate (structure below),

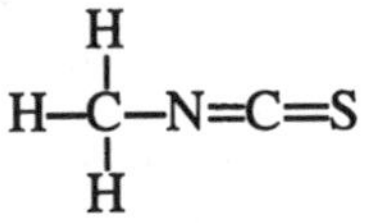

$$H_3C{-}N{=}C{=}S \quad \text{Methylisothiocyanate}$$

known as methyl mustard oil, and its ethyl analog have been developed as military poisons. Both are powerful irritants to eyes, skin, and respiratory tract. When decomposed by heat, these compounds emit sulfur oxides and hydrogen cyanide. Other common compounds in this class are allyl and phenyl isothiocyanates.

Sulfoxides and Sulfones

Sulfoxides and **sulfones** (Figure 11.9) contain both sulfur and oxygen. **Dimethylsulfoxide** (DMSO) is a liquid with numerous uses and some very interesting properties. It is used to remove paint and varnish, as a hydraulic fluid, mixed with water as an antifreeze solution, and in pharmaceutical applications as an antiinflammatory and bacteriostatic agent. A polar aprotic (no ionizable H) solvent with a relatively high dielectric constant, **sulfolane** dissolves both organic and inorganic solutes. It is the most widely produced sulfone because of its use in an industrial process called BTX processing in which it selectively extracts benzene, toluene, and xylene from aliphatic hydrocarbons; as the solvent in the Sulfinol process by which thiols and acidic compounds are removed from natural gas; as a solvent for polymerization reactions; and as a polymer plasticizer.

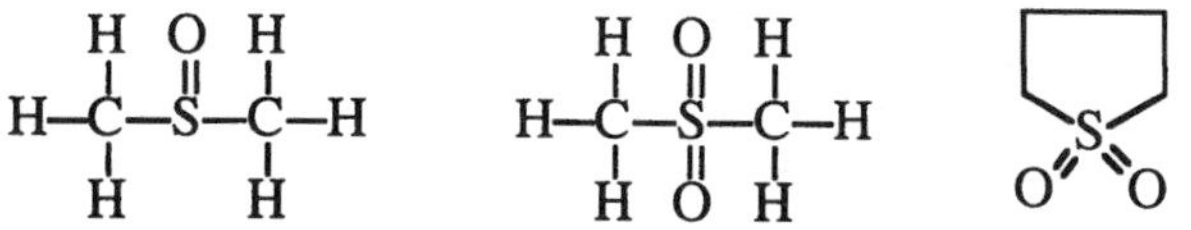

Dimethylsulfoxide (DMSO) Dimethylsulfone Sulfolane

Figure 11.9. Sulfoxides and sulfones.

Sulfonic Acids, Salts, and Esters

Sulfonic acids and sulfonate salts contain the $-SO_3H$ and $-SO_3^-$ groups, respectively, attached to a hydrocarbon moiety. The structural formula of two sulfonic acids and of sodium 1-(*p*-sulfophenyl)decane, a biodegradable detergent surfactant, are shown in Figure 11.10. The common sulfonic acids are water-soluble strong acids that lose virtually all ionizable H^+ in aqueous solution. They are used commercially to hydrolyze fat and oil esters to fatty acids and glycerol. Sulfonic acids form esters, such as methylmethane sulfonate.

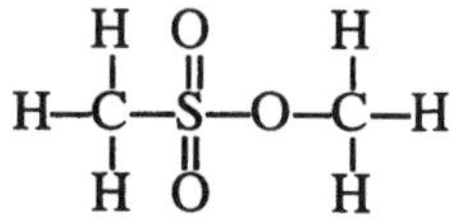

Methylmethane sulfonate

This compound is especially dangerous because it is a primary or direct-acting carcinogen that does not require metabolic conversion to cause cancer. The related compound ethyl methanesulfonate is an EPA listed waste (U119).

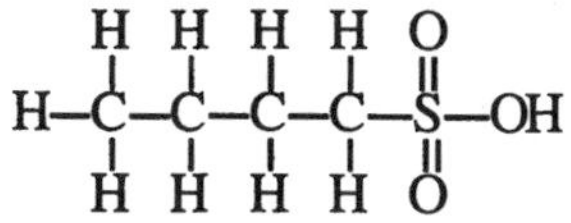

Butanesulfonic acid 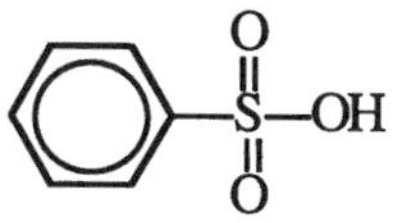 Benzenesulfonic acid

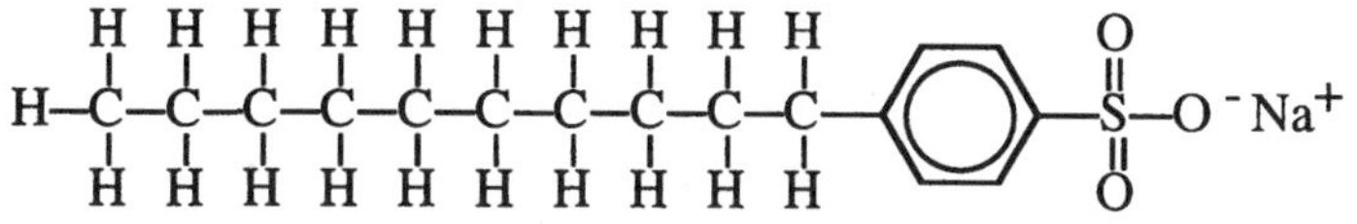

Sodium 1–(*p*–sulfophenyl)decane

Figure 11.10. Sulfonic acids and a sulfonate salt.

Organic Esters of Sulfuric Acid

Replacement of 1 H on sulfuric acid, H_2SO_4, yields an acid ester and replacement of both yields an ester. Examples of these esters are shown in Figure 11.11 in which methylsulfuric acid and ethylsulfuric acid are acid esters and dimethylsulfuric acid is an ester. As illustrated by sodium ethylsulfate in Figure 11.11, replacement of a H^+ ion by a metal ion gives a salt of an acid ester of sulfuric acid.

Sulfuric acid esters are used as alkylating agents, which act to attach alkyl groups (such as methyl) to organic molecules in the manufacture of agricultural chemicals, dyes, and drugs. **Methylsulfuric acid** and **ethylsulfuric acid** are oily water-soluble liquids that are strong irritants to skin, eyes, and mucous tissue. Liquid **dimethylsulfate** is colorless, odorless, and — like methylmethane sulfonate — a primary carcinogen.

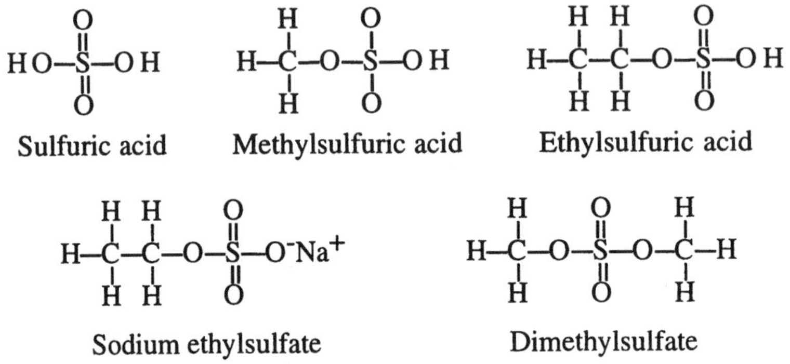

Sulfuric acid Methylsulfuric acid Ethylsulfuric acid

Sodium ethylsulfate Dimethylsulfate

Figure 11.11. Sulfuric acid and organosulfate esters.

11.7. ORGANOPHOSPHORUS COMPOUNDS

Alkyl and Aryl Phosphines

As shown in Figure 11.12, the structural formulas of alkyl and aryl phosphine compounds may be derived by substituting organic groups for the H atoms in phosphine (PH_3), the hydride of phosphorus discussed as a toxic inorganic compound in Chapter 10, Section 10.4. **Methylphosphine** is a colorless, chemically reactive gas

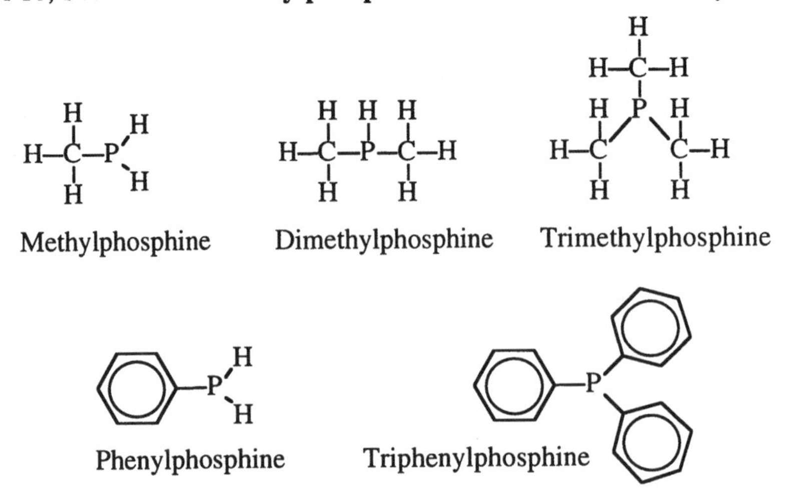

Figure 11.12. Some of the more significant alkyl and aryl phosphines.

and **dimethylphosphine** is a colorless, reactive, volatile liquid (bp 25°C). **Trimethylphosphine** is a colorless, volatile, reactive, spontaneously ignitable liquid (bp 42°C). **Phenylphosphine** (phosphaniline) is a reactive, moderately flammable liquid (bp 16°C). All of these compounds should be regarded as having high toxicities. Crystalline, solid **triphenylphosphine** has a low reactivity and moderate toxicity when inhaled or ingested.

As shown by the reaction,

$$4C_3H_9P + 26O_2 \rightarrow 12CO_2 + 18H_2O + P_4O_{10} \qquad (11.7.1)$$

combustion of aryl and alkyl phosphines produces P_4O_{10}, a corrosive irritant toxic substance, or droplets of corrosive orthophosphoric acid, H_3PO_4.

Phosphine Oxides and Sulfides

Structural formulas of a typical common phosphine oxide and a typical phosphine sulfide are the following:

$$C_2H_5-\overset{\overset{\displaystyle O}{\|}}{\underset{\underset{\displaystyle C_2H_5}{|}}{P}}-C_2H_5 \qquad\qquad C_4H_9-\overset{\overset{\displaystyle S}{\|}}{\underset{\underset{\displaystyle C_4H_9}{|}}{P}}-C_4H_9$$

Triethylphosphine oxide   Tributylphosphine oxide

The phosphine oxides and sulfides are toxic. When burned, they give off dangerous phosphorus oxide fumes and, in the case of phosphine sulfides, sulfur oxides.

Organophosphate Esters

The structural formulas of three esters of orthophosphoric acid (H_3PO_4) and an ester of pyrophosphoric acid ($H_4P_2O_6$) are shown in Figure 11.13. Although **trimethylphosphate** and **triphenylphosphate** are considered to be only moderately toxic, **tri-*o*-cresyl-phosphate, TOCP,** has a notorious record of poisonings. **Tetraethylpyrophosphate, TEPP**, was developed in Germany during World War II as a substitute for insecticidal nicotine. Although it is a very effective insecticide, its use in that application was of very short duration because it kills almost everything else, too.

Phosphorothionate and Phosphorodithioate Esters

Various kinds of thiophosphate esters have been used as the active ingredients in insecticide formulations. The insecticidal action of these compounds results from them being inhibitors to acetylcholinesterase, which is an essential enzyme in nerve function.[2] The general formulas of insecticidal **phosphorothionate** and **phosphorodithioate** esters are shown in Figure 11.14, where R is usually a methyl ($-CH_3$) or ethyl ($-C_2H_5$) group and Ar is a moiety of more complex structure, fre-

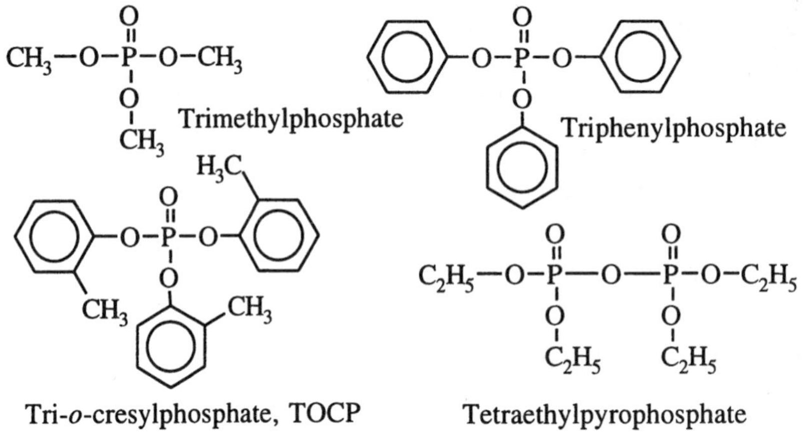

Trimethylphosphate

Triphenylphosphate

Tri-*o*-cresylphosphate, TOCP

Tetraethylpyrophosphate

Figure 11.13. Phosphate esters.

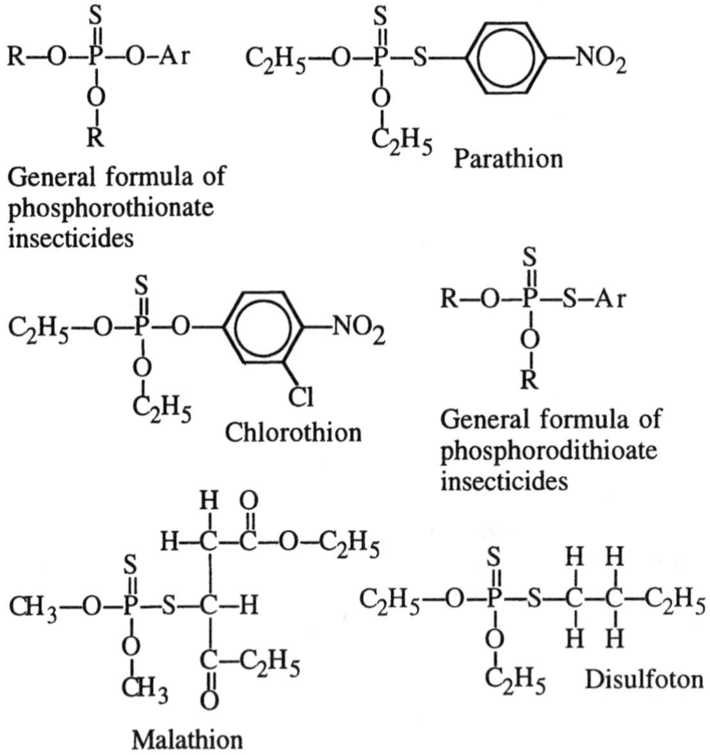

General formula of phosphorothionate insecticides

Parathion

Chlorothion

General formula of phosphorodithioate insecticides

Malathion

Disulfoton

Figure 11.14. General formulas and specific examples of phosphorothionate and phosphorodithioate organophosphate insecticides.

quently aryl. Phosphorothionate and phosphorodithioate esters contain the P=S (thiono) group, which increases their insect:mammal toxicity ratios and decreases their tendency to undergo nonenzymatic hydrolysis compared to their analogous compounds that contain the P=O functional group. The metabolic oxidative desulfuration conversion of P=S to P=O in organisms converts the phosphorothionate and phosphorodithioate esters to species that have insecticidal activity.

Since the first organophosphate insecticides were developed in Germany during the 1930s and 1940s, many insecticidal organophosphate compounds have been synthesized. One of the earliest and most successful of these was **parathion**, *O,O*-diethyl-*O*-*p*-nitrophenylphosphorothionate. This substance is now banned from use in the United States and is an EPA listed waste (P089). From a long-term environmental standpoint, organophosphate insecticides are superior to the organohalide insecticides that they largely displaced because the organophosphates readily undergo biodegradation and do not bioaccumulate.

11.8. POLYCHLORINATED BIPHENYLS

Polychlorinated biphenyls (PCBs) constitute an important class of special wastes.[3] These compounds are made by substituting from 1 to 10 Cl atoms onto the biphenyl aromatic structure as shown on the left in Figure 11.15. This substitution can produce 209 different compounds (congeners), of which one example is shown on the right in Figure 11.15.

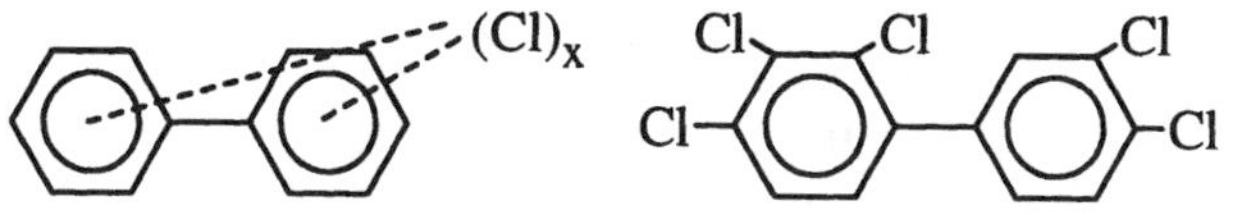

Figure 11.15. General formula of polychlorinated biphenyls (left, where X may range from 1 to 10) and a specific 5-chlorine congener (right).

Polychlorinated biphenyls have very high chemical, thermal, and biological stability; low vapor pressure; and high dielectric constants. These properties have led to the use of PCBs as coolant-insulation fluids in transformers and capacitors; for the impregnation of cotton and asbestos; as plasticizers; and as additives to some epoxy paints. The same properties that made extraordinarily stable PCBs so useful also contributed to their widespread dispersion and accumulation in the environment. By regulations issued under the authority of the Toxic Substances Control Act passed in 1976, the manufacture of PCBs was discontinued in the U.S. and their uses and disposal were strictly controlled.

Askarel

Askarel is the generic name of PCB-containing dielectric fluids that formerly were widely used in transformers. These fluids are 50–70 percent PCBs and may contain 30–50 percent trichlorobenzenes (TCBs). As of 1989 an estimated 100,000 askarel-containing transformers were still in use in the U.S., although presumably by now essentially all of them have been taken out of commission. Although the dielectric fluid may be replaced in these transformers enabling their continued use, PCBs tend to leach into the replacement fluid from the transformer core and other parts of the transformer over a several-month period.

11.9. DIOXINS IN HAZARDOUS WASTES

While PCBs discussed in the preceding section are derived from the two-ring hydrocarbon biphenyl, chlorinated dibenzo-*p*-dioxins are derived from dibenzo-*p*-dioxin, the structural formula of which is shown in Figure 11.16. The chlorinated derivatives are commonly referred to as "dioxins." They have a high environmental and toxicological significance.

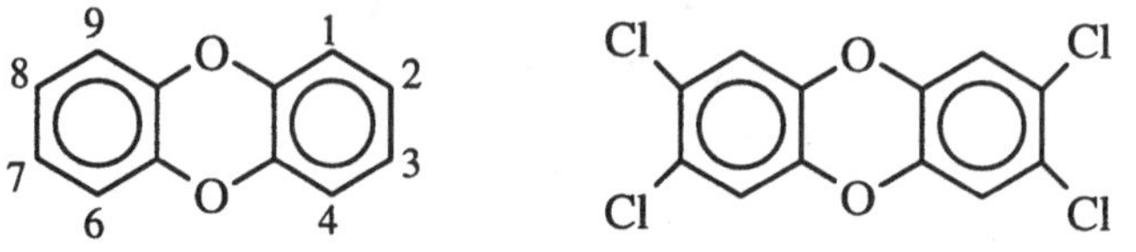

Dibenzo-*p*-dioxin 2,3,7,8-Tetrachlorodibenzo-*p*-dioxin

Figure 11.16. Dibenzo-*p*-dioxin and 2,3,7,8-tetrachlorodibenzo-*p*-dioxin (TCDD), often called simply "dioxin." In the structure of dibenzo-*p*-dioxin each number refers to a numbered carbon atom to which an H atom is bound and the names of derivatives are based upon the carbon atoms where another group has been substituted for the H atoms, as is seen by the structure and name of 2,3,7,8-tetrachlorodibenzo-*p*-dioxin.

From 1 to 8 Cl atoms may be substituted for H atoms on dibenzo-*p*-dioxin, giving a total of 75 possible chlorinated derivatives. Of these, the most notable hazardous waste compound is 2,3,7,8-tetrachlorodibenzo-*p*-dioxin (TCDD), often referred to simply as "dioxin." One of the most toxic of all synthetic substances to some animals, TCDD has never been produced commercially,[4] but was produced as a low-level contaminant in the manufacture of some aryl, oxygen-containing organohalide compounds such as chlorophenoxy herbicides and hexachlorophene (Figure 11.17) manufactured by processes used until the 1960s.

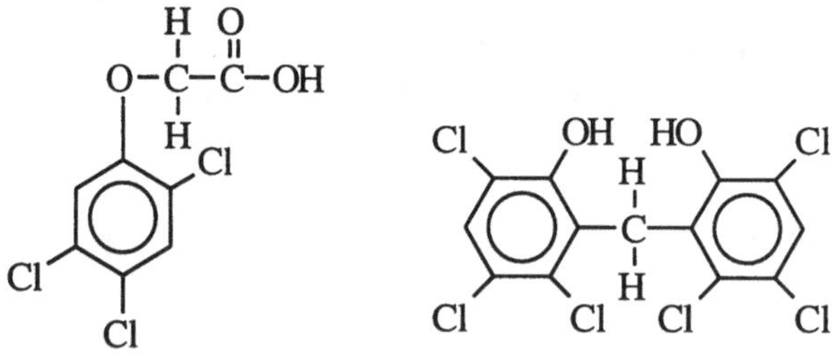

2,4,5–Trichlorophenoxy-acetic acid (and esters) Hexachlorophene

Figure 11.17. Two chemicals whose manufacture resulted in the production of byproduct TCDD contaminant.

The chlorophenoxy herbicides, such as 2,4,5-trichlorophenoxyacetic acid (2,4,5-T) shown in Figure 11.17, were manufactured on a large scale for weed and brush control and as military defoliants. Fungicidal and bactericidal hexachlorophene was once widely applied to crops in the production of vegetables and cotton and was used as an antibacterial agent in personal care products, an application that was discontinued because of toxic effects and possible TCDD contamination.

TCDD has a very low vapor pressure of only 1.7×10^{-6} mm Hg at 25°C, a high melting point of 305°C, and a water solubility of only 0.2 μg/L. It is thermally stable up to about 700°C, has a high degree of chemical stability, and is poorly biodegradable. It is very toxic to some animals, with an LD_{50} of only about 0.6 μg/kg body mass in male guinea pigs. (The type and degree of its toxicity to humans is largely unknown; it is known to cause a severe skin condition called chloracne.) Because of its properties, TCDD is a stable, persistent environmental pollutant and hazardous waste constituent of considerable concern. It has been the subject of several widely publicized environmental incidents including the contamination of the city of Times Beach, Missouri, by TCDD-containing waste oil in the early 1970s and release from a massive industrial accident at the Givaudan-La Roche Icmesa manufacturing plant near Seveso, Italy, in 1976.

LITERATURE CITED

1. McCabe, Mark M. and William Newton, "Waste Oil," Section 4.1 in *Standard Handbook of Hazardous Waste Treatment and Disposal*, 2nd ed., Harry M. Freeman, Ed., McGraw-Hill, New York, NY, 1998, pp. 4.3–4.13.
2. Ecobichon, Donald J., "Toxic Effects of Pesticides," Chapter 22 in *Casarett and Doull's Toxicology*, 5th ed., Curtis D. Klaassen, Ed., McGraw-Hill, New York, NY, 1996, pp. 643–689.
3. MacNeil, Janet C. and Drew E. McCoy, "PCB Wastes," Section 4.2 in *Standard Handbook of Hazardous Waste Treatment and Disposal*, 2nd ed., Harry M. Freeman, Ed., McGraw-Hill, New York, NY, 1998, pp. 4.14–4.24.
4. Kilgore, James D. and Chun Wai Lee, "Dioxins and Furans," Section 4.3 in *Standard Handbook of Hazardous Waste Treatment and Disposal*, 2nd ed., Harry M. Freeman, Ed., McGraw-Hill, New York, NY, 1998, pp. 4.25–4.44.

SUPPLEMENTARY REFERENCES

Cheremisinoff, Nicholas P. and Paul N. Cheremisinoff, *Hazardous Materials and Waste Management*, Noyes Publications, Park Ridge, New Jersey, 1995.

Karnofsky, Brian, Ed., *Hazardous Waste Management Compliance Handbook*, 2nd ed., Van Nostrand Reinhold, New York, NY, 1996.

12 TOXICOLGICAL AND BIOLOGICAL HAZARDS

12.1. INTRODUCTION

A major goal of industrial ecology is to reduce or eliminate the production and use of toxic materials because the ultimate concern with hazardous substances is their effect upon living organisms, especially humans. In addition, a wide variety of toxic substances are produced by organisms of various kinds. For example, aflatoxins produced by molds growing on nuts or grain are of concern in the peanut and corn products industries. Biological agents such as bacterial anthrax have the potential to be used in warfare and in terrorist acts.

This chapter addresses toxicological and biological hazards. It begins with a brief introduction to the science of toxicology, then addresses toxicological chemistry, the toxicological chemistry of various classes of substances, and finally, toxic aspects of toxins and other materials produced by biological processes and in biomedical facilities.

12.2. TOXICOLOGY

Toxicology is the science of poisons, where a **poison** or **toxicant** is a substance that is harmful to living organisms because of its detrimental effects on tissues, organs, or biological processes. Both **xenobiotic** substances, which are those that are foreign to a living system, and substances **endogenous** to living organisms may be toxic.

The degree of exposure to a toxic substances is commonly expressed by **dose**, discussed later in this chapter. The exposure **site** and **route** affect toxicity. Exposures may be **acute**, occurring at high levels for a brief period of time, or **chronic**, occurring over very long time periods. They may be **local**, where a toxicant acts at a specific site of exposure, or **systemic**, in which a toxicant is taken into an organism, distributed by means such as the blood circulatory system and metabolized in various places in the organism. In discussing exposure sites for toxicants it is useful to consider the routes and sites of exposure, distribution, and elimination of toxicants in the body as shown in Figure 12.1. The major routes of accidental or intentional exposure to toxicants by humans and other animals are the skin (percutaneous route), the lungs (inhalation, respiration, pulmonary route), and the mouth (oral route).

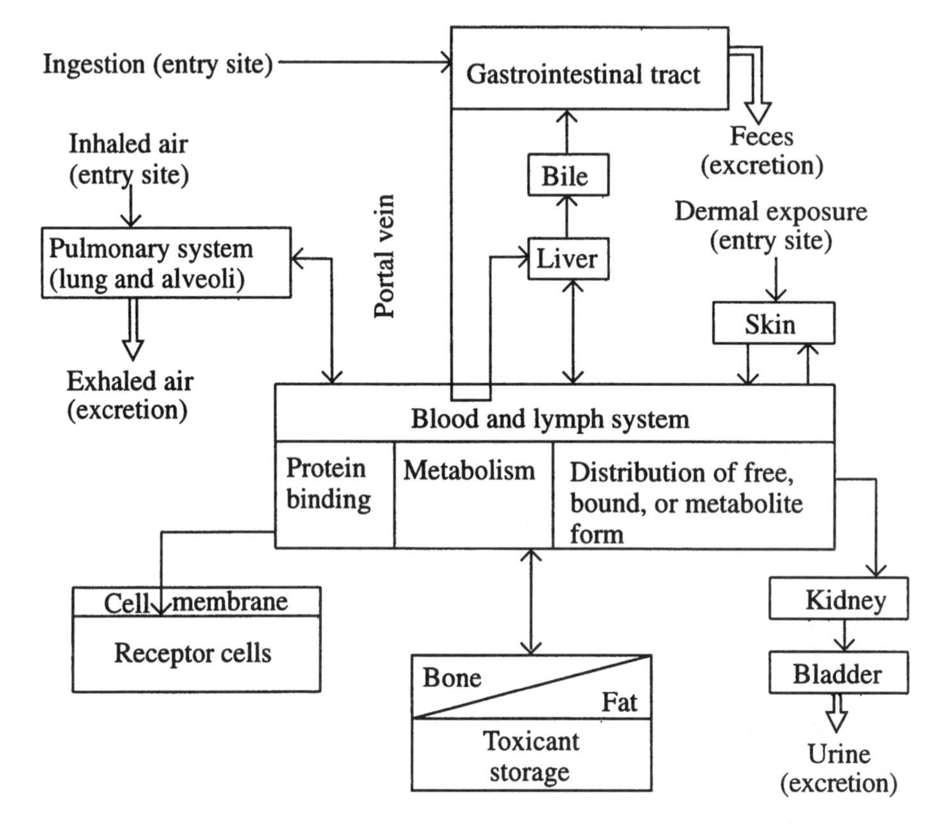

Figure 12.1. Major sites of exposure, metabolism, and storage, routes of distribution and elimination of toxic substances in the body.

Dose-Response Relationships

Factors such as minimum levels at which the onset of an effect of a toxicant is observed, the sensitivity of the organism to small increments of toxicant, and levels at which the ultimate effect (particularly death) occurs in most exposed organisms are taken into account by the **dose-response** relationship, which is one of the key concepts of toxicology. **Dose** is the amount, usually per unit body mass, of a toxicant to which an organism is exposed. **Response** is the effect upon an organism resulting from exposure to a toxicant. Figure 12.2 shows a generalized dose-response curve in which death is the response of the test organisms. The dose corresponding to the mid-point (inflection point) of this S-shaped curve is the statistical estimate of the dose that would kill 50 percent of the subjects and is designated as LD_{50}.

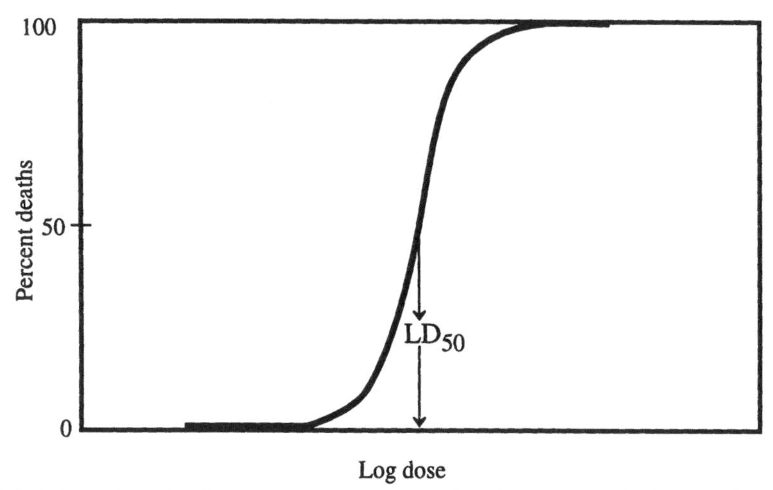

Figure 12.2. Illustration of a dose-response curve in which the response is death of the organism. The cumulative percentage of deaths of organisms is plotted on the Y axis.

Relative Toxicities

Table 12.1 illustrates standard **toxicity ratings** that are used to describe estimated toxicities of various substances to humans. In terms of fatal doses to an adult human of average size, a "taste" of a supertoxic substances (just a few drops or less) is fatal. A teaspoonful of a very toxic substance could be fatal to a human. However, as much as a liter of a slightly toxic substance might be required to kill an adult human.

As shown in Table 12.1, there is a tremendous range of toxicities of substances. For example, it is estimated that more than 100 times as much malathion is required to kill a human compared to the much more toxic parathion. Another striking figure is the astoundingly high toxicity of botulinus toxin, a natural product from bacteria.

Lethality, Reversibility, and Sensitivity

Figure 12.3 shows several important aspects of toxicants. **Lethality** refers to the degree to which a toxic substance is likely to be fatal to a subject. Sublethal doses of most toxic substances are eventually eliminated from an organism's system. If there is no lasting effect from the exposure, it is said to be **reversible**; if the effect is permanent, it is termed **irreversible**. Irreversible effects of exposure remain after the toxic substance is eliminated from the organism. The dose-response curve shown in Figure 12.3 shows that some subjects are very sensitive to a particular poison (for example, those killed at a dose corresponding to LD_5), whereas others are very resistant to the same substance (for example, those surviving a dose corresponding to LD_{95}). These two kinds of responses illustrate **hypersensitivity** and **hyposensitivity**, respectively; subjects in the mid-range of the dose-response curve are termed **normals**. Hypersensitivity can be induced, and one of the greater problems with industrial exposure to chemicals is the development of hypersensitivity as manifested by allergic responses.

Table 12.1. Toxicity Scale with Example Substances[1]

Substance	Approximate LD_{50}	Toxicity rating
	-10^5	1. Practically nontoxic $> 1.5 \times 10^4$ mg/kg
DEHP[2] →	–	
Ethanol →	-10^4	2. Slightly toxic, 5×10^3 to 1.5×10^4 mg/kg
Sodium chloride →	–	
Malathion →	-10^3	3. Moderately toxic, 500 to 5000 mg/kg
Chlordane →	–	
Heptachlor →	-10^2	4. Very toxic, 50 to 500 mg/kg
	–	
Parathion →	-10	
	–	5. Extremely toxic, 5 to 50 mg/kg
TEPP[3] →	– 1	
	–	
Tetrodotoxin[4] →	-10^{-1}	
	–	
	-10^{-2}	6. Supertoxic, <5 mg/kg
	–	
TCDD[5] →	-10^{-3}	
	–	
	-10^{-4}	
	–	
Botulinus toxin →	-10^{-5}	

[1] Doses are in units of mg of toxicant per kg of body mass. Toxicity ratings on the right are given as numbers ranging from 1 (practically nontoxic) through 6 (supertoxic) along with estimated lethal oral doses for humans in mg/kg. Estimated LD_{50} values for substances on the left have been measured in test animals, usually rats, and apply to oral doses.

[2] Bis(2-ethylhexyl)phthalate

[3] Tetraethylpyrophosphate (see Chapter 11, Section 11.7)

[4] Toxin from pufferfish

[5] TCDD represents 2,3,7,8,-tetrachlorodibenzodioxin, commonly called "dioxin."

12.3. TOXICOLOGICAL CHEMISTRY

Toxicological chemistry is the science that deals with the chemical nature and reactions of toxic substances, including their origins, uses, and chemical aspects of exposure, fates, and disposal.[1] Toxicological chemistry addresses the relationships between the chemical properties and molecular structures of molecules and their toxicological effects. Figure 12.4 outlines these terms and the relationships among them.

Toxicants in the Body

The processes by which organisms metabolize xenobiotic species are enzyme-catalyzed Phase I and Phase II reactions, which are described briefly here.

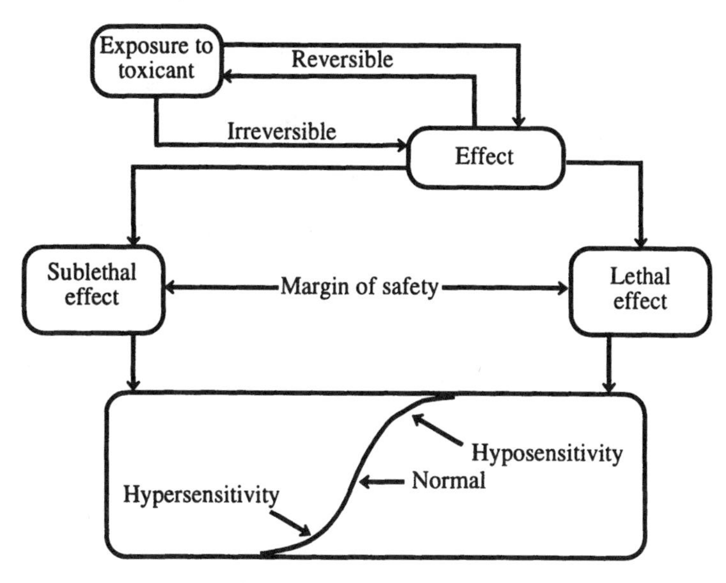

Figure 12.3. Effects of and responses to toxic substances.

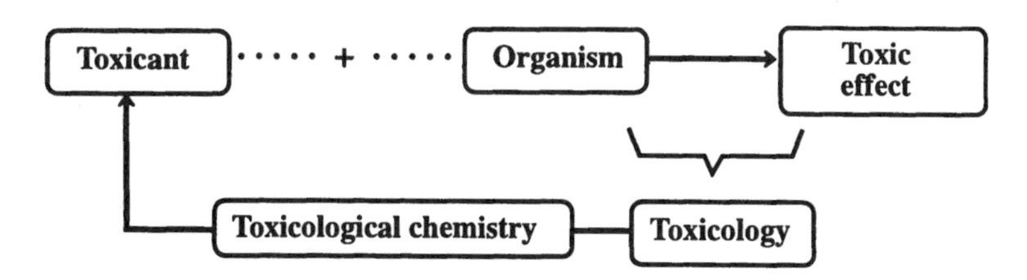

Figure 12.4. Toxicology is the science of poisons. Toxicological chemistry relates toxicology to the chemical nature of toxicants.

Lipophilic ("fat-seeking") xenobiotic species in the body tend to undergo **Phase I reactions** that make them more water-soluble and reactive by means of the attachment of polar functional groups, such as –OH (Figure 12.5). Most Phase I processes are "microsomal mixed-function oxidase" reactions catalyzed by the cytochrome P-450 enzyme system associated with the **endoplasmic reticulum** of the cell and occurring most abundantly in the liver of vertebrates.

The polar functional groups attached to a xenobiotic compound in a Phase I reaction provide reaction sites for **Phase II reactions**. Phase II reactions are **conjugation reactions** in which enzymes act to attach **conjugating agents** to xenobiotics, their phase I reaction products, and nonxenobiotic compounds (Figure 12.6). The **conjugation product** of such a reaction is usually less toxic than the original xenobiotic compound, less lipid-soluble, more water-soluble, and more readily eliminated from the body. The major conjugating agents and the enzymes that catalyze their Phase II reactions are glucuronide (UDP glucuronyltransferase enzyme), glutathione (glutathionetransferase enzyme), sulfate (sulfotransferase enzyme), and acetyl (acetylation by acetyltransferase enzymes). The conjugating agents are species that are endogenous to the organism, the most abundant of which are glucuronides.

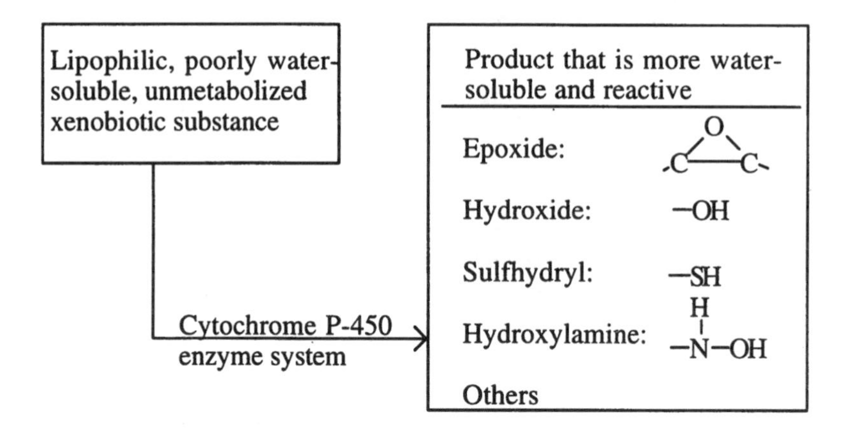

Figure 12.5. Illustration of Phase I reactions.

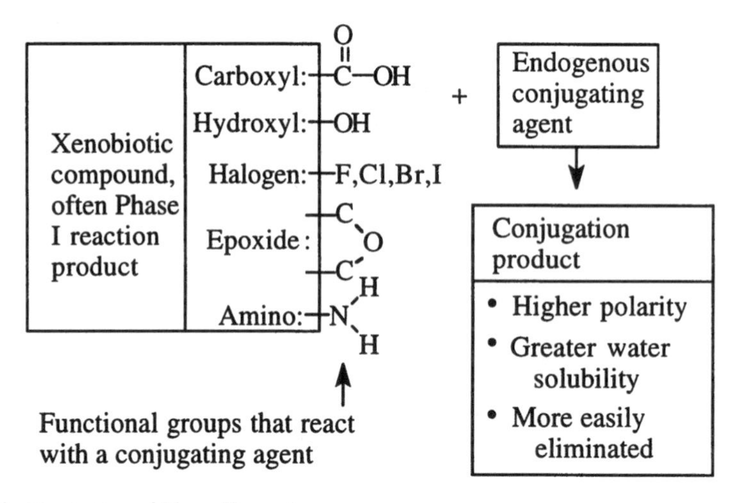

Figure 12.6. Illustration of Phase II reactions.

Kinetic Phase and Dynamic Phase

The major routes and sites of absorption, metabolism, binding, and excretion of toxic substances in the body are illustrated in Figures 12.1 and 12.7. Toxicants in the body are metabolized, transported, and excreted; they have adverse biochemical effects; and they cause manifestations of poisoning. It is convenient to divide these processes into a kinetic phase and a dynamic phase. In the **kinetic phase**, a toxicant or the metabolic precursor of a toxic substance (**protoxicant**) may undergo absorption, metabolism, temporary storage, distribution, and excretion, as illustrated in Figure 12.7. A toxicant that is absorbed may be passed through the kinetic phase unchanged as an **active parent compound**, metabolized to a **detoxified metabolite** that is excreted, or converted to a toxic **active metabolite**. These processes occur through Phase I and Phase II reactions discussed above.

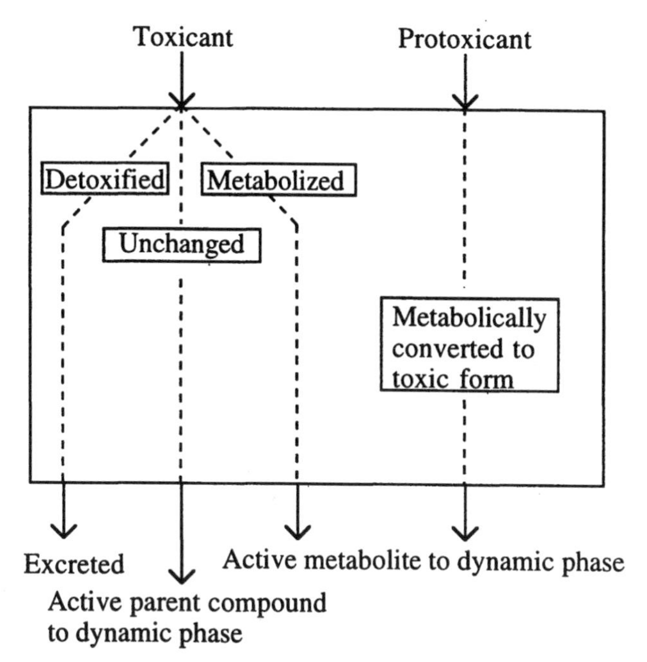

Figure 12.7. Processes involving toxicants or protoxicants in the kinetic phase.

In the **dynamic phase** (Figure 12.8), a toxicant or toxic metabolite interacts with cells, tissues, or organs in the body to cause some toxic response. The three major subdivisions of the dynamic phase are the following:

- **Primary reaction** with a receptor or target organ
- A **biochemical response**
- **Observable effects**

A toxicant or an active metabolite reacts with a receptor, such as a cell membrane or an enzyme system. The process leading to a toxic response is initiated when such a reaction occurs. A typical example is when benzene epoxide produced by the metabolic oxidation of benzene,

Benzene —Metabolic oxidation→ Benzene epoxide (12.3.1)

forms an adduct with a nucleic acid unit in DNA (receptor) resulting in alteration of the DNA. This reaction is an **irreversible** reaction between a toxicant and a receptor. A **reversible** reaction that can result in a toxic response is illustrated by the binding between carbon monoxide and oxygen-transporting hemoglobin (Hb) in blood,

$$O_2Hb + CO \leftrightarrow COHb + O_2 \qquad (12.3.2)$$

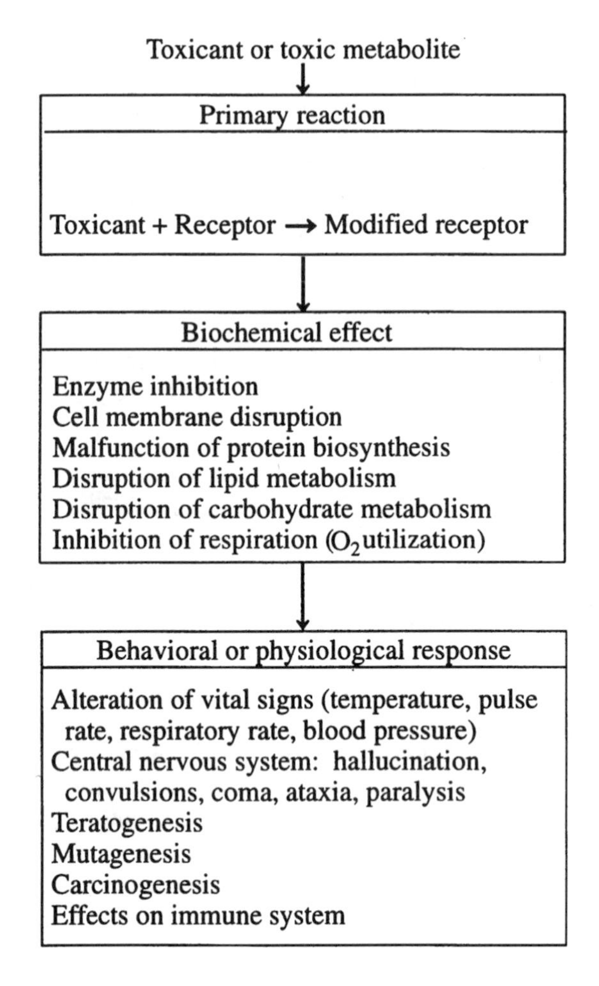

Figure 12.8. The dynamic phase of toxicant action.

The binding of a toxicant to a receptor may result in some kind of biochemical effect. Important examples of such effects are impairment of enzyme function, alteration of cell membrane, interference with carbohydrate metabolism, interference with lipid metabolism resulting in excess lipid accumulation ("fatty liver"), interference with respiration (the overall process by which electrons are transferred to molecular oxygen in the biological oxidation of energy-yielding substrates), stopping or interfering with protein biosynthesis by the action of toxicants on DNA, or interference with regulatory processes mediated by hormones or enzymes.

Among the more immediate and readily observed manifestations of poisoning are alterations in the **vital signs** of **temperature**, **pulse rate**, **respiratory rate**, and **blood pressure**. Other manifestations of poisoning include abnormal skin color, unnatural **odors**, central nervous system symptoms (convulsions, paralysis, hallucinations, lack of control of body movements, and coma). Additional observable effects, some of which may occur soon after exposure, include gastrointestinal illness, cardiovascular disease, hepatic (liver) disease, renal (kidney) malfunction, neurologic symptoms (central and peripheral nervous systems), or skin abnormalities (rash, dermatitis). Often the effects of toxicant exposure are subclinical in nature. These include damage to the immune system, chromosomal abnormalities, modification of functions of liver enzymes, and slowing of conduction of nerve impulses.

Chronic responses to toxicant exposure include mutations, cancer, birth defects, and effects on the immune system. **Teratogens** are chemical species that cause birth defects. These usually arise from damage to embryonic or fetal cells. **Mutagens** alter DNA to produce inheritable traits. The mechanisms of mutagenicity are similar to those of carcinogenicity, and mutagens often cause birth defects as well. Therefore, mutagenic hazardous substances are of major toxicological concern. An important mechanism of mutagenesis is alkylation of the nitrogenous bases present in DNA, the fundamental molecule of heredity. Figure 12.9 shows the alkylation (specifically methylation) of the nitrogenous base guanine bound in DNA. When DNA is modified by the binding of alkyl groups, it no longer accurately transmits directions for protein synthesis and mutations or cancer can result.

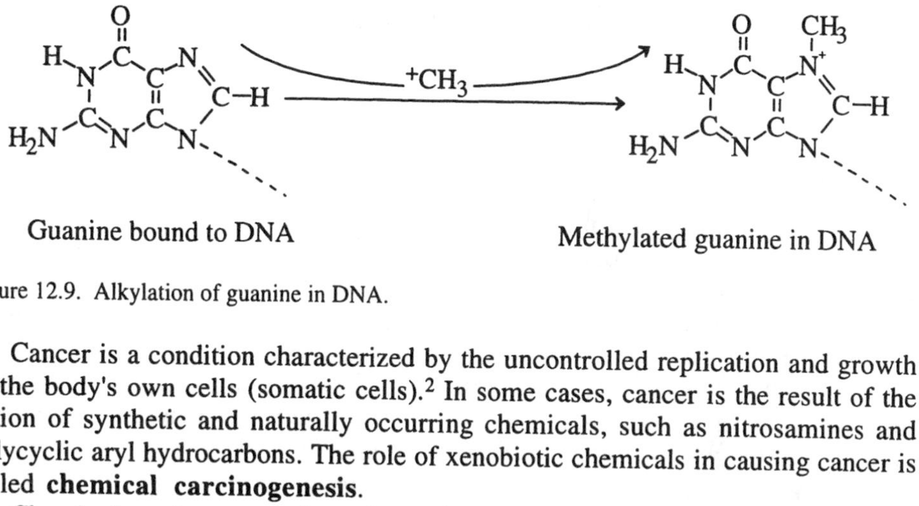

$+CH_3$

Guanine bound to DNA

Methylated guanine in DNA

Figure 12.9. Alkylation of guanine in DNA.

Cancer is a condition characterized by the uncontrolled replication and growth of the body's own cells (somatic cells).[2] In some cases, cancer is the result of the action of synthetic and naturally occurring chemicals, such as nitrosamines and polycyclic aryl hydrocarbons. The role of xenobiotic chemicals in causing cancer is called **chemical carcinogenesis**.

Chemical carcinogenesis has a long history. In 1775 Sir Percival Pott, Surgeon General serving under King George III of England, observed that chimney sweeps in London had a very high incidence of cancer of the scrotum, which he related to their exposure to soot and tar from the burning of bituminous coal. Around 1900 a German surgeon, Ludwig Rehn, reported elevated incidences of bladder cancer in dye workers exposed to chemicals extracted from coal tar; 2-naphthylamine,

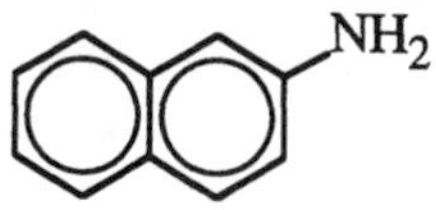

NH_2

was shown to be largely responsible. Other historical examples of carcinogenesis include observations of cancer from tobacco juice (1915), oral exposure to radium from painting luminescent watch dials (1929), tobacco smoke (1939), and asbestos (1960). Figure 12.10 shows the major steps involved in chemical carcinogenesis. It should be emphasized that most known chemical carcinogens require metabolic activation to cause cancer and are called **procarcinogens**.

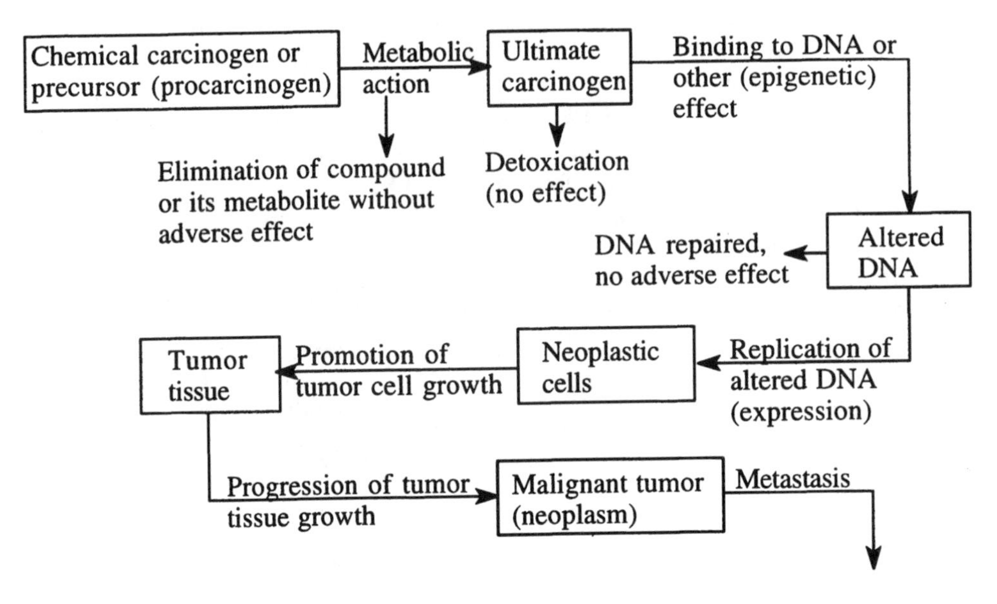

Figure 12.10. Outline of the process by which a carcinogen or procarcinogen may cause cancer.

The **immune system** acts as the body's natural defense system to protect it from xenobiotic chemicals; infectious agents, such as viruses or bacteria; and neoplastic cells, which give rise to cancerous tissue. Adverse effects on the body's immune system are being increasingly recognized as important consequences of exposure to hazardous substances. Toxicants can cause **immunosuppression**, which is the impairment of the body's natural defense mechanisms. Xenobiotics can also cause the immune system to lose its ability to control cell proliferation, resulting in leukemia or lymphoma. Another major toxic response of the immune system is **allergy** or **hypersensitivity**. This kind of condition results when the immune system overreacts to the presence of a foreign agent or its metabolites in a self-destructive manner. Among the xenobiotic materials that can cause such reactions are beryllium, chromium, nickel, formaldehyde, pesticides, resins, and plasticizers.

12.4. ASSESSMENT OF TOXICOLOGICAL HAZARDS

In recent years attention in toxicology has shifted away from readily recognized, usually severe, acute maladies that have developed on a short time scale as a result of brief, intense exposure to toxicants, toward delayed, chronic, often less severe illnesses caused by long-term exposure to low levels of toxicants. Although the total impact of the latter kinds of health effects may be substantial, their assessment is very difficult because of factors such as uncertainties in exposure, low occurrence above background levels of disease, and long latency periods.

Assessment of Potential Exposure

A critical step in assessing exposure to toxic substances, such as those from hazardous waste sites, is evaluation of potentially exposed populations. The most direct approach to this is to determine chemicals or their metabolic products in organisms. For inorganic species this is most readily done for heavy metals, radionuclides, and some minerals, such as asbestos. Symptoms associated with exposure to particular chemicals may also be evaluated. Examples of such effects include skin rashes or subclinical effects, such as chromosomal damage.

Epidemiological Evidence

Epidemiological studies applied to toxic environmental pollutants, such as those from hazardous wastes, attempt to correlate observations of particular illnesses with probable exposure to such wastes. There are two major approaches to such studies. One approach is to look for diseases known to be caused by particular agents in areas where exposure is likely from such agents. A second approach is to look for **clusters** consisting of an abnormally large number of cases of a particular disease in a limited geographic area, then attempt to locate sources of exposure to hazardous wastes that may be responsible. The most common types of maladies observed in clusters are spontaneous abortions, birth defects, and specific types of cancer.

Epidemiologic studies are complicated by long latency periods from exposure to onset of disease (which, in the case of cancer can be 20 years or more); lack of specificity in the correlation between exposure to a particular waste, pollutant, or substance to which exposure has taken place in the workplace; and the occurrence of a disease, and background levels of a disease in the absence of exposure to a hazardous waste capable of causing the disease.

Estimation of Health Effect Risks

An important part of estimating the risk of adverse health effects from exposure to toxicants involves extrapolation from experimentally observable data. Usually the end result needed is an estimate of a low occurrence of a disease in humans after a long latency period resulting from low-level exposure to a toxicant for a long period of time. The data available are almost always taken from animals exposed to high levels of the substance for a relatively short period of time. Extrapolation is then made using linear or curvilinear projections to estimate the risk to human populations. There are, of course, very substantial uncertainties in this kind of approach.

Risk Assessment

Toxicological considerations are very important in estimating potential dangers of pollutants and hazardous waste chemicals. Therefore, consideration of toxic substances plays a strong role in any system of industrial ecology. One of the major ways in which toxicology interfaces with the area of hazardous wastes is in **health risk assessment**, providing guidance for risk management, cleanup, or regulation needed at a hazardous waste site based upon knowledge about the site and the chemical and toxicological properties of wastes in it. Risk assessment includes the factors of site characteristics; substances present, including indicator species; potential receptors; potential exposure pathways; and uncertainty analysis. It may be divided into the following components:

- Identification of hazard
- Dose-response assessment
- Exposure assessment
- Risk characterization.

12.5. TOXIC ELEMENTS AND ELEMENTAL FORMS

This section discusses toxicological aspects of elements (particularly heavy metals) whose presence in a compound frequently means that the compound is toxic, as well as the toxicities of some commonly used elemental forms, such as the chemically uncombined elemental halogens.

Ozone (O_3) is a reactive and toxic form of elemental oxygen produced by ultraviolet light or electrical discharges passing through air. Air containing 1 ppm by volume ozone has a distinct odor. Inhalation of ozone at this level causes severe irritation and headache. Ozone irritates the eyes, upper respiratory system, and lungs. Inhalation of ozone can cause sometimes fatal pulmonary edema. Chromosomal damage has been observed in subjects exposed to ozone. Ozone generates free radicals (see Section 7.8) in tissue. These reactive species can cause lipid peroxidation, oxidation of sulfhydryl (–SH) groups, and other destructive oxidation processes.

Elemental **white phosphorus** can enter the body by inhalation, by skin contact, or orally. It is a systemic poison that causes anemia, gastrointestinal system dysfunction, bone brittleness, and eye damage. Exposure causes **phossy jaw**, a condition in which the jawbone deteriorates and becomes fractured.

The elemental halogens are all toxic. **Fluorine** (F_2) is a pale yellow highly reactive gas that is a strong oxidant. It is a toxic irritant and attacks skin and the mucous membranes of the nose and eyes. **Chlorine** (Cl_2) is a greenish-yellow strongly oxidizing gas that is a strong oxidant. In water, chlorine reacts to produce a strongly oxidizing solution. This reaction is responsible for some of the damage caused to the moist tissue lining the respiratory tract when the tissue is exposed to chlorine. The respiratory tract is rapidly irritated by exposure to 10–20 ppm of chlorine gas in air, causing acute discomfort that warns of the presence of the toxicant. Even brief exposure to 1,000 ppm of Cl_2 can be fatal. **Bromine** (Br_2) is a volatile dark red liquid that is toxic when inhaled or ingested. Like chlorine and fluorine, it is strongly irritating to the mucous tissue of the respiratory tract and eyes and may cause pulmonary edema. Elemental **iodine** (I_2), a solid consisting of lustrous violet-black rhombic crystals, is irritating to the lungs much like bromine or chlorine and its general effects are similar to these elements. However, the relatively low vapor pressure of iodine limits exposure to I_2 vapor.

Heavy Metals

Some metals, commonly known as **heavy metals**, are particularly toxic in their chemically combined forms and — notably in the case of mercury — some are toxic in the elemental form. The term, heavy metal, is used loosely to refer to almost any metal with an atomic number higher than that of calcium (20); some metalloids, such as arsenic and antimony, are classified as heavy metals for discussion of their toxicities. The toxic properties of some of the most hazardous heavy metals are discussed here.

Although not truly a heavy metal, **beryllium** (atomic mass 9.01) is one of the more hazardous toxic elements. Skin exposed to beryllium compounds may become ulcerated and develop granulomas. The body can become hypersensitive to beryllium, resulting in skin dermatitis, acute conjunctivitis, and corneal laceration. The most serious toxic effect of beryllium is berylliosis, a condition manifested by lung fibrosis and pneumonitis, which may develop after a latency period of 5–20 years.

Cadmium adversely affects several important enzymes and it can cause painful osteomalacia (bone disease) and kidney damage. Inhalation of cadmium oxide dusts and fumes results in cadmium pneumonitis characterized by edema and pulmonary epithelium necrosis.

The danger of exposure to **lead** is higher than that for many other toxicants because it is widely distributed as metallic lead, inorganic compounds, and organometallic compounds. Lead has a number of toxic effects, including inhibition of the synthesis of hemoglobin. It also adversely affects the central and peripheral nervous systems and the kidneys. The dispersal of, and environmental exposure to both lead and cadmium are important considerations in designing more environmentally friendly systems of industrial ecology.

Arsenic is a metalloid which forms a number of toxic compounds. The toxic +3 oxide, As_2O_3, is absorbed through the lungs and intestines. Biochemically, arsenic acts to coagulate proteins, forms complexes with coenzymes, and inhibits the production of adenosine triphosphate (ATP) in essential metabolic processes.

Elemental **mercury** vapor can enter the body through inhalation and be carried by the bloodstream to the brain where it penetrates the blood-brain barrier. It disrupts metabolic processes in the brain causing tremor and psychopathological symptoms such as shyness, insomnia, depression, and irritability. Divalent ionic mercury, Hg^{2+}, damages the kidney. Organometallic mercury compounds are also very toxic, dimethylmercury, $Hg(CH_3)_2$, extraordinarily so with at least one fatal poisoning having occurred by exposure to a single drop of the compound *through a rubber glove.*

12.6. TOXIC INORGANIC COMPOUNDS

Both **hydrogen cyanide** (HCN) and **cyanide salts** (which contain CN^- ion) are rapidly acting poisons. Metabolically, cyanide bonds to iron(III) in iron-containing ferricytochrome oxidase enzyme, preventing its reduction to iron(II) in the oxidative phosphorylation process by which the body utilizes O_2. The crucial enzyme is inhibited because ferrouscytochrome oxidase, which is required to react with O_2, is not formed and utilization of oxygen in cells is prevented so that metabolic processes cease.

Carbon monoxide, CO, is a common cause of accidental poisoning. At CO levels in air of 10 parts per million (ppm), impairment of judgement and visual perception occur; exposure to 100 ppm causes dizziness, headache, and weariness; loss of consciousness occurs at 250 ppm; and inhalation of 1,000 ppm results in rapid death. Chronic long-term exposures to low levels of carbon monoxide are suspected of causing disorders of the respiratory system and the heart.

After entering the blood stream through the lungs, carbon monoxide reacts with hemoglobin (Hb) to convert oxyhemoglobin (O_2Hb) to carboxyhemoglobin (COHb), as shown in Reaction 12.3.2. Carboxyhemoglobin is much more stable than oxyhemoglobin so that its formation prevents hemoglobin from carrying oxygen to body tissues.

Nitrogen Oxides

The two most common toxic oxides of nitrogen are **nitric oxide** (NO) and **nitrogen dioxide** (NO_2), designated collectively as NO_x. Nitric oxide is released in large quantities from the exhausts of internal combustion and turbine engines, and both oxides are common air pollutants. Regarded as the more toxic of the two gases, NO_2 causes severe irritation of the innermost parts of the lungs resulting in pulmonary edema. Approximately three weeks after severe exposures to NO_2, fatal bronchiolitis fibrosa obliterans may develop. Fatalities may result from even brief periods of inhalation of air containing 200–700 ppm of NO_2. Biochemically, NO_2 disrupts lactic dehydrogenase and some other enzyme systems, possibly acting much like ozone, a stronger oxidant discussed earlier in this chapter.

Nitrous oxide, N_2O, is used as an oxidant gas and in dental surgery as a general anesthetic. This gas was once known as "laughing gas," and was used in the late 1800s as a "recreational gas" at parties held by some of our not-so-staid Victorian ancestors. Nitrous oxide is a central nervous system depressant and can act as an asphyxiant.

Hydrogen Halides

Hydrogen halides (general formula HX, where X is F, Cl, Br, or I) are relatively toxic gases. The most widely used of these gases are HF and HCl; their toxicities are discussed here. **Hydrogen fluoride**, (HF, mp –83.1°C, bp 19.5°C) is used as a clear, colorless liquid or gas or as a 30–60% aqueous solution of **hydrofluoric acid**. Hydrofluoric acid is so reactive that it is used to etch glass and clean stone. It must be kept in plastic containers because it vigorously attacks glass and other materials containing silica (SiO_2), producing gaseous silicon tetrafluoride, SiF_4. Both gaseous and aqueous HF are extreme irritants to any part of the body that they contact, causing ulcers in affected areas of the upper respiratory tract. Lesions caused by contact with HF heal poorly over long time periods and tend to develop gangrene.

Fluoride ion, F^-, is toxic in soluble fluoride salts, such as NaF, causing **fluorosis**, a condition characterized by bone abnormalities and mottled, soft teeth. Livestock is especially susceptible to poisoning from fluoride fallout on grazing land; severely afflicted animals become lame and even die. Industrial pollution has been a common source of toxic levels of fluoride. Low levels of fluoride can have beneficial effects, however. About 1 ppm of fluoride used in some drinking water supplies prevents tooth decay. Because of the known toxic effects of higher levels of fluoride, fluoridation of water is often a very controversial and contentious issue in communities where it is practiced.

Colorless **hydrogen chloride** (HCl) is widely produced as a gas, pressurized liquid, or as a saturated aqueous solution containing 36% HCl called **hydrochloric acid** and commonly denoted simply as HCl. Hydrochloric acid is a major industrial chemical with U.S. production of about 2.3 million tons per year. Anhydrous HCl is also used in relatively large quantities. Much less toxic than HF, hydrochloric acid is a natural physiological fluid present as a dilute solution in the stomachs of humans and other animals. However, inhalation of HCl vapor can cause spasms of the larynx as well as pulmonary edema and even death at high levels. Hydrogen chloride vapor has such a high affinity for water that it tends to dehydrate eye and respiratory tract tissue.

Interhalogen Compounds and Halogen Oxides

Compounds formed from different halogens are called **interhalogen compounds**, examples of which are chlorine trifluoride, ClF_3, bromine monochloride, BrCl, and iodine pentachloride, ICl_5. Most interhalogen compounds exhibit extreme reactivity and are potent oxidizing agents for organic matter and oxidizable inorganic compounds. Interhalogen compounds react with water or steam to produce hydrohalic acid solutions (HF, HCl) and nascent oxygen {O}. Too reactive to enter biological systems in their original chemical state, interhalogen compounds tend to be powerful corrosive irritants that acidify, oxidize, and dehydrate tissue. Because of these effects, skin is readily damaged by interhalogen compounds; the eyes and mucous membranes of the mouth, throat, and pulmonary systems are especially susceptible to attack. The corrosive effects of the interhalogen compounds are much like those of the elemental forms of the elements from which they are composed.

Compounds of halogens and oxygen are **halogen oxides**, examples of which include fluorine monoxide, OF_2; chlorine dioxide, ClO_2; bromine dioxide, BrO_2; and iodine pentoxide, I_2O_5. These compounds are toxicologically similar to the interhalogen compounds discussed above, acting as corrosive substances that acidify, oxidize, and dehydrate tissue.

Oxyacids of the Halogens and their Salts

The most important of the oxyacids and their salts formed by halogens are hypochlorous acid, HOCl, and hypochlorites, such as NaOCl, used for bleaching and disinfection. The hypochlorites irritate eye, skin, and mucous membrane tissue because they react to produce active (nascent) oxygen ({O}) and acid as shown by the reaction below:

$$HClO \rightarrow H^+ + Cl^- + \{O\} \qquad (12.6.1)$$

Halogen Azides and Nitrogen Halides

The halogen azides are compounds with the general formula of XN_3, where X is a halogen. These compounds are extremely reactive and spontaneously explosive; their vapors are irritating to tissue. The halogen azides react with water to produce toxic fumes of the elemental halogen, HX, and NO_X.

Colorless, gaseous nitrogen trifluoride, NF_3, and nitrogen trichloride, NCl_3, a volatile yellow oil, are examples of nitrogen halides having the general formula N_nX_x, where X is F, Cl, Br or I. The nitrogen halides are toxic because they are irritants to eyes, skin, and mucous membranes; although because of their high reactivity they are often destroyed before exposure can occur.

Monochloramine and Dichloramine

Monochloramine (NH_2Cl), dichloramine ($NHCl_2$), and nitrogen trichloride (NCl_3) are formed deliberately in the purification of drinking water to provide **combined available chlorine**. These forms of chlorine last longer in the water distribution system than Cl_2, HOCl, and OCl^- and act to retain disinfection throughout the water distribution system.

Inorganic Compounds of Silicon

Silica, SiO_2, is a hard mineral substance known as quartz in the pure form and occurs in a variety of minerals such as sand, sandstone, and diatomaceous earth. **Silicosis** resulting from human exposure to silica dust from construction materials, sand blasting, and other sources has been a common occupational disease and is, therefore, a consideration in developing systems of industrial ecology. A type of pulmonary fibrosis that causes lung nodules and makes victims more susceptible to pneumonia and other lung diseases, silicosis is one of the most common disabling conditions resulting from industrial exposure to hazardous substances. It can cause death from insufficient oxygen or from heart failure in severe cases.

Asbestos is the name given to a group of fibrous silicate minerals, typically those of the serpentine group, for which the approximate formula is $Mg_3P(Si_2O_5)(OH)_4$. Asbestos has been widely used in structural materials, brake linings, insulation, and pipe manufacture. Inhalation of asbestos may cause asbestosis (a pneumonia condition), mesothelioma (tumor of the mesothelial tissue lining the chest cavity adjacent to the lungs), and bronchogenic carcinoma (cancer originating with the air passages in the lungs), so uses of asbestos have been severely curtailed and widespread programs have been undertaken to remove the material from buildings. The adverse health effects of asbestos have also been the subject of extensive litigation on behalf of exposed workers and their families with settlements involving billions of dollars.

Silicon tetrachloride, $SiCl_4$, is the only industrially significant of the **silicon tetrahalides**, a group of compounds with the general formula SiX_4, where X is a halogen. The two commercially produced **silicon halohydrides**, general formula $H_{4-x}SiX_x$, are dichlorosilane (SiH_2Cl_2) and trichlorosilane, ($SiHCl_3$). These compounds are used as intermediates in the synthesis of organosilicon compounds and in the production of high-purity silicon for semiconductors. Silicon tetrachloride and trichlorosilane, fuming liquids which react with water to give off HCl vapor, have suffocating odors and are irritants to eye, nasal, and lung tissue.

Inorganic Phosphorus Compounds

Phosphine (PH_3), a colorless gas that undergoes autoignition at 100°C, is used for the synthesis of organophosphorus compounds and is sometimes inadvertently produced in chemical syntheses involving other phosphorus compounds. It is a potential hazard in industrial processes and in the laboratory. Symptoms of poisoning from potentially fatal phosphine gas include pulmonary tract irritation, central nervous system depression, fatigue, vomiting, and difficult, painful breathing.

Because of its dehydrating action and formation of acid from the reaction

$$P_4O_{10} + 6H_2O \rightarrow 4H_3PO_4 \quad (12.6.2)$$

tetraphosphorus decoxide, P_4O_{10}, (formerly called phosphorus pentoxide) is a corrosive irritant to skin, eyes, and mucous membranes. This compound is produced as a fluffy white powder from the combustion of elemental phosphorus and reacts with water from air to form syrupy orthophosphoric acid, H_3PO_4.

Phosphorus halides have the general formulas PX_3 and PX_5, where X is a halogen. The most commercially important phosphorus halide is phosphorus pentachloride used as a catalyst in organic synthesis, as a chlorinating agent, and as a raw material to make phosphorus oxychloride ($POCl_3$). Because they react violently with water to produce the corresponding hydrogen halides and oxo phosphorus acids,

$$PCl_5 + 4H_2O \rightarrow H_3PO_4 + 5HCl \quad (12.6.3)$$

the phosphorus halides are strong irritants, especially to eyes, skin, and mucous membranes.

The major phosphorus oxyhalide in commercial use is **phosphorus oxychloride** ($POCl_3$), a faintly yellow fuming liquid. Reacting with water to form toxic vapors of hydrochloric acid and phosphonic acid (H_3PO_3), phosphorus oxyhalide is a strong irritant to the eyes, skin, and mucous membranes.

Inorganic Compounds of Sulfur

Sulfur forms several widely encountered toxic inorganic compounds. The toxicity of yellow crystalline or powdered elemental sulfur, S_8, is low, although chronic inhalation of it can irritate mucous membranes.

A colorless gas with a foul rotten-egg odor, **hydrogen sulfide** (H_2S, hazardous waste No. U135) occurs in large quantities as a byproduct of coal coking and petroleum refining and as a constituent of sour natural gas. In some cases inhalation of hydrogen sulfide kills faster than even hydrogen cyanide; rapid death ensues from exposure to air containing more than about 1000 ppm H_2S due to asphyxiation from respiratory system paralysis. Lower doses cause symptoms that include headache, dizziness, and excitement because of damage to the central nervous system. General debility is one of the numerous effects of chronic H_2S poisoning.

Sulfur dioxide, SO_2, dissolves in water to produce sulfurous acid, H_2SO_3; hydrogen sulfite ion, HSO_3^-; and sulfite ion, SO_3^{2-}. Because of its water solubility, sulfur dioxide is largely removed in the upper respiratory tract. It is an irritant to the eyes, skin, mucous membranes, and respiratory tract. Some individuals are hypersensitive to sodium sulfite (Na_2SO_3), which has been used as a chemical food preservative. These uses were further severely restricted in the U.S. in early 1990.

Number one in synthetic chemical production, **sulfuric acid** (H_2SO_4) is a severely corrosive poison and dehydrating agent in the concentrated liquid form; it readily penetrates skin to reach subcutaneous tissue, causing tissue necrosis with effects resembling those of severe thermal burns. Sulfuric acid fumes and mists irritate eye and respiratory tract tissue and industrial exposure has caused tooth erosion in workers.

12.7. ORGANOMETALLIC COMPOUNDS

The toxicological properties of some organometallic compounds — pharmaceutical organoarsenicals, organomercury fungicides, and tetraethyllead antiknock gasoline additives — that have been used for many years are well known. However, toxicological experience is lacking for many relatively new organometallic compounds that are now being used in semiconductors, as catalysis, and for chemical synthesis, so they should be treated with great caution until proven safe.

Organometallic compounds often behave in the body in ways totally unlike the inorganic forms of the metals that they contain. This is due in large part to the fact that, compared to inorganic forms, organometallic compounds have an organic nature and higher lipid solubility.

Perhaps the most notable hazardous organometallic compound is **tetraethyllead**, $Pb(C_2H_5)_4$ (hazardous waste No. P110), a colorless oil. Although its use in gasoline has essentially been phased out, this compound was employed for several decades as an octane-boosting gasoline additive, so there were many opportunities for exposure in its manufacture and blending and from leaded fuels. Tetraethyllead has a strong affinity for lipids and can enter the body by inhalation, ingestion, and absorption through the skin. It is highly toxic and does not act like inorganic lead compounds in the body. It affects the central nervous system with symptoms such as fatigue, weakness, restlessness, ataxia, psychosis, and convulsions. Recovery from severe lead poisoning tends to be slow. In cases of fatal tetraethyllead poisoning, death has occurred as soon as one or two days after exposure.

The greatest number of organometallic compounds in commercial use are those of tin. Organotin compounds are used as pesticides, antifouling paints, and catalysts and are of particular environmental significance because of their applications as industrial biocides. Tributyltin chloride and related tributyltin (TBT) compounds have bactericidal, fungicidal, and insecticidal properties. Organotin compounds are readily absorbed through the skin, sometimes causing a skin rash. They probably bind with sulfur groups on proteins and appear to interfere with mitochondrial function.

Metal carbonyls regarded as extremely hazardous because of their toxicities include nickel carbonyl ($Ni(CO)_4$), cobalt carbonyl, and iron pentacarbonyl. Some of the hazardous carbonyls are volatile and readily taken into the body through the respiratory tract or through the skin. The carbonyls directly affect tissue and they break down to toxic carbon monoxide and products of the metal, which have additional toxic effects.

12.8. HYDROCARBONS

The alkanes have relatively low toxicities. The lighter ones in the vapor phase, such as methane, ethane, or *n*-butane, are regarded as **simple asphyxiants**; air containing high levels of simple asphyxiants does not contain sufficient oxygen to support respiration. Inhalation of volatile liquid 5–8 carbon *n*-alkanes and branched-chain alkanes may cause central nervous system depression manifested by dizziness and loss of coordination. Exposure to *n*-hexane results in loss of myelin (a fatty substance constituting a sheath around certain nerve fibers) and degeneration of axons (part of a nerve cell through which nerve impulses are transferred out of the cell). This has resulted in **polyneuropathy**, the symptoms of which are muscle weakness and impaired sensory function of the hands and feet. Exposure to hydrocarbon liquids in the workplace results in dermatitis caused by dissolution of the fat portions of the skin and characterized by inflamed, dry, scaly skin.

Ethylene (ethene),

$$\begin{matrix} H & & H \\ & C{=}C & \\ H & & H \end{matrix}$$

a colorless gas with a somewhat sweet odor, the most widely used organic chemical, acts as a simple asphyxiant and anesthetic to animals. Ethylene is phytotoxic (toxic to plants). The toxicological properties of propylene (C_3H_6) are very similar to those of ethylene. Colorless, odorless gaseous 1,3-butadiene is an irritant to eyes and respiratory system mucous membranes; at higher levels it can cause unconsciousness and even death.

Benzene and Aryl Hydrocarbons

Benzene, C_6H_6, the simplest cyclic aryl (aromatic) hydrocarbon, is readily inhaled as the vapor and absorbed by blood, from which it is strongly taken up by fatty tissues. Some of the benzene undergoes Phase I oxidation (see Section 12.3) in the liver, producing phenol, as shown in Figure 12.11. The phenol is converted by a Phase II conjugation reaction to water-soluble glucuronide or sulfate, both of which are readily eliminated through the kidneys. The benzene epoxide intermediate in the oxidative metabolism of benzene is probably responsible for the unique toxicity of benzene, which involves damage to bone marrow.

Benzene is a skin irritant, and local exposures can cause skin redness (erythema), burning sensations, fluid accumulation (edema), and blistering. Chronic benzene poisoning causes blood abnormalities, including a lowered white cell count, an abnormal increase in blood lymphocytes (colorless corpuscles introduced to the blood from the lymph glands), anemia, a decrease in the number of blood platelets required for clotting (thrombocytopenia), and damage to bone marrow. It is thought that preleukemia, leukemia, or cancer may result.

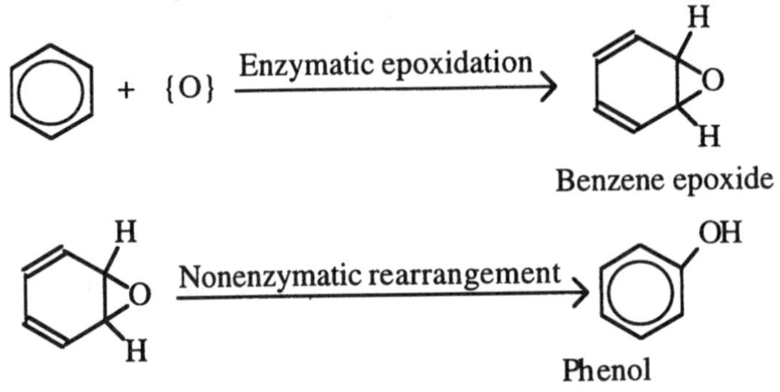

+ {O} —Enzymatic epoxidation→ Benzene epoxide

Nonenzymatic rearrangement → Phenol

Figure 12.11. Conversion of benzene to phenol in the body.

Benzo(a)pyrene is the most studied of the **polycyclic aromatic hydrocarbons** (PAHs). Some PAH metabolites, particularly the 7,8-diol-9,10 epoxide of benzo(a)pyrene shown in Figure 12.12 are known to cause cancer.

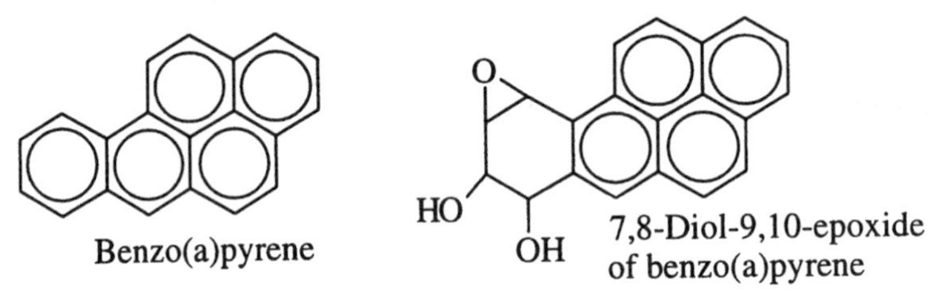

Figure 12.12. Benzo(a)pyrene and its carcinogenic metabolic product.

12.9. OXYGEN-CONTAINING ORGANIC COMPOUNDS

Ethylene oxide,

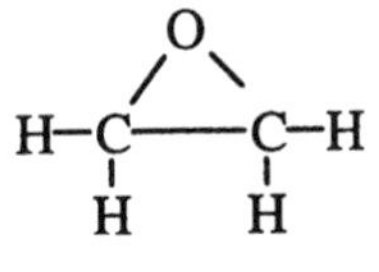

H–C(H)(–O–)C(H)–H Ethylene oxide

a gaseous colorless, sweet-smelling, flammable, explosive gas used as a chemical intermediate, sterilant, and fumigant, has a moderate to high toxicity, is a mutagen, and is carcinogenic to experimental animals. Inhalation of relatively low levels of ethylene oxide results in respiratory tract irritation, headache, drowsiness, and dyspnea, whereas exposure to higher levels causes cyanosis, pulmonary edema, kidney damage, peripheral nerve damage, and even death. Propylene oxide has similar toxic properties and 1,2,3,4-butadiene epoxide, the oxidation product of 1,3-butadiene, is notable in that it is a direct-acting (primary) carcinogen.

The common ethers have relatively low toxicities because of the low reactivity of the C–O–C functional group which has very strong carbon–oxygen bonds.

Exposure to volatile diethyl ether (structural formula below) is usually by inhalation and about 80% of this compound that gets into the body is eliminated unmetabolized as vapor through the lungs.

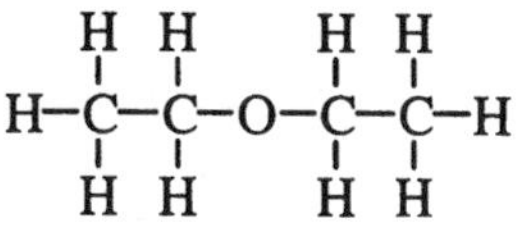

```
  H H   H H
  | |   | |
H-C-C-O-C-C-H   Diethyl ether
  | |   | |
  H H   H H
```

Diethyl ether depresses the central nervous system and is a depressant widely used as an anesthetic for surgery.

Human exposure to the three light **alcohols** shown in Figure 12.13 is common because they are widely used industrially and in consumer products.

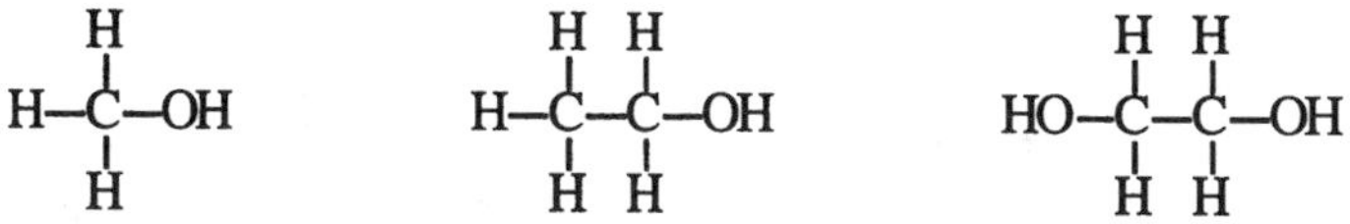

```
  H              H H              H H
  |              | |              | |
H-C-OH         H-C-C-OH        HO-C-C-OH
  |              | |              | |
  H              H H              H H
Methanol        Ethanol        Ethylene glycol
```

Figure 12.13. Three lighter alcohols with particular toxicological significance.

Metabolically, **methanol** is oxidized to formaldehyde and formic acid, causing acidosis, and affecting the central nervous system and the optic nerve. Sublethal exposures can cause blindness from deterioration of the optic nerve and retinal ganglion cells. **Ethanol** has a variety of acute effects ranging from decreased inhibitions and slowed reaction times at around 0.05% blood ethanol to coma and death around 0.5% blood ethanol. Ethanol is oxidized metabolically more rapidly than methanol, first to acetaldehyde, then to CO_2. Chronic exposure to ethanol results in chemical dependency, a condition called alcoholism. **Ethylene glycol** is metabolically oxidized to oxalate. Kidney damage occurs in later stages of ethylene glycol poisoning because of the deposition of insoluble solid calcium oxalate, CaC_2O_4, precipitated by the reaction of endogenous calcium ion with oxalic acid produced from the metabolic oxidation of ethylene glycol.

Although it was the original antiseptic used on wounds and in surgery, starting with the work of Lord Lister in 1885, **phenol,**

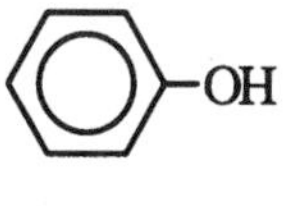

(benzene ring)-OH

is a protoplasmic poison that damages all kinds of cells and is alleged to have caused "an astonishing number of poisonings" since it came into general use.[3] The acute toxicological effects of phenol are predominantly upon the central nervous system and death can occur as soon as one-half hour after exposure. Fatal doses of phenol may be absorbed through the skin. Key organs damaged by chronic exposure to phenol include the spleen, pancreas, and kidneys.

Aldehydes and ketones are compounds that contain the carbonyl (C=O) group. The simplest of the carbonyl compounds and the one with the lowest molecular mass is **formaldehyde** (U122),

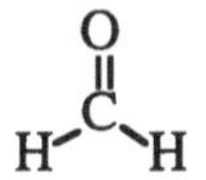

$$\mathrm{H{-}C({=}O){-}H}$$ Formaldehyde

which is uniquely important in industrial ecology because of its widespread use and toxicity. In the pure form formaldehyde is a colorless gas with a pungent, suffocating odor. **Formalin**, employed in antiseptics, fumigants, tissue and biological specimen preservatives, and embalming fluid, is marketed as a 37–50% aqueous solution of formaldehyde containing some methanol. Exposure to inhaled formaldehyde via the respiratory tract is usually to molecular formaldehyde vapor, whereas exposure by other routes is usually to formalin. Prolonged, continuous exposure to formaldehyde can cause hypersensitivity. A severe irritant to the mucous membrane linings of both the respiratory and alimentary tracts, formaldehyde reacts strongly with functional groups in molecules. Humans may be exposed to formaldehyde in the manufacture and use of phenol, urea, and melamine resin plastics and from formaldehyde-containing adhesives in pressed-wood products, such as particle board, used in especially large quantities in mobile home construction.

Carboxylic acids, oxidation products of aldehydes, contain the –C(O)OH functional group bound to an alkyl, alkenyl, or aryl hydrocarbon moiety as shown by the examples in Figure 12.14. Simple carboxylic acids are common natural products. Some of the higher carboxylic acids are constituents of oil, fat, and wax esters and can be prepared by hydrolysis of these esters. Formic acid is a relatively strong acid; it is corrosive to tissue, much like strong mineral acids. In Europe decalcifier formulations for removing mineral scale that contain about 75% formic acid are sold, and children ingesting these solutions have suffered corrosive lesions to mouth and esophageal tissue.

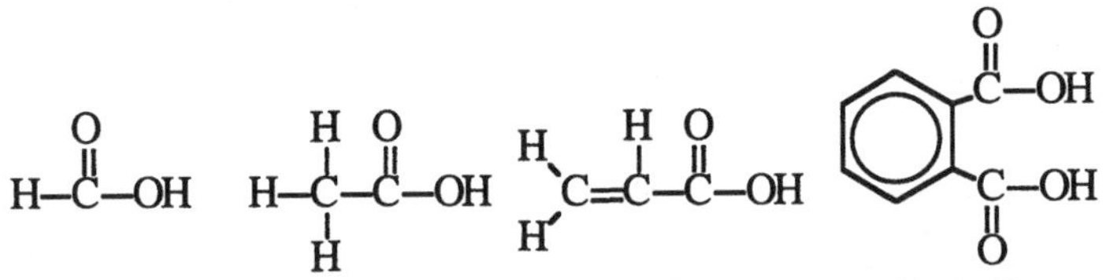

$$\mathrm{H{-}C({=}O){-}OH}\quad \mathrm{H{-}CH_2{-}C({=}O){-}OH}\quad \mathrm{H_2C{=}CH{-}C({=}O){-}OH}\quad \mathrm{C_6H_4(C({=}O){-}OH)_2}$$

Formic Acetic Acrylic Phthalic

Figure 12.14. Some common carboxylic acids. Phthalic acid is significant because phthalate esters, which are widespread environmental pollutants, are synthesized from it.

From the standpoint of industrial ecology, the most significant of the esters are those of phthalic acid. These compounds, such as *bis*(2-ethylhexyl) phthalate,

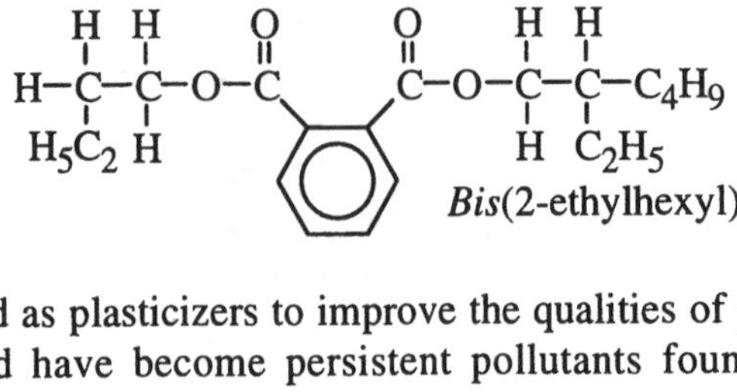

$$\mathrm{H{-}CH(C_2H_5){-}CH_2{-}O{-}C({=}O){-}C_6H_4{-}C({=}O){-}O{-}CH_2{-}CH(C_2H_5){-}C_4H_9}$$

Bis(2-ethylhexyl)phthalate

are widely used as plasticizers to improve the qualities of plastics. They are poorly degradable and have become persistent pollutants found in many parts of the environment. Although the acute toxicities of the phthalate esters are low, they are suspected of causing some chronic ill effects. There is some evidence to suggest that phthalate esters belong to a class of synthetic toxicants capable of mimicking the effects of female sex hormones in exposed children.

12.10. ORGANONITROGEN COMPOUNDS

Among the organonitrogen compounds are many toxic substances. The simple, low molecular mass amines, such as dimethylamine,

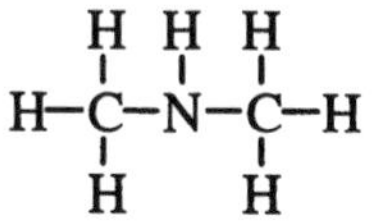

```
  H  H  H
  |  |  |
H-C--N--C-H   Dimethylamine
  |     |
  H     H
```

are rapidly and easily taken into the body by all common exposure routes. They are basic and react with water in tissue, raising the pH of the tissue to harmful levels, acting as corrosive poisons (especially to sensitive eye tissue), and causing tissue necrosis (tissue death) at the point of contact. Among the systemic effects of amines are necrosis of the liver and kidneys, lung hemorrhage and edema, and sensitization of the immune system. The lower amines are among the more toxic substances in routine, large-scale use.

Aniline,

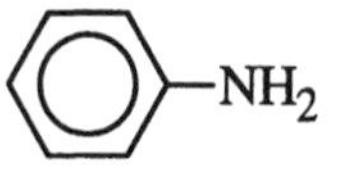

$C_6H_5{-}NH_2$ Aniline

is a widely used industrial chemical and is the simplest of the **carbocyclic aryl amines**, an industrially important class of compounds in which at least one substituent group is an aryl hydrocarbon ring bonded directly to the amino group. Some of them, such as 1-, and 2-naphthylamine have been shown to cause cancer in the human bladder, ureter, and pelvis, and are suspected of being lung, liver, and prostate carcinogens. A very toxic colorless liquid with an oily consistency and distinct odor, aniline readily enters the body by inhalation, ingestion, and through the skin. Metabolically, aniline converts iron(II) in hemoglobin to iron(III). This causes a condition called methemoglobinemia, characterized by cyanosis and a brown-black color of the blood, in which the hemoglobin can no longer transport oxygen in the body.

Nitro compounds contain the $-NO_2$ functional group. Nitrobenzene (U169), a pale yellow oily liquid with an odor of bitter almonds or shoe polish, has a toxicity rating of 5. It can enter the body through all routes and has a toxic action much like that of aniline (see above), causing methemoglobinemia.

N-nitroso compounds or **nitrosamines**, such as N-nitrosodimethylamine,

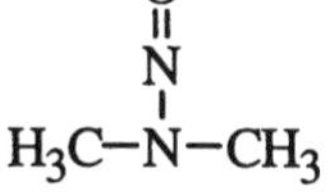

```
     O
     ||
     N        N-nitrosodimethylamine
     |        (Dimethylnitrosamine)
H3C--N--CH3
```

contain the N–N=O functional group and have been found in a variety of materials to which humans may be exposed, including beer, whiskey, and cutting oils used in machining. Cancer may result from exposure to a single large dose or from chronic exposure to relatively small doses of some nitrosamines. The carcinogenicity of these compounds was first revealed for dimethylnitrosamine. This compound, once widely used as an industrial solvent, was known to cause liver damage and jaundice in exposed workers. Its carcinogenicity was demonstrated by studies starting in the 1950s. Different nitrosamines cause cancer in different organs.

Compounds with the general formula R–N=C=O, **isocyanates** are noted for the high chemical and metabolic reactivity of their characteristic functional group. They have numerous uses in chemical synthesis, particularly in the manufacture of specialty polymers with carefully tuned properties. **Methyl isocyanate**,

```
     H
     |
  H–C–N=C=O    Methyl isocyanate
     |
     H
```

was the toxic agent involved in the catastrophic industrial poisoning in Bhopal, India on December 2, 1984, the worst industrial accident in history. In this incident several tons of methyl isocyanate were released, killing at least 2,000 people and affecting about 100,000. The lungs of victims were attacked; survivors suffered long-term shortness of breath and weakness from lung damage as well as numerous other toxic effects including nausea and bodily pain.

12.11. ORGANOHALIDE COMPOUNDS

Alkyl Halides

The toxicities of alkyl halides, such as those shown in Chapter 11, Figure 11.4, vary a great deal with the compound. Once considered almost completely safe, alkyl halides are now regarded with much more caution as additional health and animal toxicity study data have become available. Most of these compounds cause depression of the central nervous system, and individual compounds exhibit specific toxic effects. The most notably toxic alkyl halide is carbon tetrachloride, CCl_4, used for many years as a degreasing solvent, in home fire extinguishers, and for other industrial and consumer product applications. Carbon tetrachloride (U211) compiled a grim record of toxic effects which led the U.S. Food and Drug Administration (FDA) to prohibit its household use in 1970. It is a systemic poison that affects the nervous system when inhaled and the gastrointestinal tract, liver, and kidneys when ingested. The biochemical mechanism of carbon tetrachloride toxicity involves reactive radical species that react with biomolecules, such as proteins and DNA. The most damaging such reaction occurs in the liver as **lipid peroxidation**, consisting of the attack of free radicals on unsaturated lipid molecules, followed by oxidation of the lipids through a free radical mechanism.

The most significant **alkenyl** or **olefinic organohalides** are the lighter chlorinated compounds, as shown in Chapter 11, Figure 11.5. Because of their widespread use and disposal in the environment, the numerous acute and chronic toxic effects of the alkenyl halides are of considerable concern.

The central nervous system, respiratory system, liver, and blood and lymph systems are all affected by exposure to vinyl chloride (U043),

$$\overset{H}{\underset{H}{}}\!\!>C=C<\!\!\overset{Cl}{\underset{H}{}} \quad \text{Vinyl chloride}$$

which has been widespread because of this compound's use in polyvinylchloride manufacture. Most notably, vinyl chloride is carcinogenic, causing a rare and deadly angiosarcoma of the liver that has been observed in workers chronically exposed to vinyl chloride while cleaning autoclaves in the polyvinylchloride fabrication industry. The alkenyl organohalide, 1,1-dichloroethylene, is a suspect human carcinogen based upon animal studies and its structural similarity to vinyl chloride The toxicities of both *cis* and *trans* 1,2-dichloroethylene isomers are relatively low. These compounds act in different ways in that the *cis* isomer is an irritant and narcotic, whereas the *trans* isomer affects both the central nervous system and the gastrointestinal tract, causing weakness, tremors, cramps, and nausea. A suspect human carcinogen, trichloroethylene (U228) has caused liver carcinoma in experimental animals and is known to affect numerous body organs. Like other organohalide solvents, trichloroethylene causes skin dermatitis from dissolution of skin lipids and it can affect the central nervous and respiratory systems, liver, kidneys, and heart. Symptoms of exposure include disturbed vision, headaches, nausea, cardiac arrhythmias, and burning/tingling sensations in the nerves (paresthesia). Tetrachloroethylene (U210) damages the liver, kidneys, and central nervous system. It is a suspect human carcinogen based upon studies with mice.

A number of aryl halides are used industrially. Structural formulas of the most common aryl halides are shown in Chapter 11, Table 11.1. Individuals exposed to irritant monochlorobenzene by inhalation or skin contact suffer symptoms to the respiratory system, liver, skin, and eyes. Ingestion of this compound causes effects similar to those of toxic aniline, including incoordination, pallor, cyanosis, and eventual collapse. The dichlorobenzenes are irritants that affect the same organs as monochlorobenzene; the 1,4-isomer has been known to cause profuse rhinitis (running nose), nausea, jaundice, liver cirrhosis, and weight loss associated with anorexia.

The general formula of polychlorinated biphenyls (PCBs) is shown in Chapter 11, Figure 11.6. Because of their once widespread use in electrical equipment, as hydraulic fluids, and in many other applications, PCBs became widespread, extremely persistent environmental pollutants. Polybrominated biphenyl analogs (PBBs) were much less widely used and distributed. However, PBBs were involved in one major incident that resulted in catastrophic agricultural losses when livestock feed contaminated with PBB flame retardant caused massive livestock poisoning in Michigan in 1973.

12.12. ORGANOSULFUR COMPOUNDS

The structural formulas of examples of the more important organosulfur compounds are given in Chapter 11, Figure 11.7. Despite the high toxicity of H_2S, not all organosulfur compounds are particularly toxic. Their hazards are often reduced by their strong, offensive odors that warn of their presence.

Inhalation of even very low concentrations of the alkyl **thiols** (mercaptans) can cause nausea and headaches; higher levels can cause increased pulse rate, cold hands and feet, and cyanosis and, in extreme cases, unconsciousness, coma, and death. Like H_2S, the alkyl thiols are precursors to cytochrome oxidase poisons. A typical alkenyl thiol, 2-propene-1-thiol (allyl mercaptan) is an odorous volatile liquid that is highly toxic and strongly irritating to mucous membranes. Alpha-toluenethiol (benzyl mercaptan) is very toxic orally and it is an experimental carcinogen. The foul-smelling aryl thiol, benzenethiol (phenyl mercaptan), causes headache and dizziness; skin exposure results in severe contact dermatitis.

Dimethyl sulfide, an alkyl sulfide or thioether, is a volatile liquid that is moderately toxic by ingestion. Organic disulfides, such as *n*-butyldisulfide and diphenyldisulfide, may act as allergens that produce dermatitis in contact with skin. Animal studies suggest that these compounds may have several toxic effects, including homolytic anemia.

Structural formulas of **thiourea** (U219), the sulfur analog of urea, and some of its derivatives are shown in Chapter 11, Figure 11.8. Thioureas have been used as rodenticides. Thiourea, which has been shown to cause liver and thyroid cancers in experimental animals, has a moderate-to-high toxicity in humans, affecting bone marrow and causing anemia. The most effective rodenticidal thiourea compound is ANTU, **1-naphthylthiourea**, a virtually tasteless compound with a very high rodent:human toxicity ratio.

Sulfoxides and **sulfones** (Chapter 11, Figure 11.9) contain both sulfur and oxygen in their molecular structures. **Dimethylsulfoxide** (DMSO) is of toxicological concern because it has the ability to carry solutes into the skin's stratum corneum from which they are slowly released into the blood and lymph system. However, this compound's acute toxicity is remarkably low, with an LD_{50} of 10–20 *grams* per kg in several kinds of experimental animals. Applied to the skin, DMSO rapidly spreads throughout the body, giving the subject a garlic-like taste in the mouth and a garlic breath odor. Some of it undergoes partial metabolism to dimethylsulfide and dimethylsulfone and some is directly excreted in the urine. **Sulfolane** is the most widely used sulfone. Although it can cause eye and skin irritation, its overall toxicity is relatively low.

Sulfonic Acids, Salts, and Esters

Sulfonic acids and their salts, such as sodium 1-(*p*-sulfophenyl)decane contain the $-SO_3H$ or $-SO_3^-$ groups, respectively, attached to a hydrocarbon moiety (Chapter 11, Figure 11.11). Strongly acidic sulfonic acids should be treated with the precautions due strong acids. Skin, eyes, and mucous membranes are subject to strong irritation when exposed to benzenesulfonic acid; similar symptoms result from exposure to *p*-toluenesulfonic acid. Methylmethane sulfonate ester is particularly dangerous because it is a primary or direct-acting carcinogen that does not require metabolic conversion to cause cancer.

Replacement of one of the H atoms on H_2SO_4 by a hydrocarbon substituent (for example, $-CH_3$) produces an acid ester of sulfuric acid and replacement of both Hs yields an ester (see example structures in Chapter 11, Figure 11.11). Water-soluble, readily eliminated acid ester products of xenobiotic compounds (such as phenol) are produced by phase II reactions in the body. An oily water-soluble liquid, **methylsulfuric acid** is a strong irritant to skin, eyes, and mucous tissue. Colorless, odorless **dimethylsulfate** is highly toxic and like methylmethane sulfonate, a primary carcinogen. Skin or mucous membranes exposed to dimethylsulfate develop conjunctivitis and inflammation of nasal tissue and respiratory tract mucous membranes following an initial latent period during which few symptoms are observed. Damage to the liver and kidney, pulmonary edema, cloudiness of the cornea, and death within 3–4 days can result from heavier exposures.

12.13. ORGANOPHOSPHORUS COMPOUNDS

Organophosphorus compounds, examples of which are shown in the figures in Chapter 11, Section 11.7, have varying degrees of toxicity. The organophosphorus compounds of most concern are the organophosphates, among which are the "nerve gases" produced as military poisons that are deadly in minute quantities. **Trimethylphosphate** is probably moderately toxic when ingested or absorbed through the skin, whereas moderately toxic **triethylphosphate**, $(C_2H_5O)_3PO$, damages nerves and inhibits acetylcholinesterase, an enzyme crucial to the proper function of nerve impulses. Notoriously toxic **tri-*o*-cresylphosphate, TOCP,** apparently is metabolized to products that inhibit acetylcholinesterase. Exposure to TOCP causes degeneration of the neurons in the body's central and peripheral nervous systems with early symptoms of nausea, vomiting, and diarrhea accompanied by severe abdominal pain. About 1–3 weeks after these symptoms have subsided, peripheral paralysis develops manifested by "wrist drop" and "foot drop," followed by slow recovery, which may be complete or leave a permanent partial paralysis. Briefly used in Germany as a substitute for insecticidal nicotine, **tetraethylpyrophosphate, TEPP**, is a very potent acetylcholinesterase inhibitor. With a toxicity rating of 6 (supertoxic), TEPP is deadly to humans and other mammals.

Because esters containing the P=S (thiono) group are resistant to nonenzymatic hydrolysis and are not as effective as P=O compounds in inhibiting acetylcholinesterase, they exhibit higher insect:mammal toxicity ratios than their nonsulfur analogs. Therefore, **phosphorothionate** and **phosphorodithioate** esters (Figure 11.4) were once widely used as insecticides. The insecticidal activity of these compounds requires metabolic conversion of P=S to P=O (oxidative desulfuration). Environmentally, organophosphate insecticides are superior to many of the organochlorine insecticides because the organophosphates readily undergo biodegradation and do not bioaccumulate. The first commercially successful phosphorothionate and phosphorodithioate ester insecticide was **parathion**, *O,O*-diethyl-*O*-*p*-nitrophenylphosphorothionate, first licensed for use in 1944 and now banned in the U.S. This insecticide has a toxicity rating of 6 (supertoxic). Since its use began, several hundred people have been killed by parathion. Humans poisoned by parathion exhibit skin twitching and respiratory distress. In fatal cases, respiratory failure occurs due to central nervous system paralysis.

Malathion is the best known of the phosphorodithioate insecticides. It has a relatively high insect:mammal toxicity ratio because of its two carboxyester linkages which are hydrolyzable by carboxylase enzymes (possessed by mammals, but not insects) to relatively nontoxic products. For example, although malathion is a very effective insecticide, its LD_{50} for adult male rats is about 100 times that of parathion.

Powerful inhibitors of acetylcholinesterase enzyme, organophosphorus "nerve gas" military poisons, such as **Sarin** and **VX** (Figure 12.15), are among the most toxic synthetic compounds ever made. A systemic poison to the central nervous system that is readily absorbed as a liquid through the skin, Sarin may be lethal at doses as low as about 0.01 mg/kg; a single drop can kill a human. Other examples of these deadly poisons are **Tabun** (O-ethyl N,N-dimethylphosphoramidocyanidate), **Soman** (*o*-pinacolyl methylphosphonofluoridate), and "**DF**" (methylphosphonyldifluoride).

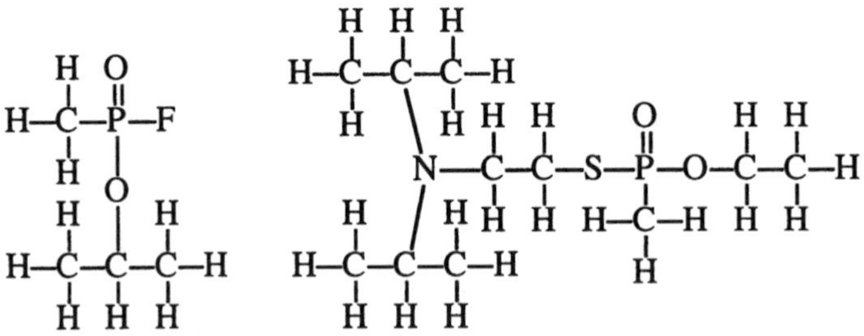

Sarin VX

Figure 12.15. Two examples of organophosphate military poisons.

12.14. BIOHAZARDS

Biohazard is used as a general term to refer to potentially harmful substances produced by organisms or by facilities that deal with organisms. For example, a toxin produced by bacteria or waste materials produced by laboratories that deal with research animals are both discussed as "biohazards."

Biomedical Waste

Biomedical waste is a term that has come into use to describe wastes that originate from sources involved with the treatment of, or research on, humans or animals. **Infectious wastes** are biomedical wastes that are capable of causing disease in individuals that come into contact with the wastes. Facilities that generate biomedical wastes obviously include hospitals and clinics, as well as pathology laboratories, nursing homes, dialysis centers, veterinary facilities, research laboratories, and pharmaceutical, cosmetics, and food industries.

The disposal of biomedical wastes has become more of a problem in recent years with the increase of single-use disposable medical items. For example, diabetics in the U.S. may use as many as 1 billion disposable syringes each year. Increased use of disposable syringes, "sharps," and other single-use materials has increased the proportion of biomedical wastes designated as infectious. Another factor leading to an increase in quantities of biomedical wastes has been the growth of facilities outside of hospitals, including nursing homes, kidney dialysis centers, blood banks, home-care sickrooms, and walk-in medical and dental clinics.

Disposable plastics account for about 1/3 of hospital wastes. Many of the plastics in biomedical wastes are organochlorine polymers, which complicates their destruction by burning because of emissions of acidic HCl and perhaps minute amounts of chlorinated dioxins. Nevertheless, incineration remains the major disposal option for biomedical wastes from hospitals.

Hospitals and clinics produce a large variety of waste materials. These include plastics, glass, paper, pads, swabs, gauze, disposable clothing, disinfectants, fluids, fecal matter, and even anatomical parts. Veterinary clinics and animal research facilities produce similar kinds of products as well as bedding material and shavings.

12.15. HAZARDOUS NATURAL PRODUCTS

Organisms of various kinds produce a variety of natural products that are hazardous because of their toxicities. Arguably the most acutely toxic substance known is botulism toxin produced by the anaerobic bacterium *Clostridium botulinum*. Mycotoxins generated by fungi (molds) can cause a number of human maladies; some mycotoxins (aflatoxins) are carcinogenic in experimental animals. Allergy-causing pollens rather than hazardous wastes are much more likely to inflict misery on the average citizen. Venoms from wasps, spiders, scorpions, and reptiles can be fatal to humans. Each year in the Orient, tetrodotoxin from improperly prepared puffer fish makes this dish the last delicacy consumed by some unfortunate diners. The stories of Socrates' execution from being forced to drink an extract of the deadly poisonous spotted hemlock plant and Cleopatra's suicide at the fangs of a venomous asp are rooted in antiquity.

It is beyond the scope of this book to discuss toxic and otherwise hazardous natural products in detail. However, two classes of natural products — mycotoxins and alkaloids — are briefly covered here because, in some respects, their behavior as contaminants or atmospheric pollutants (such as aflatoxins in dust from moldy grain) is similar to that of hazardous waste chemicals.

Aflatoxins

Aflatoxins are produced by fungi growing on moldy food, particularly nuts, some cereal grains, and oil seeds. The most notorious of the aflatoxins is aflatoxin B_1, for which the structural formula is shown in Figure 12.16. Produced by *Aspergillus niger*, it is a potent liver toxin and liver carcinogen in some species. It is metabolized in the liver to an epoxide (see Section 4.8). The product is electrophilic with a strong tendency to bond covalently to protein, DNA, and RNA. Other common aflatoxins produced by molds are those designated by the letters B_2, G_1, G_2, and M_1.

Trichothecenes are composed of 40 or more structurally related compounds produced by a variety of molds, including *Cephalosporium*, *Fusarium*, *Myro-*

thecium, and *Trichoderma*, which grow predominantly on grains. Much of the available information on human toxicity of trichothecenes was obtained from an outbreak of poisoning in Siberia in 1944. During the food shortages associated with World War II, the victims ate moldy barley, millet, and wheat. People who ate this grain suffered from skin inflammation; gastrointestinal tract disorders, including vomiting and diarrhea; and multiple hemorrhage. About 10 percent of those afflicted died.

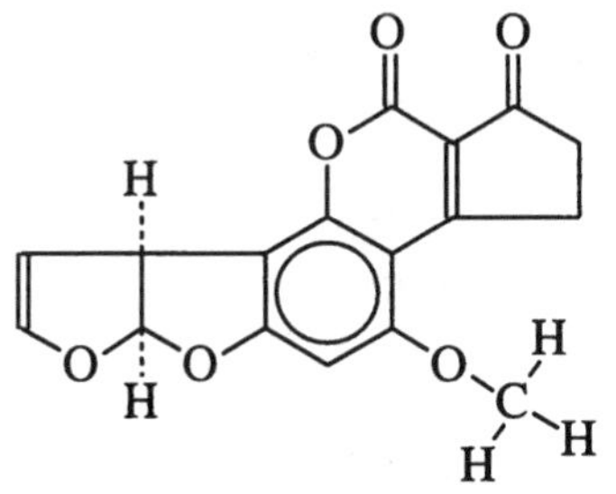

Aflatoxin B_1

Figure 12.16. Structural formula of aflatoxin B_1, a mycotoxin.

Alkaloids

Alkaloids are compounds of biosynthetic origin that contain nitrogen, usually in a heterocyclic ring. These compounds are produced by plants in which they are usually present as salts of organic acids. They tend to be basic and to have a variety of physiological effects. The structural formulas of five alkaloids are given in Figure 12.17.

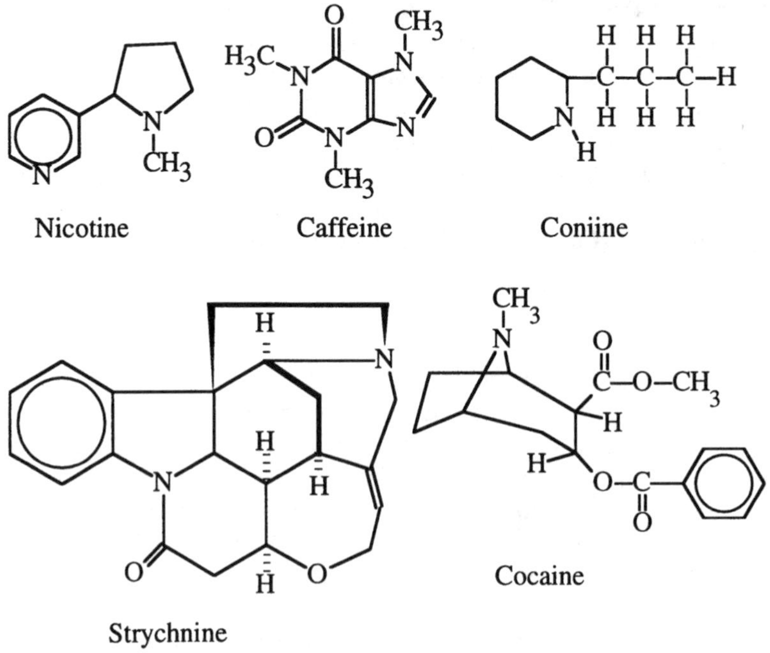

Nicotine Caffeine Coniine

Strychnine Cocaine

Figure 12.17. Structural formulas of typical alkaloids.

Among the alkaloids are some well-known (and dangerous) compounds. Cocaine is an addictive illicit drug.[4] Nicotine is an agent in tobacco, toxic enough to be used as an insecticide, that has been described as "one of the most toxic of all poisons and (it) acts with great rapidity."[5] In 1988 the U.S. Surgeon General declared nicotine to be an addictive substance. Coniine is the major toxic agent in poison hemlock. Alkaloidal strychnine is a powerful, fast-acting convulsant. Cocaine in the concentrated form of "crack" is currently the illicit drug of greatest concern. Quinine and stereoisomeric quinidine are alkaloids that are effective antimalarial agents. Like some other alkaloids, caffeine contains oxygen. It is a stimulant that can be fatal to humans in a relatively high dose of about 10 grams.

LITERATURE CITED

1. Manahan, Stanley E., *Toxicological Chemistry*, 2nd ed., CRC Press/Lewis Publishers, Boca Raton, FL, 1992.

2. Lemoine, Nicholas R., John Neoptolemos, and Timothy Cooke, *Cancer: A Molecular Approach*, Blackwell Science Inc., Cambridge, MA, 1994.

3. Gosselin, Robert E., Roger P. Smith, and Harold C. Hodge, "Phenol," in *Clinical Toxicology of Commercial Products*, 5th ed., Williams and Wilkins, Baltimore/London, 1984, pp. III-344–III-348.

4. Poklis, Alphonse, "Analytical/Forensic Toxicology," in *Casarett and Doull's Toxicology: The Basic Science of Poisons*, 5th ed., Curtis D. Klaassen, Ed., McGraw-Hill, New York, NY, 1996, pp. 951–967.

5. Gosselin, Robert E., Roger P. Smith, and Harold C. Hodge, "Nicotine," in *Clinical Toxicology of Commercial Products*, 5th ed., Williams and Wilkins, Baltimore/London, 1984, pp. III-311–III-314.

SUPPLEMENTARY REFERENCES

Allwood, Michael, Andrew Stanley, and Patricia Wright, Eds., *The Cytotoxics Handbook*, American Pharmaceutical Association, Washington, D.C., 1997.

Ballantyne, Bryan, Timothy Marrs, and Paul Turner, *General and Applied Toxicology: College Edition*, Stockton Press, New York, NY, 1995.

Chang, Louis W., *Toxicology of Metals*, CRC Press/Lewis Publishers, Boca Raton, FL, 1996.

Cockerham, Lorris G. and Barbara S. Shane, *Basic Environmental Toxicology*, CRC Press/Lewis Publishers, Boca Raton, FL, 1994.

Cooper, Andre R., *Cooper's Toxic Exposures Desk Reference*, CRC Press/Lewis Publishers, Boca Raton, FL, 1997.

Draper, William M., Ed., *Environmental Epidemiology*, American Chemical Society, Washington, D.C., 1994.

Hall, Steven K., Joanna Chakraborty, and Randall Ruch, *Chemical Exposure and Toxic Responses*, CRC Press/Lewis Publishers, Boca Raton, FL, 1997.

Klaassen, Curtis D., Ed.,*Casarett and Doull's Toxicology: The Basic Science of Poisons*, 5th ed., McGraw-Hill, New York, NY, 1996.

Landis, Wayne G. and Ming-Ho Yu, *Introduction to Environmental Toxicology*, CRC Press/Lewis Publishers, Boca Raton, FL, 1995.

Malachowski, M. J., *Health Effects of Toxic Substances*, Government Institutes, Rockville, MD, 1995.

McClellan, Roger O., Ed., *Critical Reviews in Toxicology* (journal), CRC Press/Lewis Publishers, Boca Raton, FL.

Rea, William H., *Chemical Sensitivity*, Volume 1, *Principles and Mechanisms*, CRC Press/Lewis Publishers, Boca Raton, FL, 1992.

Rea, William H., *Chemical Sensitivity*, Volume 2, *Sources of Total Body Load*, CRC Press/Lewis Publishers, Boca Raton, FL, 1994.

Rea, William H., *Chemical Sensitivity*, Volume 3, *Clinical Manifestations of Pollutant Overload*, CRC Press/Lewis Publishers, Boca Raton, FL, 1995.

Rea, William H., *Chemical Sensitivity*, Volume 4, *Tools for Diagnosis and Methods of Treatment*, CRC Press/Lewis Publishers, Boca Raton, FL, 1997.

Reinhardt, Peter A. and Judith G. Gordon, *Infectious and Medical Waste Management*, The Lab Store, Milwaukee, WI, 1991.

Revoir, William H. and Ching-tsen Bien, *Respiratory Protection Handbook*, CRC Press/Lewis Publishers, Boca Raton, FL, 1997.

Zakrzewski, Sigmund F., *Principles of Environmental Toxicology*, American Chemical Society, Washington, D.C., 1997.

13 INDUSTRIAL ECOLOGY OF WASTE MINIMIZATION

13.1. INTRODUCTION

This chapter addresses the issue of waste minimization. During recent years substantial efforts have been made in reducing the quantities of wastes and, therefore, the burden of dealing with wastes. Much of this effort has been the result of legislation and regulations restricting wastes, along with the resulting concerns over possible legal actions and lawsuits. In many cases — and ideally in all — minimizing the quantities of wastes produced is simply good business. Wastes are materials, materials have value, and, therefore, all materials should be used for some beneficial purpose and not discarded as wastes, usually at a high cost for waste disposal.

Industrial ecology is all about the efficient use of materials. Therefore, by its nature a system of industrial ecology is also a system of waste reduction and minimization. In reducing quantities of wastes, it is important to take the broadest possible view. This is because dealing with one waste problem in isolation may simply create another. Early efforts to control air and water pollution resulted in problems from hazardous wastes isolated from industrial operations. A key aspect of industrial ecology is its approach based upon industrial systems as a whole, making a system of industrial ecology by far the best means of dealing with wastes by avoiding their production.

13.2 WASTE MANAGEMENT FOR RESOURCE RECOVERY

Waste Reduction and Waste Minimization

Many hazardous waste problems can be avoided at early stages by **waste reduction** and **waste minimization**. As these terms are most commonly used, waste reduction refers to source reduction — less waste-producing materials in, less waste out. Waste minimization can include treatment processes, such as incineration, which reduce the quantities of wastes requiring ultimate disposal. Reference is sometimes made to waste abatement in terms of the four Rs — reduction, reuse, reclamation, and recycling.

Like industrial ecology, waste reduction is really an old idea that "has been around as long as man has been manufacturing products."[1] An 1883 handbook on electroplating admonishes that "nothing whatever should be allowed to go to waste in well-conducted works."[2] Although, especially for its time, zero waste was an overly ambitious goal, the manual does outline a number of measures that can be taken to reduce wastes, some of which are relevant even today.

Numerous factors, both economic and regulatory, favor waste reduction practices. Costs of treating and disposing of hazardous wastes have escalated along with problems in siting and getting permits for new hazardous waste storage and treatment units. Public opinion certainly favors reduced generation of wastes and there is a growing concern with liability associated with hazardous wastes. Furthermore, HSWA (1984) requires that "Wherever feasible, the generation of hazardous waste is to be reduced or eliminated . . .," the purpose of which is ". . . to minimize the present and future threat to human health and the environment." Section 3002(b) of HSWA requires that hazardous waste generators certify that they have implemented a program to minimize the volume and toxicity of wastes, and 3005(h) requires the same of treatment, storage, and disposal (TSD) facilities.

Legislation promoting waste minimization is contained in the U.S. Pollution Prevention Act of 1990. Under this act the U.S. Environmental Protection Agency (EPA) is required to formulate and conduct strategies of pollution prevention. It contains requirements for industry to measure the efficacy of strategies for reducing pollutants at their sources in minimizing toxic chemicals release to the environment.

In the May 28, 1993 *Federal Register*, EPA outlined the basic requirements for a good program to minimize wastes. A fundamental requirement is a commitment from management to implement and expedite such a program. The processes and products that generate wastes must be characterized through waste surveys and tracking of the kinds and amounts of wastes and the hazardous constituents in the wastes. Waste minimization assessments should be performed at regular intervals to identify opportunities for using less material, recycling, or finding less toxic substitutes. Technology transfer should be facilitated. Finally, continuous monitoring and regular evaluation of the program are required.

One of two main aspects of waste reduction is **toxicity reduction**. Toxicity reduction is always listed as a major goal of the practice of industrial ecology. An example of toxicity reduction is substitution of a water-based paint free of heavy metal-containing pigments for a solvent-based paint formulated with pigments containing heavy metals.[3] Any waste paint is thus much less toxic and less flammable. There are many other examples of the reduction of toxic substances use. Cyanide is one of the more toxic substances commonly used in manufacturing, and its replacement in metal plating operations can significantly reduce hazards. In the chemical industry, toxic solvents used as reaction media have been replaced by less toxic ones; an example is the substitution of 2-propanol (a consumer product called rubbing alcohol) for more toxic, poorly degradable 1,1,1-trichloroethane.

Volatile organic compounds (VOCs) constitute a large class of troublesome industrial chemicals and their replacement is a major objective of industrial ecology. In some cases, the use of powder-based coatings in place of solvent-based paints can reduce emissions of VOCs. Adhesives dissolved in organic solvents can be replaced in some cases by adhesives that are applied in a heated applicator and solidify when cooled or by water-based adhesives. Substitutes may be employed for heavy metals, such as cadmium, in printing inks. In some cases reuse of toxic materials, such as chlorinated bleach solutions, can reduce overall quantities of toxic substances.

The second of the two main categories of waste reduction is **volume reduction**. Volume reduction begins with taking only those amounts of raw materials required so that little or no excess will have to be disposed as waste. Large volume reductions are also achieved by source segregation and avoiding the mixing of hazardous wastes with other materials and wastes. Concentration techniques, such as pressing or centrifuging excess water from sludge filter cakes, can significantly reduce volumes of wastes.

Often substantial reductions in wastes can be achieved by fundamentally altering the ways in which manufacturing is done. It has been pointed out that in many cases deficiencies in information and lack of awareness are often greater impediments to waste reduction than are technical or economic barriers.[4] The manufacturing changes can be in both materials and equipment.

Materials can be substituted and reformulated to reduce quantities and toxicities of wastes. An example is the use of safer, biodegradable biocides in cooling tower water. Operational and maintenance procedures can be changed with the view of minimizing wastes. Leaks can be plugged and measures taken to prevent spills. "Drag-out," such as that of lubricating oils remaining on machined parts, can be minimized by measures such as increasing the time that such liquids are allowed to drain from parts. Cleaning operations may produce large quantities of wastes and should not be performed more than is essential. One clever approach in the paint formulation industry involves mixing progressively darker paints so that the mixing apparatus does not need to be cleaned between each run. Use of dry cleaning techniques or squeegees that use only minimum cleaning solution and rinse water reduces wastes.

In the manufacturing process itself various parameters may be optimized to reduce wastes. In order to reduce byproduct and waste generation in chemical synthesis, for example, the type of reaction and the kinds of reactants can be re-evaluated to see if changes could be made to reduce wastes. Other potential waste-reducing measures include changes in concentrations of reactants, reaction time, temperature, and pressure.

Modifications in equipment used in manufacturing can result in significant waste reduction. Coating equipment with Teflon can enable it to be cleaned dry without a cleaning solution. Vapor degreasing units can be equipped with refrigerated collectors to collect the degreasing solvents for reuse and prevent discharge of vapor to the atmosphere. Powder coating systems can be substituted for spray guns in parts-coating systems, eliminating waste solvents. Properly designed equipment can reduce spills and leaks. An example is use of welded pipe joints in place of leak-prone flange and gasket systems.

A number of examples exist of waste reduction through recovery and reuse. Baghouses installed on metal processing operations can recover metals, such as zinc. Electrolytic recovery techniques are used to reclaim copper and other metals from electroplating operations. One clever example of recovery is the use of sugar-rich rinse water from a soft-drink bottling facility to grow yeast. In the wood pulping industry, alkaline pulping liquor used to digest wood so that the cellulose fibers in it can be separated from lignin is dewatered and burned to reclaim the strong base in the pulping liquor.

Hierarchy of Waste Minimization

There exists a hierarchy of waste minimization, ranging from simple, readily accomplished measures through those that involve relatively drastic measures. These are the following:

- Inventory management so that only those materials needed are ordered and only for the times needed
- Increased diligence in housekeeping, such as using minimal amounts of water for washing equipment
- Substitution of less hazardous materials
- Recycling and reuse including both on-site and off-site recovery of materials
- Process modification
- Disposal

There are several ways in which wastes can be minimized. These include source reduction and waste separation and concentration. Some wastes can be reused and recycled. Waste exchange between industries can be practiced. The most effective approaches to minimizing wastes center around careful control of manufacturing processes, taking into consideration discharges and the potential for waste minimization at every step. When waste minimization is considered at a very early stage of process development or redesign, modifications may be made well upstream from discharge and treatment. Viewing the process as a whole (as outlined for a generalized chemical manufacturing process in Figure 13.1) often enables crucial identification of the source of a waste, such as a raw material impurity, which may be easier to eliminate from the feedstock than to treat as waste. Similar approaches may be used with catalysts and process solvents.

Substantial waste reduction can be accomplished by careful inventory management. This requires that excess amounts of raw materials which require disposal as hazardous wastes when left over or out of date are not ordered. Feedstocks and materials used in production processes can be examined for hazardous substance content and in some cases modified to reduce production of hazardous wastes. For this purpose the manufacturer's **Material Safety Data Sheets** can be very useful. Examples of hazardous waste reduction by substitution of materials used include replacement of cadmium in inks, pigments, and plating baths; substitution of chromium(III) for toxic chromate in plating baths; replacement of toxic cyanide in plating bath media; and substitution of water-based formulations for solvent-based paints, adhesives, and degreasers.

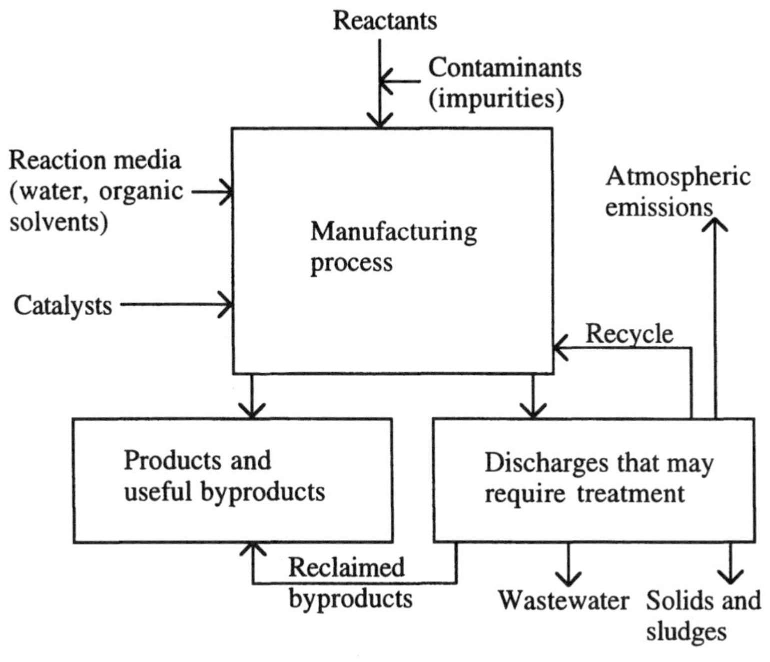

Figure 13.1. Chemical manufacturing process from the viewpoint of discharges and waste minimization.

Modification of the manufacturing process can yield substantial waste reduction. Some such modifications are of a chemical nature. Changes in chemical reaction conditions can minimize production of byproduct hazardous substances. In some cases potentially hazardous catalysts, such as those formulated from toxic substances, can be replaced by catalysts that are nonhazardous or that can be recycled rather than discarded.

Wherever possible, recycling and reuse should be accomplished on-site; a process that produces recyclable materials is often the most likely to have use for them. Metals can be recovered from waste plating solutions and recycled in the plant, waste cleaning and rinsing solvents can be used in paint formulations, and solvents can be recovered by distillation or by condensation of solvent vapor.

Some very impressive reductions in amounts of wastes produced have been described in a model study of the electronics industry, chosen for study because of its growth orientation and because it is among the top 20 waste solvent generating industries.[5] Electronics manufacture uses metals for plating (copper, nickel, tin, and lead), cyanide, chelating agents, and other chemicals. In one facility studied, the quantity of metals sludge produced was reduced from 34,250 to 500 tons per year and in another from 700 to 22 tons per year. Another facility reduced its waste methyl chloroform solvent to 6 percent of its previous production and its waste Freon to 10 percent; the solvents saved were recycled.

A Monsanto plastics plant was able to cut loss of a resin to $^{1}/_{4}$ by improved equipment maintenance and use of better gasket materials. Polymerization of excess resin converted it from a hazardous to a nonhazardous material, further reducing disposal costs.

Since 1975 Minnesota Mining and Manufacturing (3M) has been engaged in a "3P" program, standing for "Pollution Prevention Pays." The objective of this program has been to reduce the need for retrofitted add-on pollution control devices by eliminating or minimizing pollution at the source. In so doing, the 3P program has considered product development, engineering design, and manufacturing processes; it emphasizes product reformulation, changes in processes, resource recovery, and modifications of equipment design. This program has succeeded in eliminating about 1.5 billion gallons of industrial wastewater containing about 10,000 tons of water pollutants each year. Annual production of air pollutants has been cut by an estimated 100,000 tons and the production of 150,000 tons of sludge is prevented each year. Examples of measures used include burning air contaminated with solvents from tape manufacture in an industrial oven to recover fuel from the solvents, replacement of a solvent-based coating for pharmaceutical tablets with a water-based coating that produces no solvent vapors, and replacing toxic chemical cleaning sprays with pumice scrubbers to clean copper sheeting.

Waste Treatment

As illustrated in Figure 13.2, waste treatment may occur at the three major levels of **primary**, **secondary**, and **polishing**, somewhat analogous to the treatment of wastewater. Primary treatment is generally regarded as preparation for further treatment, although it can result in the removal of byproducts and reduction of the quantity and hazard of the waste. Secondary treatment detoxifies, destroys, and removes hazardous constituents. Polishing usually refers to treatment of water that is removed from wastes so that it may be safely discharged. However, the term can be broadened to apply to the treatment of other products as well so that they may be safely discharged or recycled.

Options for waste recovery during treatment may be divided among the following major categories:

- Physical separation
- Separation by phase transition of wastes
- Separation by transfer between phases
- Molecular separation
- Chemical separation

Many separation processes involve combinations of the above methods. Physical separation, which is discussed further in Chapter 14, includes settling and decanting, centrifugation, filtration, and flotation. Separation by phase transition refers to nonchemical processes by which the material separated enters a different

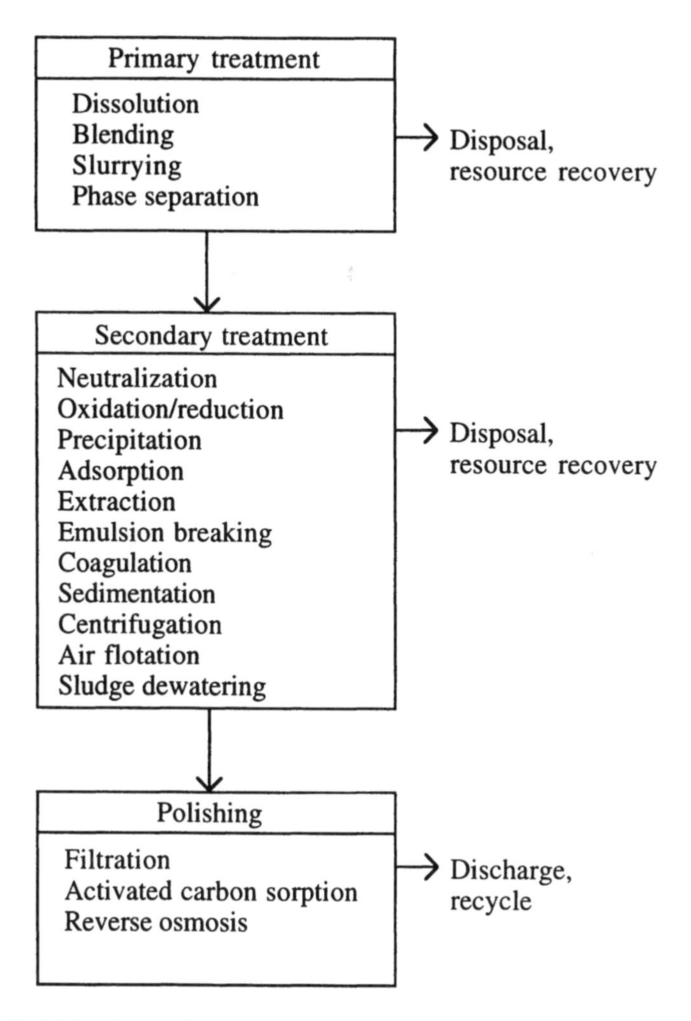

Figure 13.2. Major phases of waste treatment.

phase; it can be evaporated, sublimed, condensed from the vapor phase, distilled, or caused to precipitate by cooling a solution or evaporating solvent from it. A different approach is that of separation between phases, in which a substance in solution or bound to a solid is transferred to a similar state in another phase. An example would be a liquid-liquid extraction in which a heavy metal ion dissolved in aqueous solution is transferred to a separate organic solvent phase as a chelated metal. Other separations between phases are sorption onto solids (activated carbon or resins), ion exchange, and supercritical fluid extractions. Molecular separations occur across membranes that may be selective for particular types or sizes of species. These separations include reverse osmosis, electrodialysis, and ultrafiltration.

Chemical separations result from chemical reactions, such as the precipitation of cadmium ion in solution by hydrogen sulfide gas.

$$H_2S(g) + Cd^{2+} \rightarrow CdS(s) + 2H^+ \tag{13.2.1}$$

Other types of chemical reactions used for separations are reduction, oxidation (including electrolytic reduction and oxidation reactions), and cementation (replacement of a metal in the elemental state with a relatively less active metal than originally present as an ion in solution). Chemical separations are discussed in detail in Chapter 14.

13.3. RECYCLING

As costs of hazardous waste disposal have mounted, the economics of putting wastes to some good use have improved. This is commonly done through recycling. Several classes of substances are commonly recycled. Prominent among these are metals and compounds of metals. Recycled inorganic substances include alkaline compounds (such as sodium hydroxide used to remove sulfur compounds from petroleum products), acids (steel-pickling liquor where impurities permit reuse), and salts (for example, ammonium sulfate from coal coking used as fertilizer). The greatest quantities of recycled organic substances consist of solvents and oils, such as hydraulic and lubricating oils. In the chemical and petroleum industry, catalysts are recycled. Commercially, chemicals that are surplus, are off-specification, or have expired shelf lives may be recycled. Some recycled substances have agricultural uses, such as waste lime or phosphate-containing sludges used to treat and fertilize acidic soils.

There are four very broad areas in which something of value may be obtained from wastes. They are the following:

- Direct recycle as raw material to the generator
- Transfer as a raw material to another process
- Utilization for pollution control or waste treatment
- Recovery of energy

Direct recycle is frequently utilized in the chemical industry when raw materials are not completely consumed in a synthesis process and are simply returned as feedstock. A substance that is a waste product from one process may serve as a raw material for another, sometimes in an entirely different industry. Some process wastes can be used to treat other wastes or for pollution control. For example, waste lime from the collection of dust in lime processing can be used to neutralize waste acid or to precipitate metals from wastewater. Whenever it is desirable to incinerate wastes, energy recovery should be considered, assuming that the quantities are large enough and the production of waste is uniform enough to make energy recovery viable.

Recycling of materials, including hazardous substances, has a long history. This is especially true of economies stressed by poverty, wartime shortages, and, most recently, restrictions on the disposal of hazardous substances. Great Britain, a country that suffered severe material shortages due to destruction of shipping by enemy submarines during World War II, set up the National Industrial Recovery Association in 1942 to expedite recycling of materials in short supply. In some respects this organization resembled modern day “waste exchanges.”

Various organizations have been established to provide information about hazardous waste exchange and recycle. Several specific accomplishments of one such organization, the Minnesota Technical Assistance Program, have been described.[6] Among the examples cited were the following: (1) A user for waste redistilled dichloromethane produced by a process used to attach rubber gaskets to sewer pipes was located. (2) A process that produced large quantities of metal hydroxide sludge from the treatment of metals plating rinse water was replaced by an electrolytic and ion exchange metal recovery system that reduced sludge production by 90 percent. (3) Waste solvent from paint-thinning operations was reclaimed with a small still and reused for thinning paint and for cleaning equipment.

Numerous factors are involved in determining the feasibility of waste recycle and exchange. First of all, the material has to be good for something! It must not contain any undesirable impurities that cannot be dealt with safely and economically. For example, uranium mine tailings make good construction fill material except for the problem of radioactive radon gas emissions. Toxic lead may have to be removed from zinc oxide collected from air pollution control devices in zinc smelters if the zinc oxide is to be used to make zinc sulfate fertilizer. Transportation of the waste to an industry that can use it has to be feasible; use on site is most desirable.

Examples of Recycling

Recycling of scrap industrial impurities and products occurs on a large scale with a number of different materials. Most of these materials are not hazardous, but, as with most large-scale industrial operations, their recycle may involve the use or production of hazardous substances. Some of the more important examples are discussed briefly here.

Ferrous metals are those that are primarily composed of iron. Ferrous metal scrap is recycled from steel mills, scrap steel (such as junked automobiles), and from municipal waste. An advantage of iron for recycling is that it can be magnetically separated from other wastes. Most ferrous scrap is recycled to steel making. In recent years electric-arc furnaces have become popular for high-quality steel manufacture, which has increased the demand for ferrous scrap because the furnaces use it exclusively for feedstock.

The commonly used nonferrous metals are produced in smaller quantities than iron, are more expensive, and are in shorter supply. A number of nonferrous metals are toxic in some form (lead and cadmium compounds, mercury metal vapor). All of these factors combined make nonferrous metal recycling highly desirable, both economically and environmentally. Aluminum ranks next to iron in terms of quantities recycled. Major quantities of copper and copper alloys, zinc, and lead are recycled. Lesser amounts of cadmium, tin, and mercury are recycled. Silver is recycled from X-ray film and electronic applications using highly toxic cyanide solutions. Scrap copper is usually processed in a furnace, then electrolytically refined. Most scrap lead comes from spent automobile batteries. Recycled metal amounts to over $^1/_3$ of the U.S. supply of aluminum and over half of domestic supplies of copper and lead.

Glass, which makes up about 10 percent of municipal refuse, is a popular material to recycle. A major hurdle with glass recycling is the sorting of glass by color after it has been separated from municipal refuse. Some municipal recycling systems attempt to accomplish this by having consumers sort when they put glass out for recycling. Few municipal glass reclamation systems accomplish more than 50 percent recycle of waste glass.

Paper, along with metals and glass, is one of the items most commonly recycled from municipal refuse. The cellulose fibers in recycled paper tend to be shorter, more flattened, less strong, and drier than fibers freshly produced from wood pulp. In addition, used paper is contaminated with constituents such as adhesives, inks, clay, and coatings, as well as dirt, grease, and other impurities. There have also been reports of paper contaminated with hazardous materials, such as PCBs. All of these factors tend to complicate paper recycling. Careful sorting of scrap paper at the source is important for successful paper recycling. New developments in paper recycling have made it more practical and applicable to a wider variety of raw materials. Consideration of recycling in the composition of new paper products would also be helpful.

Plastic is a term that describes a variety of polymeric materials that can be shaped or molded while in a liquid (plastic) state and formed as desired for containers and other objects. Thermosetting plastics are those that are synthesized and formed as part of the manufacturing process These plastics cannot be melted and reprocessed, which limits the possibilities for recycling. Thermoplastic materials can be melted by heat and refabricated, which enables their recycle. Recycling of scrap thermoplastic produced in plastic fabrication is accomplished by grinding, adding more heat stabilizers, and mixing it with the feed of raw plastic.

Since World War II, plastic has become a major constituent of municipal wastes. Most plastics do not biodegrade well or at all and they contribute a significant fraction of residual solid wastes remaining in landfills after biodegradation has occurred. Because of variable composition, impurities, pigment constituents, and other factors, recycling of plastics in municipal solid wastes has proven difficult. As with paper, formulation of plastics with recycling as a major goal would prove helpful.

Rubber is a recyclable material as was demonstrated by the rubber-short major powers in World War II. Now synthetic polymers provide an abundance of rubber, but the need to recycle rubber remains because of the literally mountains of used rubber tires that have accumulated in numerous locations. The magnitude of the problem was illustrated by a massive fire involving 14 million tires at a tire disposal site in Hagersville, Ontario, Canada in February, 1990.[7] Rubber is a hydrocarbon polymer that may contain other materials, such as carbon black filler; the Hagersville fire yielded about 160,000 gallons of oil byproduct that was collected and recycled in a nearby petroleum refinery. Rubber has a good fuel value of about 14,000 Btu/lb, which is comparable to the best grades of coal. Some kinds of rubber can be ground and recycled through the rubber fabrication process. A major complication in the burning and recycling of rubber in tires is the presence of strong steel wires in many radial tires. The steel can jam grinding equipment and handling mechanisms in both recycling and incineration operations.

13.4. WASTE OIL UTILIZATION AND RECOVERY

Annual production of **waste oil** in the U.S. is of the order of 4 billion liters per year. Around half of this amount is burned as fuel and lesser quantities are recycled or disposed as waste. Much of the improper disposal of waste oil is from individuals who change oil in their own vehicles.

Properties of Waste Oil

Although waste oil is regarded as an organic material, it contains inorganic substances, such as metals. Waste oil is generated from lubricants for vehicle engines, lubricants in industrial processes, and spent hydraulic fluids. The collection, recycling, treatment, and disposal of waste oil are all complicated by the fact that it comes from diverse, widely dispersed sources.

Waste oil contains several classes of potentially hazardous contaminants divided between organic constituents and inorganic constituents. Among the organic constituents are polycyclic aromatic hydrocarbons consisting of two rings (naphthalene) or more (including the procarcinogen, benzo(a)pyrene). Some of these come from the petroleum base stock used to make the oil and others are generated by pyrolysis during the exposure of the oil to high temperatures. Chlorinated hydrocarbons, such as trichloroethylene, get into waste oil as contaminants and possibly from chemical processes associated with use of the oil.

Metals are the predominant contaminants of waste oil that are of concern. Aluminum, chromium, and iron may get into the oil from wear of metal parts. Barium and zinc are often present from oil additives. Lead from leaded gasoline can contaminate motor oil, but this source of lead in oil has diminished significantly with the phaseout of leaded gasoline. Limits for arsenic and cadmium are specified in used, recycled oil.

Recycling Waste Oil

Several processes are used to convert waste oil to a feedstock hydrocarbon liquid for lubricant formulation. These processes may be divided into the three major steps shown in Figure 13.3.

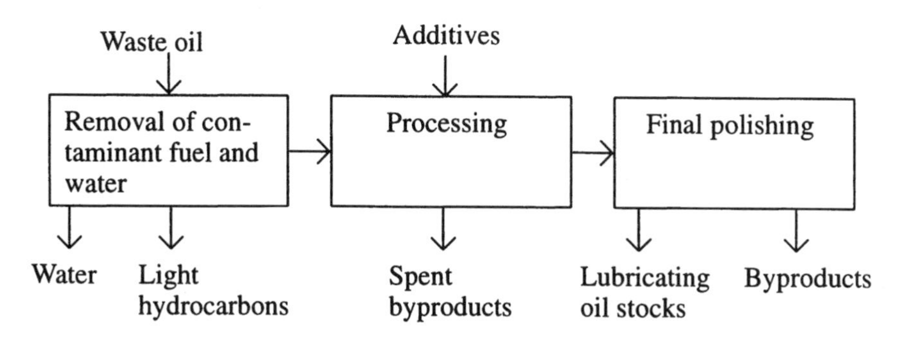

Figure 13.3. Major steps in reprocessing waste oil.

The first phase of waste oil treatment involves distillation to remove water and light ends that have come from condensation and contaminant fuel. The second phase, designated processing in Figure 13.3, varies considerably. It may be as simple as a vacuum distillation in which the three products are oil for further processing, a fuel oil cut, and a heavy residue. Another process uses a mixture of solvents including isopropyl and butyl alcohols and methylethyl ketone to dissolve the oil and leave contaminants as a sludge. Treatment with sulfuric acid followed by treatment with clay has been widely used. The sulfuric acid reacts with inorganic contaminants, which settle from the mixture as a sludge. Contact of the oil with clay removes acid and contaminants that cause odor and color. Treatment with clay has lost favor because of the large amount of waste byproduct generated.

The third step shown in Figure 13.3 employs vacuum distillation to separate lubricating oil stocks from a fuel fraction and heavy residue. This phase of treatment may also involve hydrofinishing, treatment with clay, and filtration.

Waste Oil Fuel

For economic reasons waste oil that is to be used for fuel is given minimal treatment of a physical nature, including settling, removal of water, and filtration. Metals in waste fuel oil become highly concentrated in its fly ash, which may be hazardous.

13.5. RECOVERY OF SOLUTES FROM WASTEWATER

Wastewater produced by a variety of processes is the most abundant hazardous waste material. Often such water contains contaminants that have economic value if they are reclaimed from the water. The value of the reclaimed materials may pay part or, in favorable cases, all of the costs of recovery. Several examples are cited in this section.

Recovery of Acids

Acidic solutions containing dissolved metals are common and troublesome wastes from a number of metal processing operations. Some examples of these solutions are listed below:

- HCl and Zn^{2+} from zinc stripping
- HNO_3 and Ni^{2+} from nickel stripping
- H_2SO_4 and Al^{3+} from aluminum anodizing
- H_2SO_4, HCl and Zn^{2+} from steel pickling
- HNO_3, HF, Fe^{2+}, and Cr^{3+} from stainless steel pickling
- HNO_3, H_2SO_4, Cu^{2+}, and Zn^{2+} from brass etching

These waste solutions must be treated to remove both metals and acids. Addition of lime works for this purpose as shown by the following reactions:

$$2HNO_3 + Ca(OH)_2 \rightarrow Ca^{2+} + 2NO_3^- + 2H_2O \quad (13.5.1)$$

(Acid neutralization)

$$Cu^{2+} + Ca(OH)_2 \rightarrow Ca^{2+} + Cu(OH)_2 \quad (13.5.2)$$

(Metal precipitation)

$$2HF + Ca(OH)_2 \rightarrow CaF_2(s) + 2H_2O \quad (13.5.3)$$

(Precipitation of fluoride)

$$H_2SO_4 + Ca(OH)_2 \rightarrow CaSO_4(s) + 2H_2O \quad (13.5.4)$$

(Acid neutralization, sulfate precipitation)

However, lime produces large quantities of precipitate from which it is not possible to economically recover metals.

A system has been described in which acid is absorbed by a resin and subsequently reclaimed by elution with water.[8] The acid recovered can be recycled to the anodizing, pickling, plating, or etching operation from which it was recovered or it can be used for other purposes.

Acetic acid,

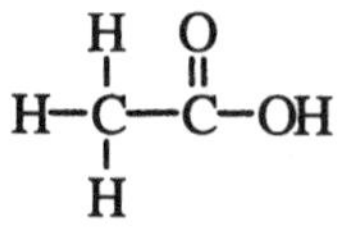

```
  H  O
  |  ||
H-C--C-OH
  |
  H
```

is a common component of wastewater. Acetic acid can be extracted into organic solvents such as ethyl acetate or a mixture of ethyl acetate and benzene. After extraction, the volatile solvents may be separated from acetic acid by distillation.

Recovery of Phenol

Phenol is an important industrial chemical with many uses. Phenol and related compounds are produced in abundance in coal gasification and liquefaction. However, these industries are small on a global basis and the major source of phenol from coal is from coal coking where it occurs in a relatively concentrated form in byproduct water. Wastewaters from phenol resin manufacture and petroleum refining may also contain recoverable phenol.

Phenol is recovered from wastewater by solvent extraction, which involves transfer of a solute from water or an organic solvent and vice versa. Phenol is extracted from wastewater by a variety of water-immiscible organic solvents. Some of the solvents used for phenol extraction are hydrocarbons, including fractions distilled from crude oil, benzene, and toluene. Chlorinated hydrocarbons, alcohols (1-octanol), ethers (diisopropyl ether), esters (*n*-butyl acetate, tricresyl phosphate), and ketones (methylisobutyl ketone) will also extract phenol. In practice, mixtures of solvents may be most effective. One of the criteria in choosing a solvent is its selectivity for extraction of phenol in preference to other organic contaminants that are present in relatively high concentrations in water produced in petroleum refining and coal coking. Phenols extracted into the organic liquid are back-extracted to aqueous alkali, a process that occurs because of the ionization of phenol in contact with base,

$$C_6H_5\text{—}OH(org) + OH^- \rightarrow C_6H_5\text{—}O^-(aq) + H_2O \quad (11.5.5)$$

Phenol   Phenolate anion

Acidification of the basic aqueous extract of phenolic compounds precipitates them in the solid form,

$$C_6H_5\text{—}O^-(aq) \quad H^+ \rightarrow C_6H_5\text{—}OH(s) \quad (11.5.6)$$

Phenolate anion   Phenol

Solvent that dissolves in water during extraction is recovered by steam distillation under vacuum.

Recovery of Metals

Metals from electroplating wastes and wastewater from other metal processing operations may be recovered for their economic value as part of the wastewater treatment process. In some cases metals may be recovered by electrochemical reduction (see Chapter 14) in which a direct current is applied between electrodes immersed in the wastewater. Metals plate out on the cathode,

$$Cu^{2+} + 2e^- \rightarrow Cu \quad (13.5.7)$$

An electrolytic system has been described for the removal of zinc and cadmium from waste electroplating water containing cyanide.[9] The metals may be recycled to the electoplating process. This system was designed to remove metals from dilute solution with a high surface area cathode composed of carbon fibers. Cyanide was effectively oxidized at the anode and removed from the water.

Ion exchange (see Chapter 14) is widely used for the recovery of metals from wastewater. The metal-containing water is passed over a solid cation exchange resin where metal is exchanged for H^+ ion.

$$2H^+\{^-CatExRes\} + Zn^{2+} \rightarrow Zn^{2+}\{^-CatExRes\}_2 + 2H^+ \quad (13.5.8)$$

The resin is regenerated and metal reclaimed in a concentrated form by treatment with a relatively concentrated solution of strong acid.

$$Zn^{2+}\{^-CatExRes\}_2 + 2H^+ \rightarrow 2H^+\{^-CatExRes\} + Zn^{2+} \quad (13.5.9)$$

Some wastewater solutions, such as those from various kinds of electroplating baths, contain cyanide. Cyanide forms negatively charged complex ions with various metals. For example, the cyanide complex of copper(I), $Cu(CN)_4^{3-}$, is very stable. Metal cyanide complexes may be removed from water by anion exchange resins. Regeneration of these resins may be difficult because of their strong affinity for the metal complex ions.

Chromium(VI) in water is in the form of a negatively charged species such as chromate ion, CrO_4^{2-}. It is recovered by passing the chromate-containing solution over an anion exchange resin in the base form. If other cationic metals are present, they must first be removed by cation exchange (see above) or other means to prevent precipitation of hydroxides in contact with the basic anion exchange resin. The reaction for the removal of chromate ion is the following:

$$2\{AnExRes^+\}OH^- + CrO_4^{2-} \rightarrow \{AnExRes^+\}_2CrO_4^{2-} + 2OH \qquad (13.5.10)$$

The chromate retained by the anion exchange resin is eluted with strong base (NaOH) and reclaimed. This process regenerates the anion exchange column for further use.

13.6. RECOVERY OF WATER FROM WASTEWATER

It is often desirable to reclaim water from wastewater. This is especially true in regions where water is in short supply. Even where water is abundant, water recycling is desirable to minimize the amount of water that is discharged.

A little more than half of the water used in the U.S. is consumed by agriculture, primarily for irrigation. Steam-generating power plants consume about one-fourth of the water, and other uses, including manufacturing and domestic uses, account for the remainder.

The three major manufacturing consumers of water are chemicals and allied products, paper and allied products, and primary metals. These industries use water for cooling, processing, and boilers. Their potential for water reuse is high and their total consumption of water is projected to drop in future years as recycling becomes more common.

The degree of treatment required for reuse of wastewater depends upon its application. Water used for industrial quenching and washing usually requires the least treatment, and wastewater from some other processes may be suitable for these purposes without additional treatment. At the other end of the scale, boiler makeup water, potable (drinking) water, water used to directly recharge aquifers, and water that people will directly contact (in boating, water skiing, and similar activities) must be of very high quality.

The treatment processes applied to wastewater for reuse and recycle depend upon both the characteristics of the wastewater and its intended uses. Solids can be removed by sedimentation and filtration. Biochemical oxygen demand is reduced by biological treatment, including trickling filters and activated sludge treatment. For uses conducive to the growth of nuisance algae, nutrients may have to be removed. The easiest of these to handle is nutrient phosphate, which can be precipitated with lime. Nitrogen can be removed by denitrification processes.

Two of the major problems with industrial water recycle are heavy metals and dissolved toxic organic species. Heavy metals may be removed by ion exchange or precipitation by base or sulfide. The organic species are usually removed with activated carbon filtration. Some organic species are biologically degraded by bacteria in biological wastewater treatment.

The ultimate water quality is achieved by processes that remove solutes from water, leaving pure H_2O. A combination of activated carbon treatment to remove organics, cation exchange to remove dissolved cations, and anion exchange for dissolved anions can provide very high quality water from wastewater. Reverse osmosis (see Chapter 14) can accomplish the same objective. However, these processes generate spent activated carbon, ion exchange resins that require regeneration, and concentrated brines (from reverse osmosis) that require disposal.

13.7. SOLVENT RECOVERY

Organic solvents of various kinds are used for many purposes, such as reaction media, extraction of fats, extraction of oils, degreasing, and dry cleaning. A large number of different compounds occur in hazardous waste solvents. Among the halogenated solvents listed in hazardous wastes are dichloromethane, carbon tetrachloride, trichlorofluoromethane, tetrachloroethylene, trichloroethylene, 1,1,1-trichloroethane, 1,1,2-trichloro-1,2,2-trifluoroethane, chlorobenzene, and 1,2-dichlorobenzene. Hydrocarbon solvents that occur in hazardous wastes include benzene, xylene, ethylbenzene and liquid alkanes. Other assorted compounds in waste solvents include acetone, methylisobutyl ketone, cyclohexanone, methanol, *n*-butyl alcohol, isobutanol, pyridine, carbon disulfide, 2-nitropropane, diethyl ether, and 2-ethoxyethanol.

For reasons of both economics and pollution control, many industrial processes that use solvents are equipped for solvent recycling. The basic scheme for solvent reclamation and reuse is shown in Figure 13.4. Some loss of solvent occurs to the product and in purification, so fresh makeup solvent is added.

Solvents usually must be recovered from solids or parts that have been extracted or cleaned with solvent. This can be accomplished by heating the solids in vacuum or in a nonoxidizing atmosphere such as CO_2 or nitrogen. When solvent flammability is not a problem, liquids can be evaporated from solids with hot air. In either case solvents are recovered by condensation in a heat exchanger, adsorption with activated carbon, or scrubbing with another liquid, usually water. Solvents may also be separated from products by extraction with another solvent.

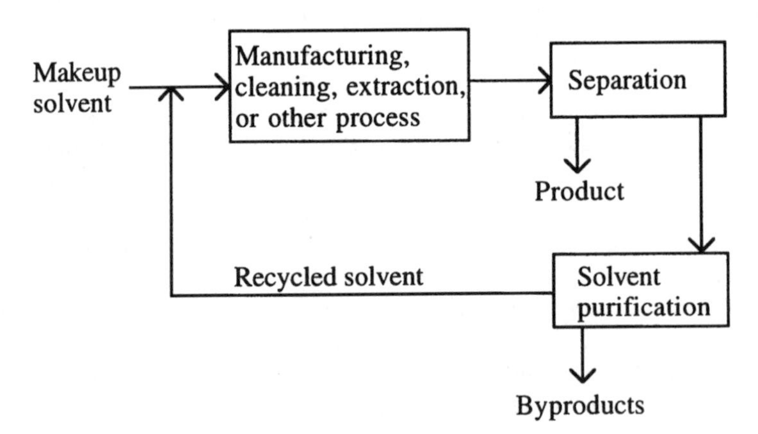

Figure 13.4. Overall process for recycling solvents.

A number of operations are used in solvent purification. Entrained solids are removed by settling, filtration, or centrifugation. Drying agents may be used to remove water from solvents and various adsorption techniques and chemical treatment may be required to free the solvent from specific impurities. Fractional distillation, often requiring several distillation steps, is the most important operation in solvent purification and recycle. It is used to separate solvents from impurities, water, and other solvents. High efficiency columns, including packed, bubble-cap, and sieve plate columns may be required. The sophistication of the distillation apparatus required increases with the complexity of the mixture to be purified. A high degree of segregation (see Section 8.4 and Figure 8.1) of solvents is desirable and is a good reason for performing the solvent purification and recycle on site rather than shipping the solvent to another facility for purification, which may result in mixing with other solvents and impurites.

LITERATURE CITED

1. Hunt, Gary E. and Greg P. Newman, "Waste Reduction: A Cost-Effective Approach to Hazardous Waste Management," Section 5.1 in *Standard Handbook of Hazardous Waste Treatment and Disposal*, Harry M. Freeman, Ed., McGraw-Hill, New York, 1998, pp. 5.3–5.21.

2. Wahl, William, *A Practical Guide for the Gold and Silver Electroplater and the Galvanoplastic Operator*, Henry Carey Baird and Co., Philadelphia, PA, 1883.

3. "Waste Minimization," Chapter 13 in *Hazardous Waste Management Compliance Handbook*, 2nd ed., Brian Karnofsky, Ed., Van Nostrand Reinhold, New York, NY, 1996, pp. 153–158.

4. "Waste Minimization," Chapter 7 in *Hazardous Materials and Waste Management*, Nicholas P. Cheremisinoff and Paul N. Cheremisinoff, Noyes Publications, Park Ridge, NJ, 1995, pp. 107–116.

5. Nunno, Thomas J. and Mark Arienti, "Waste Minimization Case Studies for Solvents and Metals Waste Streams," in *Land Disposal, Remedial Action, Incineration and Treatment of Hazardous Waste*, EPA/600/9–86/022, U.S. Environmental Protection Agency, Cincinnati, OH, 1986, pp. 278–284.

6. Thompson, Fay M. and Cindy A. McComas, "Technical Assistance for Hazardous Waste Reduction," *Environmental Science and Technology*, **21**, 1987, pp. 1154–1158.

7. Schneider, Keith, "Worst Tire Inferno Has Put Focus on Disposal Problem," *New York Times*, March 2, 1990, p. A8.

8. Brown, Craig J., "Ion Exchange," Section 6.5 in *Standard Handbook of Hazardous Waste Treatment and Disposal*, Harry M. Freeman, Ed., McGraw-Hill, New York, NY, 1998, pp. 6.57–6.72.

9. "Evaluation of the HSA Reactor for Metal Recovery and Cyanide Oxidation in Metal Plating Operations," EPA/600/S2–86/094, U.S. Environmental Protection Agency, Washington, D.C., 1986.

SUPPLEMENTARY REFERENCES

Cheremisinoff, Paul N., *Waste Minimization and Cost Reduction for the Process Industries*, Noyes Publications, Park Ridge, NJ, 1995.

Ciambrone, David F., *Waste Minimization As a Strategic Weapon*, CRC Press/Lewis Publishers, Boca Raton, FL, 1996.

Clark, J. H., Ed., *Chemistry of Waste Minimization*, Blackie Academic and Professional, New York, NY, 1995.

Frosch, Robert A., "Industrial Ecology: Adapting Technology for a Sustainable World," *Environment Magazine*, **37**, 16–37, 1995.

Graedel, Thomas E. and B. R. Allenby, "Industrial Process Residues: Composition and Minimization," Chapter 15 in *Industrial Ecology*, Prentice Hall, Englewood Cliffs, NJ, 1995, pp. 204–230.

International Atomic Energy Agency Vienna, *Minimization of Radioactive Waste from Nuclear Power Plants and the Back End of the Nuclear Fuel*, International Atomic Energy Agency, Vienna, 1995.

Nemorow, Nelson, *Zero Pollution for Industry : Waste Minimization Through Industrial Complexes*, John Wiley & Sons, New York, NY, 1995.

Reinhardt, Peter A., K. Leigh Leonard, and Peter C. Ashbrook, Eds., *Pollution Prevention and Waste Minimization in Laboratories*, CRC Press/Lewis Publishers, Boca Raton, FL, 1996.

Rossiter, Alan P., Ed., *Waste Minimization Through Process Design*, McGraw-Hill, New York, NY, 1995.

Thomas, Suzanne T., *Facility Manager's Guide to Pollution Prevention and Waste Minimization*, BNA Books, Castro Valley, CA, 1995.

14 INDUSTRIAL ECOLOGY OF WASTE TREATMENT

14.1. INTRODUCTION

In an ideal system of industrial ecology, waste treatment would be irrelevant because there would be no wastes. In reality, there will always be wastes to treat, so it is necessary to consider waste treatment processes. Furthermore, most of the processes used to treat wastes are also useful in preventing their production and release. For example, electrolysis to reclaim heavy metals from waste metal plating bath solutions can enable safe disposal of the remaining solution as wastewater, thus eliminating their disposal as hazardous waste.

Large quantities of hazardous wastes are dissolved in, or suspended in water. Therefore, water treatment is very important in the treatment of hazardous wastes. A variety of long-established and sophisticated, more recent methods are used to treat wastewater associated with hazardous wastes.[1]

This chapter presents an overview of waste treatment. It first covers physical processes, usually the simplest and least expensive to perform. Addressed next are chemical processes followed by thermal, then biological approaches to waste treatment. The industrial ecology of waste disposal is addressed in Chapter 15. An excellent comprehensive source of information on hazardous waste treatment and disposal is contained in a handbook on that topic.[2]

14.2. PHYSICAL METHODS OF WASTE TREATMENT

It should be kept in mind that most waste treatment measures have both physical and chemical aspects. The appropriate treatment technology for hazardous wastes obviously depends upon the nature of the wastes. These may consist of volatile wastes (gases, volatile solutes in water, gases or volatile liquids held by solids, such as catalysts), liquid wastes (wastewater, organic solvents), dissolved or soluble wastes (water-soluble inorganic species, water-soluble organic species, compounds soluble in organic solvents), semisolids (sludges, greases), and solids (dry solids, including granular solids with a significant water content, such as dewatered sludges, as well as solids suspended in liquids). One of the most common physical forms of waste treated consists of a semisolid, generally water-containing material called sludge.[3] Different kinds of sludge come from a variety of sources and include biological, inorganic chemical, and organic chemical sludge. The type of physical treatment to be applied to wastes strongly depends upon the physical properties of the material treated, including state of matter, solubility in water and organic solvents, density, volatility, boiling point, and melting point.

Methods of Physical Treatment

Knowledge of the physical behavior of wastes has been used to develop various unit operations for waste treatment that are based upon physical properties. These operations include the following:

- Phase separation
  - Filtration
  - Decantation
- Phase transition
  - Distillation
  - Evaporation
  - Physical precipitation
- Phase transfer
  - Extraction
  - Sorption
- Membrane separations
  - Reverse osmosis
  - Hyper- and ultrafiltration

Phase Separations

The most straightforward means of physical treatment involves separation of components of a mixture that are already in two different phases. **Sedimentation** and **decanting** are easily accomplished with simple equipment. In many cases the separation must be aided by mechanical means, particularly **filtration** or **centrifugation**. **Flotation** is used to bring suspended organic matter or finely divided particles to the surface of a suspension. In the process of **dissolved air flotation**, air is dissolved in the suspending medium under pressure and comes out of solution when the pressure is released as minute air bubbles attached to suspended particles, which causes the particles to float to the surface. Dissolved air flotation (DAF) float from the petroleum refining industry is a listed hazardous waste (K048).

An important and often difficult waste treatment step is **emulsion breaking** in which colloidal-sized **emulsions** are caused to aggregate and settle from suspension. Agitation, heat, acid, and the addition of **coagulants** consisting of organic polyelectrolytes or inorganic substances, such as an aluminum salt, may be used for this purpose. The chemical additive acts as a flocculating agent to cause the particles to stick together and settle out.

Phase Transition

A second major class of physical separation is that of **phase transition** in which a material changes from one physical phase to another. It is best exemplified by **distillation**, which is used in treating and recycling solvents, waste oil, aqueous phenolic wastes, xylene contaminated with paraffin from histological laboratories, and mixtures of ethylbenzene and styrene. Distillation produces **distillation bottoms** (still bottoms), which are often hazardous and polluting. These consist of unevaporated solids, semisolid tars, and sludges from distillation. Specific examples with their hazardous waste numbers are distillation bottoms from the production of acetaldehyde from ethylene (hazardous waste number K009), and still bottoms from toluene reclamation distillation in the production of disulfoton (K036). The landfill disposal of these and other hazardous distillation bottoms was widely practiced, but is now severely limited.

Evaporation is usually employed to remove water from an aqueous waste to concentrate it. A special case of this technique is **thin-film evaporation** in which volatile constituents are removed by heating a thin layer of liquid or sludge waste spread on a heated surface.

Drying — removal of solvent or water from a solid or semisolid (sludge) or the removal of solvent from a liquid or suspension — is a very important operation because water is often the major constituent of waste products, such as sludges obtained from emulsion breaking. In **freeze drying**, the solvent, usually water, is sublimed from a frozen material. Hazardous waste solids and sludges are dried to reduce the quantity of waste, to remove solvent or water that might interfere with subsequent treatment processes, and to remove hazardous volatile constituents. Dewatering can often be improved with addition of a filter aid, such as diatomaceous earth, during the filtration step.

Stripping is a means of separating volatile components from those less volatile in a liquid mixture by the partitioning of the more volatile materials to a gas phase of air or steam (steam stripping). The gas phase is introduced into the aqueous solution or suspension containing the waste in a stripping tower that is equipped with trays or packed to provide maximum turbulence and contact between the liquid and gas phases. The two major products are condensed vapor and a stripped bottoms residue. Examples of two volatile components that can be removed from water by air stripping are benzene and dichloromethane. Air stripping can also be used to remove ammonia from water that has been treated with a base to convert ammonium ion to volatile ammonia.

Physical precipitation is used here as a term to describe processes in which a solid forms from a solute in solution as a result of a physical change in the solution, as compared to chemical precipitation (see Section 14.5) in which a chemical reaction in solution produces an insoluble material. The major changes that can cause physical precipitation are cooling the solution, evaporation of solvent, or alteration of solvent composition. The most common type of physical precipitation by alteration of solvent composition occurs when a water-miscible organic solvent is added to an aqueous solution, so that the solubility of a salt is lowered below its concentration in the solution.

Phase Transfer

Phase transfer consists of the transfer of a solute in a mixture from one phase to another. An important type of phase transfer process is **solvent extraction**, a process in which a substance is transferred from solution in one solvent (usually water) to another (usually an organic solvent) without any chemical change taking place. When solvents are used to leach substances from solids or sludges, the process is called **leaching**. Solvent extraction and the major terms applicable to it are summarized in Figure 14.1. The same terms and general principles apply to leaching. The major application of solvent extraction to waste treatment has been in the removal of phenol from byproduct water produced in coal coking, petroleum refining, and chemical syntheses that involve phenol.

A relatively new approach to solvent extraction and leaching of hazardous wastes is the use of **supercritical fluids**, most commonly CO_2, as extraction solvents. A supercritical fluid is one that has characteristics of both liquid and gas and consists of a substance above its critical temperature and pressure (31.1°C and 73.8 atm, respectively, for CO_2). After a substance has been extracted from a waste into a supercritical fluid at high pressure, the pressure can be released, resulting in separation of the substance extracted. The fluid can then be compressed again and recirculated through the extraction system. Some possibilities for treatment of hazardous wastes by extraction with supercritical CO_2 include removal of organic contaminants from wastewater, extraction of organohalide pesticides from soil, extraction of oil from emulsions used in aluminum and steel processing, and regeneration of spent activated carbon. Waste oils contaminated with PCBs, metals, and water can be purified using supercritical ethane.

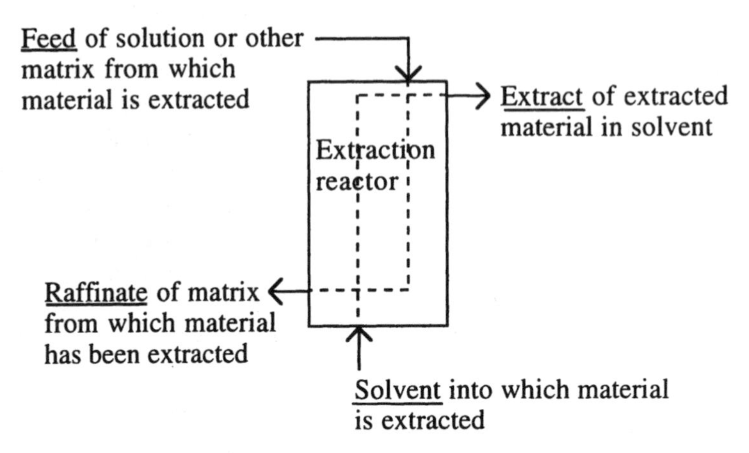

Figure 14.1. Outline of solvent extraction/leaching process with important terms underlined.

Transfer of a substance from a solution to a solid phase is called **sorption**. The most important sorbent is **activated carbon** used for several purposes and in some cases it is adequate for complete treatment. It can also be applied to pretreatment of waste streams going into processes such as reverse osmosis to improve treatment efficiency and reduce fouling. Effluents from other treatment processes, such as biological treatment of degradable organic solutes in water, can be polished with activated carbon. Activated carbon sorption is most effective for removing poorly water-soluble high-molar-mass materials, such as xylene, naphthalene (U165), cyclohexane (U056); chlorinated hydrocarbons, phenol (U188), aniline (U012), dyes, and surfactants from water. Activated carbon does not work well for organic compounds that are highly water-soluble or polar.

Solids other than activated carbon can be used for sorption of contaminants from liquid wastes. These include synthetic resins composed of organic polymers and mineral substances. Of the latter, clay has been employed to remove impurities from waste lubricating oils in some oil recycling processes.

Molecular Separation

A third major class of physical separation is **molecular separation**, often based upon **membrane processes** in which dissolved contaminants or solvent pass through a size-selective membrane under pressure. The products are a relatively pure solvent phase (usually water) and a concentrate enriched in the solute impurities. **Hyperfiltration** allows passage of species with molecular masses of about 100 to 500, whereas **ultrafiltration** is used for the separation of organic solutes with molar masses of 500 to 1,000,000. With both of these techniques, water and lower molar mass solutes under pressure pass through the membrane as a stream of purified **permeate**, leaving behind a stream of **concentrate** containing impurities in solution or suspension. Ultrafiltration and hyperfiltration are especially useful for concentrating suspended oil, grease, and fine solids in water. They also serve to concentrate solutions of large organic molecules and heavy metal ion complexes.

Reverse osmosis is the most widely used of the membrane techniques. Although superficially similar to ultrafiltration and hyperfiltration, it operates on a different principle in that the membrane is selectively permeable to water and excludes ionic solutes. Reverse osmosis uses high pressures to force permeate through the membrane, producing a concentrate containing high levels of dissolved salts (Figure 14.2).

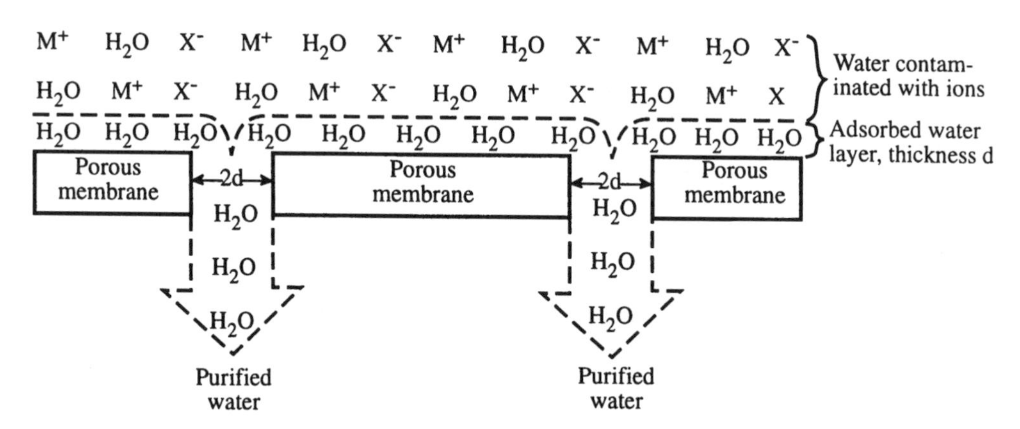

Figure 14.2. Reverse osmosis treatment of wastewater. This process is one of the most effective and least costly ways of treating water enabling its recycle.

Electrodialysis, sometimes used to concentrate plating wastes, employs membranes alternately permeable to cations and to anions. The driving force for the separation is provided by electrolysis with a direct current between two electrodes. Alternate layers between the membranes contain concentrate (brine) and purified water.

14.3. CHEMICAL TREATMENT OF WASTES

The applicability of chemical treatment to wastes depends upon the chemical properties of the waste constituents. These properties include acid-base, oxidation-reduction, precipitation, and complexation behavior; flammability/combustibility; reactivity; corrosivity; and compatibility with other wastes. The chemical behavior of wastes translates to various unit operations for waste treatment that are based upon chemical properties and reactions. These include the following:

- Acid/base neutralization
- Chemical precipitation
- Chemical flocculation
- Oxidation
- Reduction
- Chemical extraction and leaching
- Ion exchange

Chemical reactions of various types are used for the treatment and destruction of hazardous wastes as summarized in this chapter. An attractive feature of chemical treatment is the opportunity to treat wastes with other wastes. One of the largest such applications is the mutual neutralization of waste acids (such as acids from steel pickling liquor) and waste bases (for example, alkali remaining from the removal of sulfur from petroleum products). Waste ash containing high contents of calcium and magnesium oxides can be substituted for lime in waste treatment or stabilization. Combustible organic materials separated from waste oils, solvents, and sludges can be used as supplemental fuel in hazardous waste incinerators.

Methods for destroying some of the chemicals often used in laboratory procedures have been discussed in some detail.[4] Among the chemicals for which treatment procedures are given in this work are acid halides, acid anhydrides, amines, azo compounds, azoxy compounds, tetrazenes, boron trifluoride, calcium carbide, chloromethylsilanes, chromates, hydrides, cyanides, dimethyl sulfate, ethidium bromide, hydrazines, mercaptans, inorganic sulfides, nitriles, peroxides, picric acid, and sodium amide.

14.4. ACID/BASE NEUTRALIZATION

The process used to eliminate waste acids and bases is called **neutralization**, as shown by the following reaction:

$$H^+ + OH^- \rightarrow H_2O \quad (14.4.1)$$

If too much base is present, acid is added to react with OH^- and, if too much acid is present, base is added to react with H^+. Although simple in principle, neutralization can present some problems in practice, such as evolution of volatile contaminants or mobilization of soluble substances. The heat generated from the above reaction when the wastes involved are relatively concentrated can result in dangerously hot solutions and spattering. Strongly acidic or basic solutions are corrosive to pipes, containers, and mixing apparatus. There is a danger of adding too much acid or base and producing a product that is too acidic or basic.

The sources of acid or base used to treat alkaline or acidic wastes are determined by cost and safety. Ideally, wastes, such as waste acid metal pickling liquor or waste base from petroleum refining can be used. Lime, $Ca(OH)_2$, is a widely used base for treating acidic wastes, and it has the advantage of limited solubility, which prevents solutions of excess lime from reaching extremely high pH values. Sulfuric acid, H_2SO_4, is a relatively inexpensive acid for treating alkaline wastes. However, addition of too much sulfuric acid can produce highly acidic products; for some applications, acetic acid, CH_3COOH, is preferable. As noted above, acetic acid is a weak acid and an excess of it does little harm. It is also a natural product and biodegradable. These characteristics make its use desirable for soil washing and flushing where it may be difficult to control the quantity of acid needed to neutralize alkali in contaminated soil.

Neutralization, or pH adjustment, is often required prior to the application of other waste treatment processes. These include activated carbon sorption, oxidation/reduction, wet air oxidation, stripping, and ion exchange. Microorganisms usually require a pH in the range of 6–9, so neutralization may be required prior to biochemical treatment.

14.5. CHEMICAL PRECIPITATION

Precipitation is used in hazardous waste treatment primarily for the removal of heavy metal ions from water as shown below for the chemical precipitation of cadmium:

$$Cd^{2+}(aq) + HS^-(aq) \rightarrow CdS(s) + H^+(aq) \quad (14.5.1)$$

Physical precipitation in which solutes are removed by cooling the solution, evaporation, or alteration of solvent composition was mentioned in Section 14.2. This section deals with **chemical precipitation** where a chemical reaction in solution is used to form an insoluble species.

Precipitation of Metals

Hydroxides and Carbonates

The most widely used means of precipitating metal ions is by the formation of hydroxides such as chromium(III) hydroxide.

$$Cr^{3+} + 3OH^- \rightarrow Cr(OH)_3 \quad (14.5.2)$$

The source of hydroxide ion, OH^-, is a base (alkali), such as lime ($Ca(OH)_2$), sodium hydroxide (NaOH), or sodium carbonate (Na_2CO_3). The base, itself, may be a waste material, such as sodium hydroxide used to remove sulfur compounds from petroleum products. Most metal ions tend to produce basic salt precipitates, such as basic copper(II) sulfate, $CuSO_4{\bullet}3Cu(OH)_2$, formed as a solid when hydroxide is added to a solution containing Cu^{2+} and SO_4^{2-} ions. The solubilities of many heavy metal hydroxides reach a minimum value, often at a pH in the range of 9–11, then increase with increasing pH values due to the formation of soluble hydroxo complexes, as illustrated by the following reaction:

$$Zn(OH)_2(s) + OH^-(aq) \rightarrow Zn(OH)_3^-(aq) \quad (14.5.3)$$

The chemical precipitation method most used is precipitation of metals as hydroxides and basic salts with lime. Sodium carbonate can be used to precipitate hydroxides, carbonates, or basic carbonate salt precipitates. The carbonate anion produces hydroxide by virtue of its hydrolysis reaction with water,

$$CO_3^{2-} + H_2O \rightarrow HCO_3^- + OH^- \quad (14.5.4)$$

Carbonate, alone, does not give as high a pH as alkali metal hydroxides, which may have to be used to precipitate metals that form hydroxides only at relatively high pH values. Some heavy metals may be precipitated as carbonates. A typical carbonate salt formed in the presence of carbonate ion is cadmium carbonate, $CdCO_3$, and a typical basic carbonate is one formed with lead ion having the formula $2PbCO_3 \cdot Pb(OH)_2$. Some carbonate precipitates are more filterable than hydroxides.

Sulfides

The solubilities of some heavy metal sulfides are extremely low, so sulfide precipitation (see Reaction 14.5.1) can be a very effective means of treatment. Sources of sulfide ion include sodium sulfide (Na_2S), sodium hydrosulfide (NaHS), hydrogen sulfide (H_2S), and iron(II) sulfide (FeS). Hydrogen sulfide is a toxic gas that is, itself, considered to be a hazardous waste (U135). Iron(II) sulfide (ferrous sulfide) can be used as a safe source of sulfide ion to produce sulfide precipitates with other metals that are less soluble than FeS. Since iron(II) sulfide is only slightly soluble, itself, it presents a relatively low hazard due to the presence of sulfide ion in solution and the potential to form volatile hydrogen sulfide. However, H_2S can be a problem when metal sulfide wastes contact acid, which results in the production of H_2S gas by the following reaction:

$$MS + 2H^+ \rightarrow M^{2+} + H_2S \quad (14.5.5)$$

Precipitation as Metals

Some metals can be precipitated from solution in the elemental metal form by the action of a reducing agent. A reducing agent used for this purpose is sodium borohydride. This reagent can be used, for example, to precipitate copper metal from plating solutions in which the metal is stabilized as copper(I) in the form of a complex ion,

$$8Cu(I) + NaBH_4 + 2H_2O \rightarrow 8Cu + NaBO_2 + 8H^+ \quad (14.5.6)$$

The volume of sludge from sodium borohydride precipitation is usually much less than that from precipitation of metals with lime.

Metal ions can be converted to the elemental form and removed from solution by reduction to the metallic state through reaction with more active metals, a process called **cementation**. An example is shown below for the reduction of toxic cadmium with relatively harmless zinc.

$$Cd^{2+} + Zn \rightarrow Cd + Zn^{2+} \qquad (14.5.7)$$

14.6. CHEMICAL FLOCCULATION

Many solids formed by chemical reactions do not produce precipitates that readily settle. Instead, many precipitates are colloidal suspensions which do not settle or which form a very bulky precipitate with a high water content. Therefore, it is often desirable to add a chemical **flocculating agent** that binds to the colloidal particles, as bridging groups between them that enable particles to join together and settle as a relatively dense, filterable precipitate.

Flocculants are polymeric species, often **polyelectrolytes** that consist of large molecules with functional groups, such as SO_3^- or NH^+ attached. Examples of polymeric flocculating agents are illustrated in Figure 14.3.

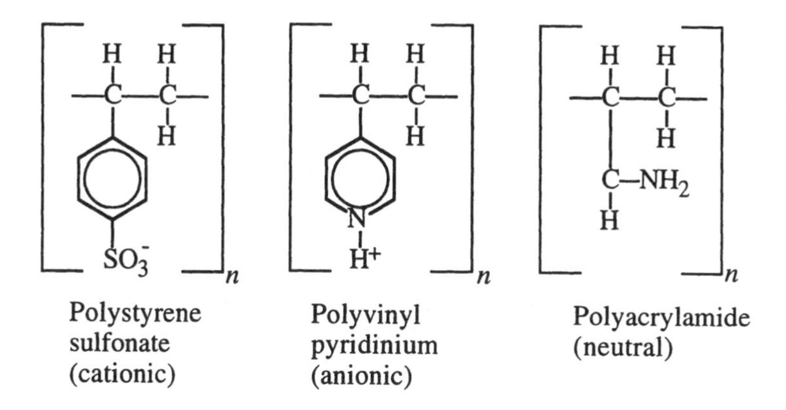

Figure 14.3. Polymeric species used as flocculants. Examples are shown of anionic and cationic polyelectrolytes and a nonionic polymer.

14.7. OXIDATION/REDUCTION

Oxidation and **reduction** can be used for the treatment and removal of a variety of inorganic and organic wastes. Some waste oxidants can be used to treat oxidizable wastes in water and cyanides. The net result of an oxidation or reduction reaction used to treat a waste constituent in aqueous solution is the conversion of the waste to a nonhazardous form or to a form that can be physically isolated. Important examples of oxidation/reduction processes for the treatment of hazardous wastes are given in Table 14.1.

Table 14.1. Oxidation/Reduction Reactions Used to Treat Water

Waste Substance	Reaction with Oxidant or Reductant
Oxidation of Organics	
Organic matter, $\{CH_2O\}$	$\{CH_2O\} + 2\{O\} \rightarrow CO_2 + H_2O$
Aldehyde	$CH_3CHO + \{O\} \rightarrow CH_3COOH$ (acid)
Oxidation of Inorganics	
Cyanide	$2CN^- + 5OCl^- + H_2O \rightarrow N_2 + 2HCO_3^- + 5Cl^-$
Iron(II)	$4Fe^{2+} + O_2 + 10H_2O \rightarrow 4Fe(OH)_3 + 8H^+$
Sulfur dioxide	$2SO_2 + O_2 + 2H_2O \rightarrow 2H_2SO_4$
Reduction of Inorganics	
Chromate	$2CrO_4^{2-} + 3SO_2 + 4H^+ \rightarrow Cr_2(SO_4)_3 + 2H_2O$
Permanganate	$MnO_4^- + 3Fe^{2+} + 7H_2O \rightarrow MnO_2(s) + 3Fe(OH)_3(s) + 5H^+$

Ozone as an Oxidant

Ozone, O_3, is a strong oxidant that can be generated on site by an electrical discharge through dry air or oxygen as shown by the following reaction:

$$3O_2 \xrightarrow[\text{discharge}]{\text{Electrical}} 2O_3 \quad (14.7.1)$$

A system for generating ozone is shown in Figure 14.4.

Ozone is employed as an oxidant gas at levels of 1–2 wt% in air and 2–5 wt% in oxygen. It has been used to treat a large variety of oxidizable contaminants, effluents, and wastes in the following categories:

- Municipal drinking water

  Disinfection; color, taste, and odor removal

- Wastewater

  Disinfects and removes oxidizable chemical contaminants from municipal and industrial wastewater. Oxidizes organic compounds, including unsaturated alcohols, phenols, aldehydes; inorganic species, including H_2S, nitrite (to less toxic nitrate), cyanide, and Fe^{2+} (to insoluble Fe(III)).

- Sludges containing oxidizable constituents
- Gas streams containing toxic gases and odor-causing organic compounds

A safety advantage with ozone is its generation on site so that oxidant does not have to be stored and shipped. Its toxicity and status as an air pollutant are disadvantages, and scrupulous measures must be employed to prevent its release.

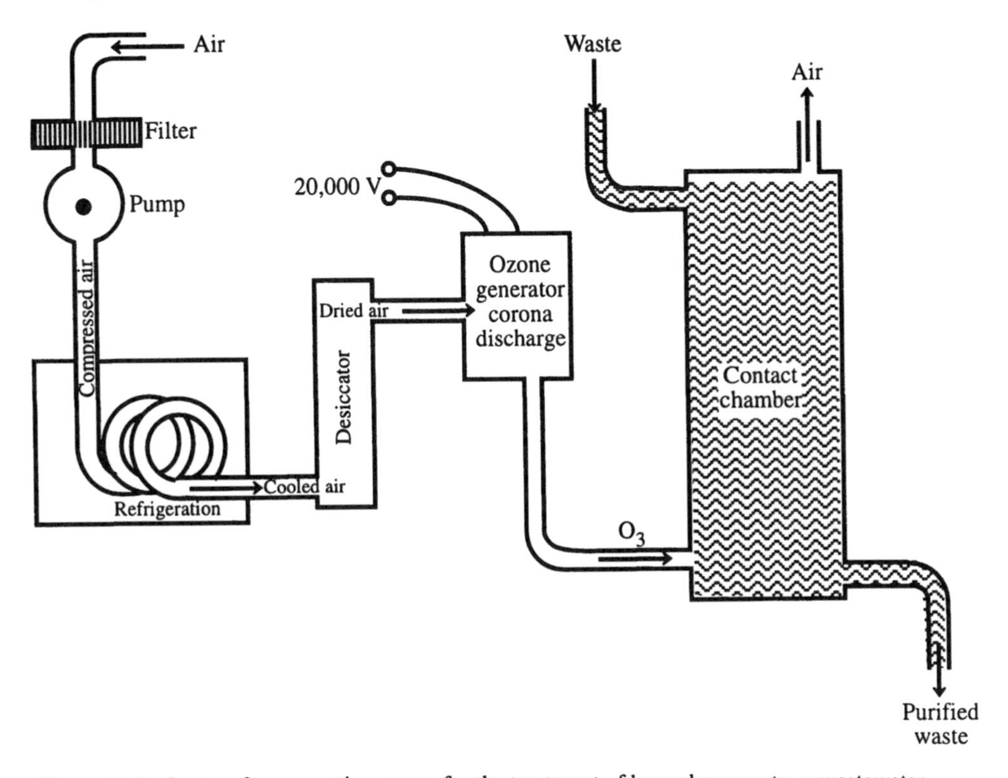

Figure 14.4. System for generating ozone for the treatment of hazardous waste or wastewater.

14.8. ELECTROLYSIS

Electrolysis consists of the electrochemical reduction and oxidation of chemical species in solution by means of electricity applied to electrodes from an external source. One species in solution (usually a metal ion) is reduced by electrons at the **cathode** and another gives up electrons to the **anode** and is oxidized there. Examples of cathodic and anodic half-reactions and the general concept of electrolysis are illustrated in Figure 14.5.

In hazardous waste applications, electrolysis is most widely used in the recovery of metals. An obvious application is metal recovery from spent electroplating media. Other types of media to which electrolysis is applicable include wastewaters and rinsewaters from the electronics industries and from metal finishing operations. The metals that are most commonly recovered by electrolysis are cadmium, copper, gold, lead, silver, and zinc. Metal recovery by electrolysis is made more difficult by the presence of cyanide ion which stabilizes metals in solution as the cyanide complexes, such as $Ni(CN)_4^{2-}$. Recovered metals are usually recycled to the process that produced the wastes.

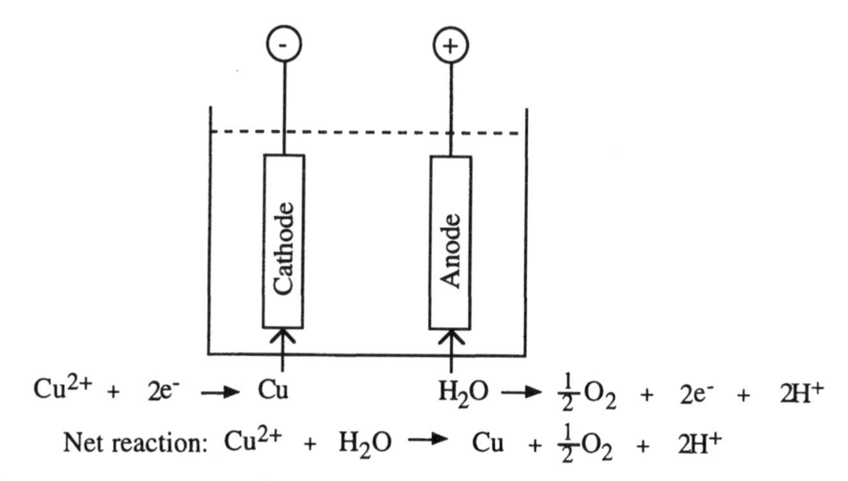

Figure 14.5. Electrolysis of copper solution.

14.9. REACTION WITH WATER

Many inorganic and organic chemicals are hazardous because of their reactions with water, commonly called **hydrolysis**. In some cases the reaction is so vigorous that a fire or even an explosion may result. In others, a hazardous gas, such as explosive H_2 or corrosive, toxic HCl, is evolved. One of the ways to dispose of chemicals that are reactive with water is to subject them to hydrolysis under controlled conditions.

Inorganic chemicals that can be treated by hydrolysis include metals that react with water; metal carbides, hydrides, amides, alkoxides, and halides; and nonmetal oxyhalides and sulfides. Examples of the treatment of these classes of inorganic species are given in Table 14.2.

Organic chemicals may also be treated by hydrolysis. The types of organic chemicals that can be hydrolyzed include the following:

$$H{-}CH_2{-}C(=O){-}O{-}C(=O){-}CH_2{-}H + H_2O \rightarrow 2H{-}CH_2{-}C(=O){-}O{-}H \quad (14.9.1)$$

Acid anydride (acetic anhydride)

$$H{-}CH_2{-}C(=O){-}Cl + H_2O \rightarrow H{-}CH_2{-}C(=O){-}OH + HCl \quad (14.9.2)$$

Acid halide (acetyl chloride)

$$H_3C{-}N{=}C{=}O + H_2O \rightarrow \text{Hydrolysis products} \quad (14.9.3)$$

Isocyanates (methyl isocyanate)

14.10. CHEMICAL EXTRACTION AND LEACHING

Chemical extraction or **leaching** in hazardous waste treatment is the removal of a hazardous constituent by reaction with an extractant in solution. Some examples of chemical phenomena in extraction are cited here.

Acidic solutions dissolve poorly soluble heavy metal salts by reaction of the salt anions with H^+ as illustrated by the following:

$$PbCO_3 + H^+ \rightarrow Pb^{2+} + HCO_3^- \quad (14.10.1)$$

Acids also dissolve basic organic compounds such as amines and aniline. Extraction with acids should be avoided if cyanides or sulfides are present to prevent formation of toxic hydrogen cyanide or hydrogen sulfide. Nontoxic weak acids are usually the safest to use. These include acetic acid, CH_3COOH, and the acid salt, NaH_2PO_4.

Table 14.2. Inorganic Chemicals that may be Treated by Hydrolysis

Waste Substance	Reaction with Water
Active metals (calcium)	$Ca + 2H_2O \rightarrow H_2 + Ca(OH)_2$
Hydrides (sodium aluminum hydride)	$NaAlH_4 + 4H_2O \rightarrow 4H_2 + NaOH + Al(OH)_3$
Carbides (calcium carbide)	$CaC_2 + 2H_2O \rightarrow Ca(OH)_2 + C_2H_2$
Amides (sodium amide)	$NaNH_2 + H_2O \rightarrow NaOH + NH_3$
Halides (silicon tetrachloride)	$SiCl_4 + 2H_2O \rightarrow SiO_2 + 4HCl$
Alkoxides (sodium ethoxide)	$NaOC_2H_5 + H_2O \rightarrow NaOH + C_2H_5OH$

Chelating agents, such as dissolved ethylenedinitrilotetraacetate (EDTA, HY^{2-}), dissolve insoluble metal salts by forming soluble species with metal ions,

$$FeS + H_2Y^{2-} \rightarrow FeY^{2-} + H_2S \quad (14.10.2)$$

Heavy metal ions in soil contaminated by hazardous wastes may be present in a coprecipitated form with insoluble iron(III) and manganese(IV) oxides, Fe_2O_3 and MnO_2, respectively. These oxides can be dissolved by reducing agents, such as sodium dithionate/citrate or hydroxylamine, which result in the production of soluble Fe^{2+} and Mn^{2+}. Heavy metal species such as Cd^{2+} or Ni^{2+} are released and removed with the water.

14.11. ION EXCHANGE

Ion exchange is a means of removing cations or anions from solution onto a solid resin as illustrated by the two following reactions showing cadmium ion in solution being taken up by a cation exchanger and sulfate ion by an anion exchanger:

$$2H^+\{^-CatExRes\} + Cd^{2+} \rightarrow Cd^{2+}\{^-CatExRes\}_2 + 2H^+ \qquad (14.11.1)$$

$$2\{AnExRes^+\}OH^- + SO_4^{\ 2-} \rightarrow \{AnExRes^+\}_2SO_4^{\ 2-} + 2OH^- \qquad (14.11.2)$$

Ions taken up by an ion exchange resin may be removed by treating the resin with concentrated solutions of acid, base, or salt (NaCl). The net result is to concentrate the ions originally removed from dilute solution in water into a much more concentrated solution.

The greatest use of ion exchange in hazardous waste treatment is for the removal of low levels of heavy metal ions from wastewater. Ion exchange is employed in the metal-plating industry to purify rinsewater and spent plating bath solutions. Cation exchangers are used to remove cationic metal species, such as Cu^{2+}, from such solutions. Anion exchangers remove anionic cyanide metal complexes (for example, $Ni(CN)_4^{\ 2-}$) and chromium(VI) species, such as $CrO_4^{\ 2-}$.

Radionuclides may be removed from radioactive wastes and mixed radioactive and hazardous waste by ion exchange resins. High levels of radioactivity cause deterioration of ion exchange resins, although in some applications inorganic ion exchangers may be used.

Several characteristics of wastes must be considered prior to application of ion exchange. These include physical form of the wastes, presence of oxidants, and contents of metals, organic consituents, and suspended solids.

14.12. CHEMICAL DESTRUCTION OF PCBs

Polychlorinated biphenyls (PCBs) and dielectric fluid made from PCBs (askarel) were discussed in Section 11.5. Because of their extreme chemical and biological stabilities, PCBs are difficult to destroy. These compounds can be burned in high-efficiency incinerators meeting stringent requirements for residence time, temperature, and excess oxygen. Proprietary processes that use metallic sodium dissolved in various solvents can be used to selectively destroy PCBs dissolved at levels up to 10 parts per thousand in transformer fluids and other media. The reaction involves attack on C–Cl bonds as shown by the following generalized reaction:

$$(PCB)\text{–}Cl + Na \rightarrow \text{Biphenyl polymer} + NaCl \qquad (14.12.1)$$

The sodium chloride solid and a sludge consisting of polymer formed from the dechlorinated biphenyl constituents and other reaction byproducts are filtered from the liquid, which can be recycled as a dielectric fluid.

14.13. PHOTOLYTIC REACTIONS

Photolytic reactions or the process of **photolysis** are those in which **photons** of electromagnetic radiation consisting of short-wavelength visible light or ultraviolet radiation are absorbed by a molecule, causing a chemical reaction to occur. The energy, E, of the photon is equal to the product of Planck's constant, h, and the frequency, ν, of the radiation and is represented in photochemical reactions as $h\nu$. An example of an important photochemical reaction is the photodecomposition of nitrogen dioxide, which occurs in the earth's atmosphere and produces reactive oxygen atoms that start the processes by which photochemical smog is formed.

$$NO_2 + h\nu \rightarrow NO + O \qquad (14.13.1)$$

Photolysis can be used to destroy a variety of hazardous wastes. In such applications it is most useful in breaking strong chemical bonds in refractory organic compounds. The irradiation of TCDD (see Section 11.9) by ultraviolet light in the presence of hydrogen atom donors {H} results in reactions such as the following:

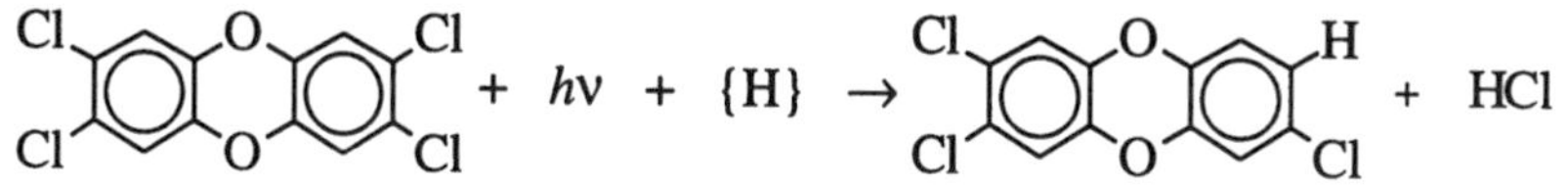

Cl, O, Cl, Cl, O, Cl + $h\nu$ + {H} → Cl, O, H, Cl, O, Cl + HCl

(14.13.2)

As photolysis proceeds, more H–C bonds are broken, the C–O bonds are broken, and the final product is a harmless organic polymer.

An initial photolyis reaction can initiate the generation of reactive intermediates that participate in **chain reactions** that lead to the destruction of a compound. One of the most important reactive intermediates is the hydroxyl radical, HO•, where the dot represents an unpaired electron. (The hydroxyl radical was mentioned in Chapter 7, Section 7.8 for its role in atmospheric photochemistry.) Molecular fragments with unpaired electrons are highly reactive species called **free radicals**. In some cases substances are added to the reaction mixture to absorb radiation and generate reactive species that destroy wastes. Such a substance is called a **sensitizer**.

In addition to TCDD, photolysis has been used to destroy several other kinds of hazardous waste substances. Included are herbicides (atrazine), 2,4,6-trinitrotoluene (TNT), and polychlorinated biphenyls (PCBs). The addition of a chemical oxidant, such as potassium peroxidisulfate, $K_2S_2O_8$, enhances destruction by oxidizing active photolytic products.

14.14. THERMAL TREATMENT METHODS

Thermal treatment of hazardous wastes can be used to accomplish most of the common objectives of waste treatment — volume reduction; removal of volatile, combustible, mobile organic matter; and destruction of toxic and pathogenic materials. The most widely applied means of thermal treatment of hazardous wastes is **incineration**. Incineration utilizes high temperatures, an oxidizing atmosphere, and often turbulent combustion conditions to destroy wastes. Methods other than incineration that make use of high temperatures to destroy or neutralize hazardous wastes are discussed briefly at the end of this section.

One of the largest contaminated soil incineration projects ever undertaken in the U.S. was concluded in 1997 at the site of the former town of Times Beach, Missouri.[5] The gravel roads of Times Beach, along with a number of other places in Missouri, were sprayed with waste oil contaminated with "dioxin," 2,3,7,8-tetrachloro-*p*-dioxin (TCDD), by a waste oil hauler in the early 1970s. Following a devastating flood in 1982, the contaminated town was bought out with Federal funds, and all the residents were required to evacuate.

As part of a cleanup of Times Beach and 26 other nearby areas contaminated with dioxin by the same waste oil hauler, about 240,000 tons of dioxin-contaminated soil were burned in a large incinerator on the former town site. Soil was fed into a rotary kiln where dioxins and other organic contaminants were driven off by heat, then burned in a secondary combustion chamber. Soil levels of 30–200 parts-per-billion dioxin were reduced to 1 part-per-billion. Emissions of dioxin in the stack gas were estimated to average 0.0036 nanograms per cubic meter, compared to an allowable limit of 1 ng/m^3.

Incineration

Hazardous waste incineration will be defined here as a process that involves exposure of the waste materials to oxidizing conditions at a high temperature, usually in excess of 900°C. Normally the heat required for incineration comes from the oxidation of organically bound carbon and hydrogen contained in the waste material or in supplemental fuel.

$$C(organic) + O_2 \rightarrow CO_2 + heat \quad (14.14.1)$$

$$4H(organic) + O_2 \rightarrow 2H_2O + heat \quad (14.14.2)$$

These reactions destroy organic matter and generate heat required for endothermic reactions, such as the breaking of C–Cl bonds in organochlorine compounds.

Incinerable Wastes

Ideally, incinerable wastes are predominantly organic materials that will burn with a heating value of at least 5,000 Btu/lb and preferably over 8,000 Btu/lb. Such heating values are readily attained with wastes having high contents of the most commonly incinerated waste organic substances, including methanol, acetonitrile, toluene, ethanol, amyl acetate, acetone, xylene, methyl ethyl ketone, adipic acid, and ethyl acetate. In some cases, however, it is desirable to incinerate wastes that will not burn alone and which require **supplemental fuel,** such as methane and petroleum liquids. Examples of such wastes are nonflammable organochlorine wastes, some aqueous wastes, or soil in which the elimination of a particularly troublesome contaminant is worth the expense and trouble of incinerating it. Inorganic matter, water, and organic hetero element contents of liquid wastes are important in determining their incinerability.

Hazardous Waste Fuel

Many industrial wastes, including hazardous wastes, are burned as **hazardous waste fuel** for energy recovery in industrial furnaces and boilers and in incinerators for nonhazardous wastes, such as sewage sludge incinerators. This process is called **coincineration**, and more combustible wastes are utilized by it than are burned solely for the purpose of waste destruction. In addition to heat recovery from combustible wastes, it is a major advantage to use an existing on-site facility for waste disposal rather than a separate hazardous waste incinerator.

Incineration Systems

The four major components of hazardous waste incineration systems are shown in Figure 14.6.

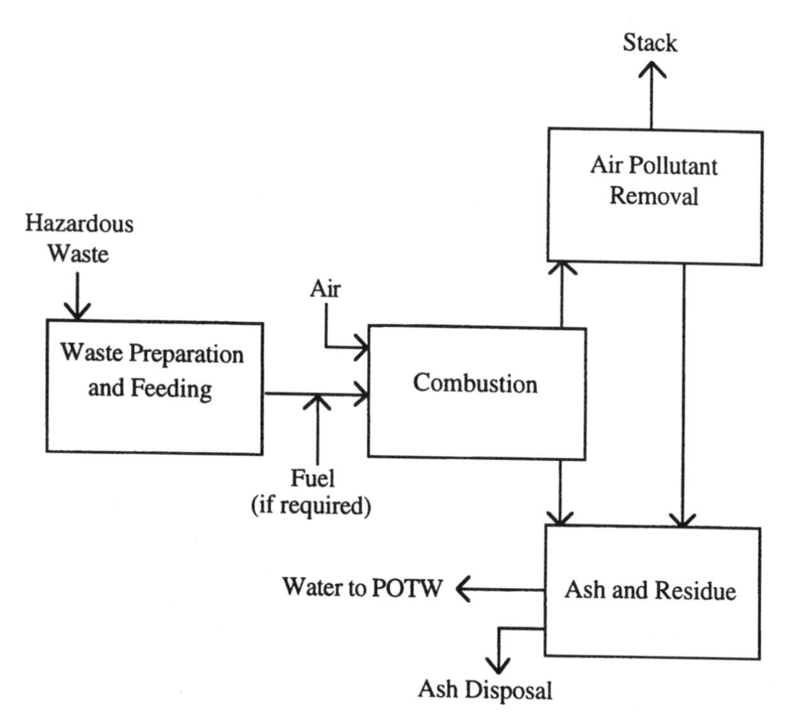

Figure 14.6. Major components of a hazardous waste incinerator system.

Waste preparation for liquid wastes may require filtration, settling to remove solid material and water, blending to obtain the optimum incinerable mixture, or heating to decrease viscosity. Solids may require shredding and screening. Atomization is commonly used to feed liquid wastes. Several mechanical devices, such as rams and augers, are used to introduce solids into the incinerator.

The most common kinds of **combustion chambers** are liquid injection, fixed hearth, rotary kiln, and fluidized bed. These types are discussed in more detail later in this section. Often the most complex part of a hazardous waste incineration system is the **air pollution control system**, which involves several operations. The most common operations in air pollution control from hazardous waste incinerators are combustion gas cooling, heat recovery, quenching, particulate matter removal, acid gas removal, and treatment and handling of byproduct solids, sludges, and liquids.

Hot ash is often quenched in water. Prior to disposal, it may require dewatering and chemical stabilization. A major consideration with hazardous waste incinerators and the types of wastes that are incinerated is the disposal problem posed by the ash, especially in respect to potential leaching of heavy metals.

Types of Incinerators

Hazardous waste incinerators may be divided among the following, based upon type of combustion chamber:

- **Rotary kiln** (about 40% of U.S. hazardous waste incinerator capacity) in which the primary combustion chamber is a rotating cylinder lined with refractory materials and an afterburner downstream from the kiln to complete destruction of the wastes
- **Liquid injection** incinerators (also about 40% of U.S. hazardous waste incinerator capacity) that burn pumpable liquid wastes dispersed as small droplets
- **Fixed-hearth incinerators** with single or multiple hearths upon which combustion of liquid or solid wastes occurs
- **Fluidized-bed incinerators** that have a bed of granular solid (such as limestone) maintained in a suspended state by injection of air to remove pollutant acid gas and ash products
- **Advanced design incinerators** including **plasma incinerators** that make use of an extremely hot plasma of ionized air injected through an electrical arc; **electric reactors** that use resistance-heated incinerator walls at around 2,200°C to heat and pyrolyze wastes by radiative heat transfer; **infrared systems**, which generate intense infrared radiation by passing electricity through silicon carbide resistance heating elements; **molten salt combustion** that use a bed of molten sodium carbonate at about 900°C to destroy the wastes and retain gaseous pollutants; and **molten glass processes** that use a pool of molten glass to transfer heat to the waste and to retain products in a poorly leachable glass form.

Combustion Conditions

The key to effective incineration of hazardous wastes lies in the combustion conditions. These require (1) sufficient free oxygen in the combustion zone; (2) turbulence for thorough mixing of waste, oxidant, and (where used) supplemental fuel; (3) high combustion temperatures above about 900°C to ensure that thermally resistant compounds do react; and (4) sufficient residence time (at least 2 seconds) to allow reactions to occur.

Effectiveness of Incineration

EPA standards for hazardous waste incineration are based upon the effectiveness of destruction of the **principal organic hazardous constituents** (POHC). Measurement of these compounds before and after incineration gives the **destruction removal efficiency** (DRE) according to the formula,

$$DRE = \frac{W_{in} - W_{out}}{W_{in}} \times 100 \qquad (14.14.3)$$

where W_{in} and W_{out} are the mass flow rates of the principal organic hazardous constituent (POHC) input and output (at the stack downstream from emission controls), respectively. United States EPA regulations call for destruction of 99.99% of POHCs and 99.9999% ("six nines") destruction of 2,3,7,8-tetrachloro-dibenzo-*p*-dioxin, commonly called TCDD or "dioxin."

Wet Air Oxidation

Organic compounds and oxidizable inorganic species can be oxidized by oxygen in aqueous solution. The source of oxygen usually is air. Rather extreme conditions of temperature and pressure are required, with a temperature range of 175–327°C and a pressure range of 300–3,000 psig (2070–20,700 kPa). The high pressures allow a high concentration of oxygen to be dissolved in the water and the high temperatures enable the reaction to occur.

Wet air oxidation has been applied to the destruction of cyanides in electroplating wastewaters. The oxidation reaction for sodium cyanide is the following:

$$2Na^+ + 2CN^- + O_2 + 4H_2O \rightarrow 2Na^+ + 2HCO_3^- + 2NH_3 \qquad (14.14.4)$$

A method has been described in which cyanide is oxidized on an aerated carbon bed with added copper(II) ion and sulfite as catalysts.[6] In addition to destroying CN^-, it also oxidizes highly stable complexed cyanide in species such as $Fe(CN)_6^{4-}$.

Organic wastes can be oxidized in supercritical water, taking advantage of the ability of supercritical fluids to dissolve organic compounds. Wastes are contacted with water and the mixture raised to a temperature and pressure required for supercritical conditions for water. Sufficient oxygen is then pumped in to oxidize the wastes. The process produces only small quantities of CO, and no SO_2 or NO_x. It has been used to degrade PCBs, dioxins, organochlorine insecticides, benzene, urea, and numerous other materials.

14.15. BIODEGRADATION OF WASTES

The latter part of this chapter covers biological treatment of hazardous wastes. It does not directly consider the widely practiced biological treatment of municipal wastewaters, food processing byproducts, and other wastes for which biological treatment has long been established. Although it has some shortcomings in the degradation of complex chemical mixtures, biological treatment offers a number of significant advantages and has considerable potential for the degradation of hazardous wastes, even *in situ.*

Biodegradation of wastes is their conversion by biological processes to simple inorganic molecules and, to a certain extent, to biological materials. The complete bioconversion of a substance to inorganic species such as CO_2, NH_3, and phosphate is called **mineralization. Detoxification** refers to the biological conversion of a toxic substance to a less toxic species, which may still be a relatively complex, or even more complex material. An example of detoxification is illustrated below for the enzymatic conversion of paraoxon (a highly toxic organophosphate insecticide) to *p*-nitrophenol, which has only about 1/200 the toxicity of the parent compound.

Paraoxon (H–C(H)(H)–C(H)(H)–O–P(=O)(–O–CH₂–CH₃)–O–C₆H₄–NO₂) $\xrightarrow[\text{Enzyme action}]{H_2O,\ \{O\}}$ HO–C₆H₄–NO₂ + Other products (14.15.1)

Usually the products of biodegradation are molecular forms that tend to occur in nature and that are in greater thermodynamic equilibrium with their surroundings. The definition of biodegradation is illustrated by an example in Figure 14.7. Biodegradation is usually carried out by the action of microorganisms, particularly bacteria and fungi.

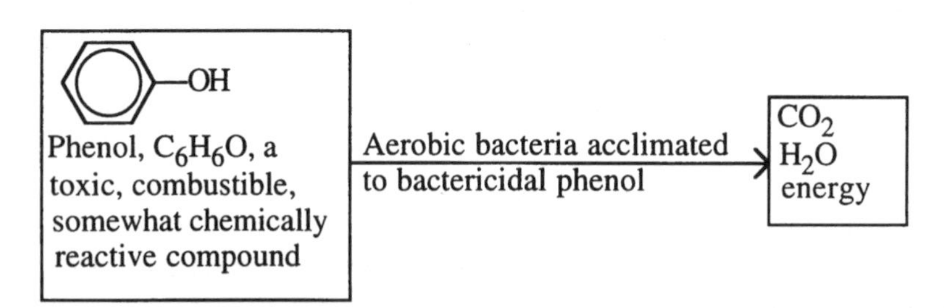

Figure 14.7. Illustration of biological treatment of a hazardous waste constituent.

Biodegradation of municipal wastewater and solid wastes in landfills occurs by design. Biodegradation of any kind of waste that can be metabolized takes place whenever the wastes are subjected to conditions conducive to biological processes.

The most common type of biodegradation is that of organic compounds in the presence of air, that is, **aerobic processes**. However, in the absence of air, **anaerobic biodegradation** may also take place. Furthermore, inorganic species are subject to both aerobic and anaerobic biological processes.

Although biological treatment of wastes is normally regarded as degradation to simple inorganic species such as carbon dioxide, water, sulfates, and phosphates, the possibility must always be considered of forming more complex or more hazardous chemical species. An example of the latter is the production of volatile, soluble, toxic methylated forms of arsenic and mercury from inorganic species of these elements by bacteria under anaerobic conditions.

Biochemical Aspects of Biodegradation

Biotransformation is what happens to any substance that is **metabolized** by the biochemical processes in an organism and is altered by these processes. **Metabolism** is divided into the two general categories of **catabolism**, which is the breaking down of more complex molecules, and **anabolism**, which is the building up of life molecules from simpler materials.The substances subjected to biotransformation may be naturally occurring or *anthropogenic* (made by human activities). They may consist of *xenobiotic* molecules that are foreign to living systems.

An important biochemical process that occurs in the biodegradation of many synthetic and hazardous waste materials is **cometabolism**. Cometabolism does not serve a useful purpose to an organism in terms of providing energy or raw material to build biomass, but occurs concurrently with normal metabolic processes. An example of cometabolism of hazardous wastes is provided by the white rot fungus, *Phanerochaete chrysosporium*, which degrades a number of kinds of organochlorine compounds, including DDT, PCBs, and chlorodioxins, under the appropriate conditions. The enzyme system responsible for this degradation is one that the fungus uses to break down lignin in plant material under normal conditions.

Enzymes in Waste Degradation

Enzyme systems hold the key to biodegradation of hazardous wastes. For most biological treatment processes currently in use, enzymes are present in living organisms in contact with the wastes. However, in some cases it is possible to extract enzymes from organisms and use cell-free extracts of enzymes to treat hazardous wastes. For this application the enzymes may be present in solution or, more commonly, immobilized in biochemical reactors.

14.16. FACTORS INVOLVED IN BIOLOGICAL TREATMENT

Biological treatment of hazardous wastes is usually carried out by microorganisms — bacteria and sometimes fungi. In considering waste biodegradation, it is useful to keep in mind that microorganisms require sources of energy, carbon, macronutrients, and micronurients. Except for algae and photosynthetic bacteria, which utilize sunlight for energy, microorganisms extract energy from the reactions that they mediate or "catalyze." Carbon can come from inorganic sources (HCO_3^-, CO_2) or from organic sources, including the substances undergoing biodegradation. Macronutrients required include sources of elements in addition to carbon used in building biomass — hydrogen, oxygen, nitrogen as well as smaller quantities of sulfur, and phosphorus. Trace amounts of micronutrients are needed to support biological processes and as constituents of enzymes. Important micronutrients are calcium, magnesium, potassium, sodium, chlorine, cobalt, iron, vanadium, and zinc. Sometimes sulfur, phosphorus, and micronutrients must be added to media in which microorganisms are used to degrade hazardous wastes in order for optimum growth to occur.

Organisms

A crucial factor in the biodegradation of hazardous waste substances is the selection of organisms that will degrade a particular type of material. Genetic engineering holds considerable promise for the development of microorganisms that can readily degrade toxic and recalcitrant wastes. A more practical approach at the present time is selection of indigenous microorganisms at a hazardous waste site that have the ability to degrade particular kinds of molecules. Often the populations of such microorganisms can be increased and conditions (such as those of nutrition) enhanced for their growth.

Biodegradability

An obvious requirement for the biological treatment of wastes is **biodegradability**. The biodegradability of a compound is influenced by its physical characteristics, such as solubility in water and vapor pressure, and by its chemical properties, including molecular mass, molecular structure, and presence of various kinds of functional groups, some of which provide a "biochemical handle" for the initiation of biodegradation. Many hazardous waste materials are biocidal and are often therefore not even considered candidates for biodegradation. However, with the appropriate organisms and under the right conditions, even substances that are biocidal to most microorganisms can undergo biodegradation. For example, as shown in Figure 14.7, normally bactericidal phenol is readily metabolized and degraded by the appropriate bacteria acclimated to its use as a carbon and energy source.

Recalcitrant or **biorefractory** substances are defined as those that resist biodegradation and tend to persist and accumulate in the environment. Such materials are not necessarily toxic to organisms, but simply resist their metabolic attack. However, even some compounds regarded as biorefractory may be degraded by microorganisms adapted to their biodegradation. Examples of such compounds and the types of microorganisms that can degrade them include endrin (*Arthrobacter*), DDT (*Hydrogenomonas*), phenylmercuric acetate (*Pseudomonas*), and raw rubber (*Actinomycetes*). Chemical pretreatment, especially by partial oxidation, can make some kinds of recalcitrant wastes much more biodegradable.

Properties of hazardous wastes can be changed to increase biodegradability. This is especially true of wastes that consist of several constituents, one or more of which inhibit biological processes. Sometimes a waste substance that is toxic to microorganisms at a relatively high concentration is well degraded in more dilute media. Biodegradation that is inhibited by extremes of pH in the waste may occur when excess acid or base is neutralized. Toxic organic and inorganic substances, such as heavy metal ions, can be removed in some cases prior to biological treatment. Substances that are precursors to formation of solid precipitates that form during biodegradation and inhibit it can be removed. For example, soluble iron(II) should be removed because it forms bacteria-inhibiting deposits of gelatinous iron(III) hydroxide during aerobic treatment of wastes.

$$2\{CH_2O\} \rightarrow CO_2 + CH_4 \qquad (14.16.1)$$

Conditions

The conditions under which biodegradation is carried out have a strong effect on the efficiency of biological treatment. Temperature affects the rate of biodegradation which, to a point, usually increases in rate with increasing temperature. The pH of the medium needs to be controlled, usually to within a range of 6–9. The need to remove toxic impurities was mentioned above. Stirring and mixing can be factors in biodegradation. The load of material is important to the process; sudden fluctuations can be detrimental. Oxygen is important; an adequate supply is needed for aerobic treatment, whereas oxygen is excluded for anaerobic processes.

Bioreactors

Biological treatment processes may take place in a **bioreactor**, which may have a continuous flow of material or may be operated in a batch mode. Efficient biodegradation occurs when a large mass of microorganisms is maintained relative to the amount of waste present at any particular time. This can be done with an attached fixed film of organisms. One such system is the **trickling filter** bioreactor in which layers of microorganisms are established on media, such as rocks, over which aqueous waste is sprayed. The microorganisms may also be suspended as a sludge in the bioreactor and recirculated as is done in the **activated sludge** process (Figure 14.8). Widely used in sewage treatment, the activated sludge process enables buildup of a large mass of microorganisms and is very efficient.

14.17. AEROBIC TREATMENT OF ORGANIC WASTES

Aerobic waste treatment processes utilize aerobic bacteria and fungi that require molecular oxygen, O_2. These processes are often favored by microorganisms, in part because of the high energy yield obtained when molecular oxygen reacts with organic matter. Aerobic waste treatment is well adapted to the use of an activated sludge process (above). It can be applied to hazardous wastes such as chemical process wastes and landfill leachates. Some systems use powdered activated carbon as an additive to absorb nonbiodegradable organic wastes.

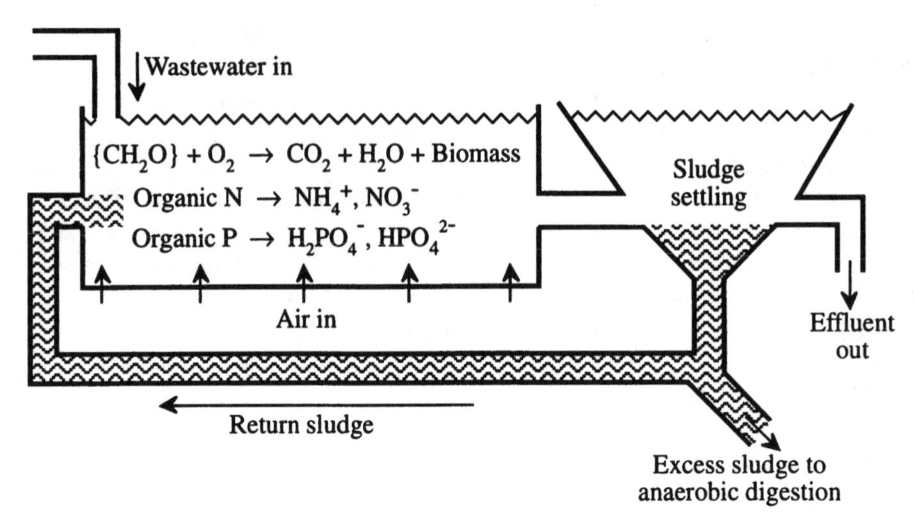

Figure 14.8. Activated sludge process.

Contaminated soils can be mixed with water and treated in a bioreactor to eliminate biodegradable contaminants in the soil. It is also possible, in principle, to use aerobic biological processes to treat contaminated soils in place. This is accomplished by pumping oxygenated, nutrient-enriched water through the soil in a recirculating system. In a sense, the method is a biochemical variation of soil flushing. Aerobic treatment of contaminated soil is relatively expensive, in part because of the high amount of energy required to pump both air and water into the system.

The off-gases from aerobic treatment of hazardous wastes often contain volatile organic compounds. Because of the potential toxicities of these compounds and their odors, the off-gases may require activated carbon filtration or other treatment.

14.18. ANAEROBIC TREATMENT OF ORGANIC WASTES

Anaerobic waste treatment in which microorganisms degrade wastes in the absence of oxygen can be practiced on a variety of organic hazardous wastes. Compared to the aerated activated sludge process, anaerobic digestion requires less energy, yields less sludge byproduct, and generates sulfide (H_2S), which precipitates toxic heavy metals.[7]

The overall process for the anaerobic digestion of biomass is a fermentation process in which organic matter is both oxidized and reduced. Where biomass is represented by the formula, $\{CH_2O\}$, the simplified reaction for anaerobic fermentation producing carbon monoxide and methane is the following:

$$2\{CH_2O\} \rightarrow CO_2 + CH_4 \qquad (14.18.1)$$

In practice, the microbial processes involved are quite complex. Most of the wastes for which anaerobic digestion is suitable consist of oxygenated compounds, including those given in Figure 14.9.

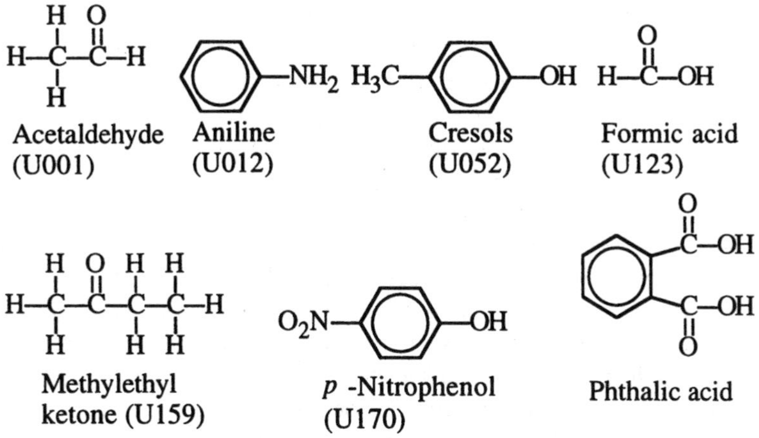

CH_3CHO — Acetaldehyde (U001)

$C_6H_5NH_2$ — Aniline (U012)

$H_3C-C_6H_4-OH$ — Cresols (U052)

$HCOOH$ — Formic acid (U123)

$CH_3COCH_2CH_3$ — Methylethyl ketone (U159)

$O_2N-C_6H_4-OH$ — *p* -Nitrophenol (U170)

$C_6H_4(COOH)_2$ — Phthalic acid

Figure 14.9. Examples of waste compounds that can be degraded by anaerobic digestion.

14.19. COMPOSTING AND LAND TREATMENT

Composting and land treatment are two means of treating solid hazardous wastes utilizing the biological activity of naturally occurring organisms. These two techniques have many similarities. In land treatment, the waste to be treated is mixed with soil, whereas composting does not involve soil as a major part of the medium. Both of these techniques are useful for solid or solidified wastes.

Land Treatment

A variety of enzyme activities are exhibited by microorganisms in soil. Even sterile soil may show enzyme activity due to extracellular enzymes secreted by microorganisms. Some of these enzymes are hydrolase enzymes which split molecules into two with the addition of water molecules, such as those that catalyze the hydrolysis of organophosphate compounds as shown by the reaction,

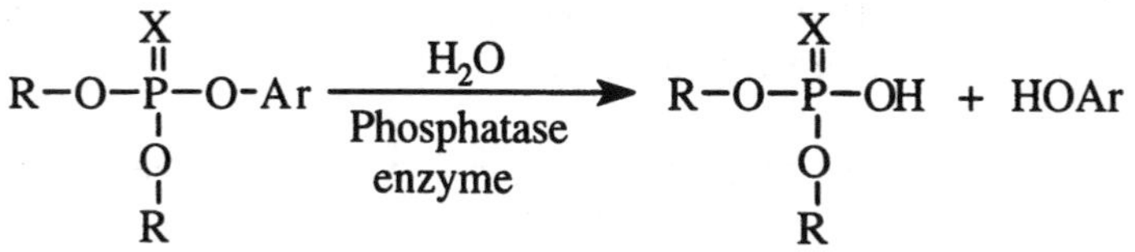

$$R-O-P(=X)(-O-R)-O-Ar \xrightarrow[\text{Phosphatase enzyme}]{H_2O} R-O-P(=X)(-O-R)-OH + HOAr \quad (14.19.1)$$

where R is an alkyl group, Ar is a substituent group that is frequently aromatic, and X is either S or O. Another example of a reaction catalyzed by soil enzymes is the oxidation of phenolic compounds by diphenol oxidase.

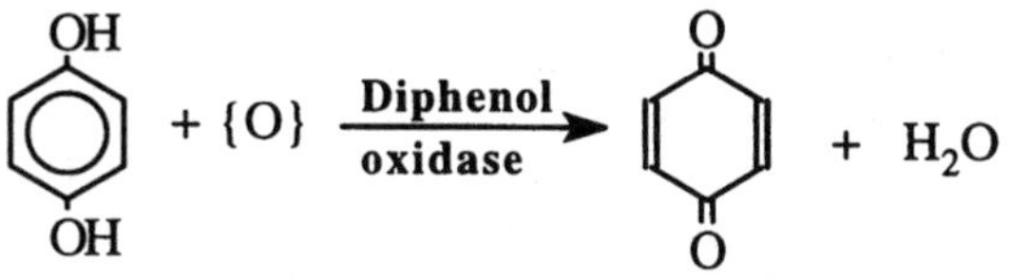

$$HO-C_6H_4-OH + \{O\} \xrightarrow[\text{oxidase}]{\text{Diphenol}} O=C_6H_4=O + H_2O \quad (14.19.2)$$

Soil may be viewed as a natural filter for wastes. Soil has physical, chemical, and biological characteristics that can enable waste detoxification, biodegradation, chemical decomposition, and physical and chemical fixation. A number of soil characteristics are important in determining its use for land treatment of wastes. These characteristics include physical form, ability to retain water, aeration, organic content, acid-base characteristics, and oxidation-reduction behavior. Soil is a natural medium for a number of living organisms that may have an effect upon biodegradation of hazardous wastes. Of these, the most important are bacteria, including those from the genera *Agrobacterium*, *Arthrobacteri*, *Bacillus*, *Flavobacterium*, and *Pseudomonas*. Actinomycetes and fungi are important in decay of vegetable matter and may be involved in biodegradation of wastes. Other unicellular organisms that may be present in or on soil are protozoa and algae. Soil animals, such as earthworms, affect soil parameters such as soil texture. The growth of plants in soil may have an influence on its waste treatment potential in areas such as uptake of soluble wastes and erosion control.

Wastes that are amenable to land treatment are biodegradable organic substances. However, soil bacterial cultures may develop that are effective in degrading normally recalcitrant compounds through acclimation over a long period of time. This occurs particularly at contaminated sites, such as those where soil has been exposed to crude oil for many years. Land treatment is most used for petroleum refining wastes and is applicable to the treatment of fuels and wastes from leaking underground storage tanks. It can also be applied to biodegradable organic chemical wastes, including some organohalide compounds. Land treatment is not suitable for the treatment of wastes containing acids, bases, toxic inorganic compounds, salts, heavy metals, and organic compounds that are excessively soluble, volatile, or flammable.

Composting

Composting of hazardous wastes is the biodegradation of solid or solidified materials in a medium other than soil. Bulking material, such as plant residue, paper, municipal refuse, or sawdust may be added to retain water and enable air to penetrate to the waste material. Composting can be carried out in a container, often with air pumped through it to aid biodegradation. Alternatively, the waste material may be placed in open piles or rows (windrows) which are reformed periodically to facilitate penetration of air.

Successful composting of hazardous waste depends upon a number of factors, including those discussed in Section 14.16. The first of these is the selection of the appropriate microorganism or **inoculum**. Once a sucessful composting operation is underway, a good inoculum is maintained by recirculating spent compost to each new batch. Other parameters that must be controlled include oxygen supply, moisture content (which should be maintained at a minimum of about 40%), pH (usually around neutral), and temperature. The composting process generates heat so, if the mass of the compost pile is sufficiently high, it can be self-heating under most conditions. Some wastes are deficient in nutrients, which must be supplied from commercial sources, such as commercial fertilizer. One of the nutrients most likely to be deficient is nitrogen, and the carbon/nitrogen level is a critical parameter for many wastes. This ratio is very high in some plant wastes such as straw or sawdust, so nitrogen should be added. Other wastes, such as manure and, particularly, urine, have a low C/N ratio reflecting a high nitrogen content, and these wastes may be used in composting to increase the nitrogen content to required levels.

As with other treatment technologies, composting requires consideration of potential environmental pollution factors. Volatile toxic or noxious organic compounds may be emitted to the air from composting operations. Leachate from composting may contain undegraded organic substances, heavy metals, and other inorganic solutes, so it must be collected, analyzed and treated.

14.20. BACTERIAL DESTRUCTION OF PCBs

Potentially, bacteria can be used for bioremediation of sites contaminated with organic wastes such as polychlorinated biphenyls (PCBs). The best place to search for bacteria capable of degrading organic wastes is in sites where the wastes have been in contact with soil for some time giving the bacteria the chance to develop strains capable of biodegrading the wastes. Because of their prevalence in soil, these bacteria are often of the *Pseudomonas* or *Alciligenes* genuses. For wastes such as PCBs that consist of many different molecular variations of the same class of compounds, bioremediation is complicated by the high selectivity of various strains of bacteria for molecular type. For example, aerobic bacteria generally are required to degrade PCBs with 3 or 4 Cl atoms per molecule, whereas anaerobic bacteria are not effective for these compounds. Some anaerobic bacteria metabolize PCBs with 5 or more Cls per molecule. It is not possible to devise a system that is hospitable for both types of bacteria at the same time. Different isomers of PCB molecules that have the same number of Cl atoms per molecule may show big differences in biodegradability by a particular strain of bacteria.

LITERATURE CITED

1. "Wastewater Treatment Systems," Chapter 4 in *Hazardous Materials and Waste Management*, Nicholas P. Cheremisinoff and Paul N. Cheremisinoff, Noyes Publications, Park Ridge, NJ, 1995, pp. 53–70.

2. Freeman, Harry M., Ed., *Standard Handbook of Hazardous Waste Treatment and Disposal*, McGraw-Hill, New York, NY, 1998.

3. "Hazardous Materials and Sludge Treatment, Disposal Methods," Chapter 5 in *Hazardous Materials and Waste Management*, Nicholas P. Cheremisinoff and Paul N. Cheremisinoff, Noyes Publications, Park Ridge, NJ, 1995, pp. 71–89.

4. Lunn, George and Eric B. Sansone, *Destruction of Hazardous Chemicals in the Laboratory*, John Wiley & Sons, Somerset, NJ, 1990.

5. Hileman, Bette, "Times Beach Detox Nears End," *Chemical and Engineering News*, January 6, 1997, pp. 20–22.

6. Chen, Yi-shen, Chung-guang You, and Wei-chi Ying, "Cyanide Destruction by Catalytic Oxidation," *46th Purdue Industrial Waste Treatment Conference Proceedings*, Lewis Publishers, Chelsea, Michigan, 1992, pp. 539–545.

7. Torpy, Michael F., "Anaerobic Digestion," Section 9.2 in *Standard Handbook of Hazardous Waste Treatment and Disposal*, Freeman, Harry M., Ed., McGraw-Hill Book Company, New York, 1989, pp. 9.19–9.28.

SUPPLEMENTARY REFERENCES

Boardman, Gregory D., Ed., *Proceedings of the Twenty-Ninth Mid-Atlantic Industrial and Hazardous Waste Conference*, Technomic Publishing Co., Lancaster, PA, 1997.

Cheremisinoff, Nicholas P., *Biotechnology for Waste and Wastewater Treatment*, Noyes Publications, Park Ridge, NJ, 1996.

Haas, Charles N. and Richard J. Vamos, *Hazardous and Industrial Waste Treatment*, Prentice Hall, Englewood Cliffs, NJ, 1995.

Hester, R. E. and R. M. Harrison, Eds., *Waste Treatment and Disposal*, Royal Society of Chemistry, Cambridge, U.K., 1995.

Hickey, Robert F. and Gretchen Smith, Eds., *Biotechnology in Industrial Waste Treatment and Bioremediation*, CRC Press/Lewis Publishers, Boca Raton, FL, 1996.

Hinchee, Robert E., Rodney S. Skeen, and Gregory D. Sayles, *Biological Unit Processes for Hazardous Waste Treatment*, Battelle Press, Columbus, OH 1995.

Karnofsky, Brian, Ed., *Hazardous Waste Management Compliance Handbook*, 2nd ed., Van Nostrand Reinhold, New York, NY, 1996.

Lewandowski, Gordon A. and Louis J. DeFilippi, *Biological Treatment of Hazardous Wastes*, Wiley, New York, NY, 1997.

Reed, Sherwood C., Ronald W. Crites, and E. Joe Middlebrooks, *Natural Systems for Waste Management and Treatment*, McGraw-Hill, New York, NY, 1995.

Tedder, D. William and Frederick G. Pohland, Eds., *Emerging Technologies in Hazardous Waste Management* **7**, Plenum, New York, NY, 1997.

Waste Characterization and Treatment, Society for Mining Metallurgy & Exploration, Littleton, CO, 1998.

Water and Residue Treatment, Volume II, Hazardous Materials Control Research Institute, Silver Spring, MD, 1987.

Williams, Paul T., *Waste Treatment and Disposal*, John Wiley & Sons, New York, NY, 1998.

Winkler, M. A., *Biological Treatment of Waste-Water*, Halsted Press (John Wiley & Sons), New York, NY, 1997.

15 INDUSTRIAL ECOLOGY OF WASTE DISPOSAL

15.1. IMMOBILIZATION

Immobilization and **stabilization** are used here as general terms to describe techniques whereby hazardous wastes are placed in a form suitable for long term disposal. **Fixation** is sometimes used to mean much the same thing.[1] **Solidification** describes a process in which a liquid or semisolid sludge waste is converted into a monolithic solid form or granular solid material.

Immobilization serves several purposes in the management and disposal of wastes. It usually improves the handling and physical characteristics of wastes. It isolates the wastes from their environment, especially groundwater, so that they have the least possible tendency to migrate. This is accomplished by physically isolating the waste, reducing its solubility, and decreasing its surface area. Immobilization includes physical and chemical processes that reduce surface areas of wastes to minimize leaching.

There are several ways in which wastes can be immobilized and effectively isolated from their surroundings. The processes used to do that include solidification with cement (Portland cement), solidification with silicate materials, sorption to a solid matrix material, imbedding in thermoplastics, imbedding in organic polymers, surface encapsulation, and vitrification (in glassy material)

Stabilization

Stabilization means the conversion of a waste to a physically and chemically more stable material. Stabilization may include chemical reactions that produce products that are less volatile, soluble, and reactive. Solidification by chemical means (formation of a precipitate by a chemical reaction) or solidification by physical means (evaporation of water from aqueous wastes or sludges) are both used for stabilization. Stabilization is required for land disposal of wastes.

Solidification

Solidification is done to improve the handling characteristics of waste and to reduce its mobility after disposal. Solidification may involve several measures, including chemical reaction of the waste with the solidification agent, mechanical isolation in a protective binding matrix, or a combination of chemical and physical processes. It can be accomplished by sorption onto solid material, reaction with cement, reaction with silicates, encapsulation, and imbedding in polymers or thermoplastic materials.

The simplest solidification process is one in which water is evaporated from an aqueous or sludge waste to leave a solid residue. As an example consider a sludge of iron(III) hydroxide containing heavy metal contaminants and some free water. Evaporation of the water results in formation of a solid product which with further heating and drying is converted to solid iron(III) oxide, Fe_2O_3, a very insoluble material that holds heavy metal ions.

Solidification of aqueous sludges is often accomplished by the addition of a substance that absorbs water and produces a solid. Portland cement is effective for this purpose for many wastes. **Pozzolanic material** consisting of hygroscopic silicates, such as those produced in fly ash, is an effective drying agent for solidification processes.

Encapsulation

As the name implies, **encapsulation** is used to coat wastes with an impervious material so that they do not contact their surroundings. For example, a water-soluble waste salt encapsulated in asphalt would not dissolve, so long as the asphalt layer remains intact preventing water from contacting the soluble waste. Coating of individual small particles with an impervious material is called **microencapsulation**, whereas imbedding larger aggregates of material in such a matrix is **macroencapsulation**. A common means of encapsulation is the use of heated, molten materials that solidify when they are cooled. Thermoplastic polymers, asphalt, and waxes can be used for this purpose. A more sophisticated approach to encapsulation is to form polymeric resins from monomeric substances in the presence of the waste. The substances that form the polymers can thoroughly penetrate the waste material, then bind it strongly in a polymer matrix. Problems can occur when the wastes adversely affect the polymerization process or the quality of the product formed.

15.2. CHEMICAL FIXATION

Chemical fixation is a process that binds a hazardous waste substance in a less mobile, less toxic form by a chemical reaction that chemically alters the waste. As an example consider the reaction of toxic, soluble chromium(VI) on a carbon surface when the carbon is partially gasified by reaction with O_2 and steam,

$$4C + 4Na_2CrO_4 + O_2 \xrightarrow[\text{Carbon matrix}]{\text{Heat}} 2Cr_2O_3 + 4Na_2CO_3 \qquad (15.2.1)$$

The chromium(III) product, Cr_2O_3, is insoluble, relatively nontoxic, and firmly attached to the carbon matrix.

Although chemical fixation and physical fixation are discussed separately here, they often occur together. Polymeric inorganic silicates containing some calcium and often some aluminum are the inorganic materials most widely used as a fixation matrix. Many kinds of heavy metals are chemically bound in such a matrix, as well as being physically held by it. Similarly, some organic wastes are bound by reactions with matrix constituents. For example, humic acid wastes react with calcium in a solidification matrix to produce insoluble calcium humates.[2]

15.3. PHYSICAL FIXATION

Solidification with Cement

Portland cement is widely used for solidification of hazardous wastes. In this application, Portland cement provides a solid matrix for isolation of the wastes, chemically binds water from sludge wastes, and may chemically react with wastes (for example, the calcium and base in Portland cement chemically react with inorganic arsenic sulfide wastes to reduce their solubilities). However, most wastes are physically held in the rigid Portland cement matrix and are subject to leaching. There are several different kinds of Portland cement; the one most widely used for waste solidification is Type I. Wastes containing sulfate (SO_4^{2-} ion) or sulfite (SO_3^{2-} ion) are more successfully solidified with Types II and V Portland cement.

When Portland cement is used for solidification, it is usually mixed directly with wastes, which may provide part or all of the water required to set the cement. Ideally, the product is a monolithic solid with a low exposed surface area, although often it is a granular material with a relatively high surface/volume ratio. Portland cement is often used in relatively small quantities as a setting agent for other immobilization reagents such as silicates.

As a solidification matrix, Portland cement is most applicable to inorganic sludges containing heavy metal ions that form insoluble hydroxides and carbonates in the basic carbonate medium provided by the cement. Addition of Portland cement to an arsenic sulfide sludge produces a concrete material in which the waste is held in a matrix of calcium silicates and aluminum hydrates. The success of solidification with Portland cement strongly depends upon whether or not the waste adversely affects the strength and stability of the concrete product. A number of substances are incompatible with Portland cement because they interfere with its set and cure and cause deterioration of the cement matrix with time.[3] These substances include organic matter such as petroleum or coal; some silts and clays; sodium salts of arsenate, borate, phosphate, iodate, and sulfide; and salts of copper, lead, magnesium, tin, and zinc.

However, a reasonably good disposal form can be obtained by absorbing organic wastes with a solid material, which in turn is set in Portland cement. This approach has been used with hydrocarbon wastes sorbed by an activated coal char matrix.[4] The product was reported to be a reasonably strong monolithic solid material. Because of its carbon content, this material can be burned in incinerators, keeping that option open if required for later, more permanent disposal.

Advantages of Portland cement for waste immobilization include its nonhazardous nature, affinity for water in wastes, ability to form solids, universal availability at low cost, and well-known procedures for working with it. However, it does add to the mass and volume of waste and often provides only minimal protection against leaching.

Solidification with Silicate Materials

The term, **silicate**, is used to denote a number of substances containing oxyanionic silicon such as SiO_3^{2-}. Water-insoluble silicates (pozzolanic substances) used for waste solidification include fly ash, flue dust, clay (an alumino silicate mineral), calcium silicates, and ground up slag from blast furnaces. Soluble silicates, such as sodium silicate, may also be used. Silicate solidification usually requires a setting agent, which may be Portland cement (see above), gypsum (hydrated $CaSO_4$), lime, or compounds of aluminum, magnesium, or iron. The product may vary from a granular material to a concrete-like solid. In some cases the product is improved by additives, such as emulsifiers, surfactants, activators, calcium chloride, clays, carbon, zeolites, and various proprietary materials.

Success has been reported for the solidification of both inorganic wastes and organic wastes (including oily sludges) with silicates. The advantages and disadvantages of silicate solidification are similar to those of Portland cement discussed above. One consideration that is especially applicable to fly ash is the presence in some silicate materials of leachable hazardous substances, which may include arsenic and selenium.

An Example of a Solidification Process

An example of a typical commercial solidification process is discussed here to illustrate the numerous parameters that must be considered. The SEALOSAFE® process uses insoluble silicates and cement to produce a solid that contains hazardous wastes. It is typical of many similar processes that use silicates and/or cement.[5]

The first step is to determine whether a waste can be solidified and effectively immobilized by solidification. This is done by chemical analysis, assessment of the materials that went into the waste (when known), and laboratory-scale solidification of samples of the waste. The major parameters analyzed include selected heavy metals, organic carbon, acidity, alkalinity, pH, organic carbon, cyanide, sulfide, and specific gravity.

A number of major kinds of wastes may not be amenable to treatment by cement and silicate. These include high levels of leachable soluble salts; wastes that produce noxious gas in reactions with water or base (carbides, phosphides, and ammonium salts); water-soluble anionic heavy metal species that are not well bound by the solid matrix (chromates, borates, sugar); substances, such as sulfate and borate salts, that prevent setting of cement; and carcinogenic, explosive, flammable, or volatile organic compounds. Wastes that are amenable to solidification by the SEALOSAFE or related processes include liquid electroplating wastes and electroplating sludges containing acids, bases, cyanide, chromium, cadmium, copper, nickel, and zinc; liquid waste from electronics manufacture containing chromium, copper, lead, selenium, and tin; solid spent catalysts containing cobalt, copper, chromium, manganese, molybdenum, nickel, and vanadium; solid residues from oil incineration containing vanadium; asbestos; chloralkali sludge containing mercury; sludge from paint manufacture containing antimony, cadmium, cobalt, and lead; and sludge from the manufacture of fungicides, insecticides, and preservatives containing arsenic, chromium, mercury, and sulfides.

Waste pretreatment is carried out to remove especially harmful constituents and to make the waste compatible with the solid matrix. Wastes exhibiting extremes of pH are neutralized to a pH in the range of approximately 6–9. Cyanides should be oxidized and toxic, soluble chromate reduced to nontoxic, insoluble chromium(III). Interference by some organic species may require addition of activated carbon to the waste mixture (see sorption below).

Following pretreatment, a slurry of the waste is prepared. It is important for the slurry to have a water content within a specified range; excess water may be removed by settling, centrifugation, or addition of dry wastes (fly ash, solids collected in air pollution control).

The final step in solidification is addition of the reagents that produce the solid, in this case cement and silicates. These materials react with water in the waste to form solids, such as hydrated silicates, which bind chemically to some of the waste constituents. After the solid product has been made and cured, it is desirable to test its properties to see if any special precautions are required in the final disposal of the wastes. Tests performed may include leaching, compressive strength, and permeability by water.

Thermal Processes in Solidification

In the solidification processes described above, water is an important ingredient of the hydrated solid matrix. Therefore, the solid should not be heated excessively or exposed to extremely dry conditions, which could result in diminished structural integrity from loss of water. In some cases, however, heating a solidified waste is an essential part of the overall solidification procedure. For example, an iron hydroxide matrix can be converted to highly insoluble, refractory iron oxide by heating. Organic constituents of solidified wastes may be converted to inert carbon by heating. Heating is an integral part of the process of vitrification (see below). Because of the cost of energy, heating of solidified wastes is not done unless it is required for special kinds of wastes.

Sorption to a Solid Matrix Material

Hazardous waste liquids, emulsions, sludges, and free liquids in contact with sludges may be sorbed by solid **sorbents**, including activated carbon (for organics), fly ash, kiln dust, clays, vermiculite, and various proprietary materials. Sorption may be done to convert liquids and semisolids to dry solids, improve waste handling, and reduce solubility of waste constituents. Sorption can also be used to improve waste compatibility with substances such as Portland cement.

Liquid waste that is incompatible with Portland cement may first be absorbed by granular carbon and the dry carbon/waste mixture set in Portland cement. Specific sorbents may also be used to stabilize pH and pE (a measure of the tendency of a medium to be oxidizing or reducing).[6]

The action of sorbents can include simple mechanical retention of wastes, physical sorption, and chemical reactions. It is important to match the sorbent to the waste. A substance with a strong affinity for water should be employed for wastes containing excess water and one with a strong affinity for organic materials should be used for wastes with excess organic solvents. The amount of sorbent required depends upon the levels of consituents to be removed.

Thermoplastics and Organic Polymers

Thermoplastics are solids or semisolids that become liquified at elevated temperatures. Hazardous waste materials may be mixed with hot thermoplastic liquids and immobilized in the cooled thermoplastic matrix, which is rigid but deformable. The thermoplastic material most used for this purpose is asphalt bitumen. Other thermoplastics, such as paraffin and polyethylene, have also been used to immobilize hazardous wastes.

Among the wastes that can be immobilized with thermoplastics are those containing heavy metals, such as electroplating wastes. Organic thermoplastics repel water and reduce the tendency toward leaching in contact with groundwater. Compared to cement, thermoplastics add relatively less material to the waste.

An immobilization technique similar to that just described uses **organic polymers** produced in contact with solid wastes to imbed the wastes in a polymer matrix. Three kinds of polymers that have been used for this purpose include polybutadiene, urea-formaldehyde, and vinyl ester-styrene polymers. The reagents used to make the polymers (monomers) are mixed with the waste and a polymerization catalyst to enable polymerization to occur. The product is a polymeric solid in which the wastes are imbedded. This procedure is more complicated than the use of thermoplastics but, in favorable cases, yields a product in which the waste is more strongly held.

Vitrification

Vitrification or **glassification** consists of imbedding wastes in a glass material. In this application, glass may be regarded as a high-melting-temperature inorganic thermoplastic. Molten glass can be used, or glass can be synthesized in contact with the waste by mixing and heating with glass constituents — silicon dioxide (SiO_2), sodium carbonate (Na_2CO_3), and calcium oxide (CaO). Other constituents may include boric oxide, B_2O_3, which yields a borosilicate glass that is especially resistant to changes in temperature and chemical attack. In some cases glass is used in conjunction with thermal waste destruction processes, serving to immobilize hazardous waste ash consituents. Some wastes are detrimental to the quality of the glass. Aluminum oxide, for example, may prevent glass from fusing.

Vitrification is relatively complicated and expensive, the latter because of the energy consumed in fusing glass. Despite these disadvantages, it is the best immobilization technique for some special wastes and has been promoted for solidification of radionuclear wastes because glass is chemically inert and resistant to leaching. However, high levels of radioactivity can cause deterioration of glass and lower its resistance to leaching.

15.4. ULTIMATE DISPOSAL

Regardless of the destruction, treatment, and immobilization techniques used, there will always remain some material from hazardous wastes that has to be put somewhere. A very favorable case is that of hydrocarbon wastes that are incinerated; the products are carbon dioxide and water that is discharged to the atmosphere. At the other extreme is contaminated soil; it can be detoxified, but little or nothing can be done to reduce its bulk. In some cases even the incineration product is still considered to be a hazardous waste (specifically, residues resulting from the incineration or thermal treatment of soil contaminated with some chlorinated benzene and phenol compounds (F028)). This section briefly addresses the ultimate disposal of ash, salts, liquids, solidified liquids, and other residues that must be placed where their potential to do harm is minimized.

Disposal Aboveground

In some important respects disposal aboveground, essentially in a pile designed to prevent erosion and water infiltration, is the best way to store solid wastes.[7] Perhaps its most important advantage is that it avoids infiltration by groundwater that can result in leaching and groundwater contamination common to storage in pits and landfills. In a properly designed aboveground disposal facility, any leachate that is produced quickly drains by gravity to the leachate collection system, where it can be detected and treated.

Aboveground disposal can be accomplished with a storage mound, which must have several features to successfully operate. Such a mound begins with a layer of compacted soil or clay lain somewhat above the original soil surface and shaped to allow leachate flow and collection. Two flexible membrane liners with a layer of low-permeability clay are placed on top of the compacted soil and an appropriate leachate collection system is installed. Solid wastes are then placed atop the liners and covered with a flexible membrane, clay cap, and topsoil layer planted with vegetation. The slopes around the edges of the storage mound are sufficiently great to allow good drainage of precipitation, but gentle enough to deter erosion.

Landfill

Landfill historically has been the most common way of disposing of solid hazardous wastes and some liquids, but it is now severely limited in the U.S. by regulations. It involves disposal that is at least partially underground in excavated cells, quarries, or natural depressions. Usually fill is continued above ground to efficiently utilize space and provide a grade for drainage of precipitation.

The greatest environmental concern with landfill of hazardous wastes is the generation of leachate from infiltrating surface water and groundwater with resultant contamination of groundwater supplies. Hazardous waste landfills constructed under current regulations provide elaborate systems to contain, collect, and control such leachate.[8]

There are several components to a modern landfill. A landfill should be placed on a compacted low-permeability medium, preferably clay, which is covered by a flexible-membrane liner consisting of water-tight impermeable material. This liner is covered with granular material in which is installed a secondary drainage system. Next is another flexible-membrane liner above which is installed a primary drainage system for the removal of leachate. This drainage system is covered with a layer of granular filter medium, upon which the wastes are placed. In the landfill, wastes of different kinds are separated by berms consisting of clay or soil covered with liner material. (In some installations only one liner is required.) Batches of wastes placed in the landfill at different times are covered with soil. When the fill is complete, the waste is capped to prevent surface water infiltration and covered with compacted soil. In addition to leachate collection, provision may be made for a system to treat evolved gases, particularly when methane-generating biodegradable materials are disposed in the landfill.

The flexible-membrane liner is a key component of state-of-the-art landfills. It controls seepage out of, and infiltration into the landfill. Obviously, liners have to meet stringent standards to serve their intended purpose. In addition to being impermeable, the liner material must be strongly resistant to biodegradation, chemical attack, and tearing. A means must be available for joining liner segments together with impermeable seams. Liner materials used for hazardous waste landfills consist of rubber (including chlorosulfonated polyethylene) or plastic (including chlorinated polyethylene, high-density polyethylene, and polyvinylchloride).

Capping is done to cover the wastes, prevent infiltration of excessive amounts of surface water, and prevent release of wastes to overlying soil and the atmosphere. In cases where the wastes may generate gases, such as methane from anaerobic biodegradation, provision should be made for gas collection. Caps come in a variety of forms and are often multilayered. Some of the problems that may occur with caps are settling, erosion, ponding, damage by rodents and penetration by plant roots.

The description above applies to an idealized landfill constructed under current regulations. Most older landfills do not meet such stringent standards, which has caused many of the current problems with hazardous waste disposal.

Surface Impoundments of Liquids

Many liquid hazardous wastes, slurries, and sludges are placed in **surface impoundments**, which usually serve as treatment areas and often are designed to eventually be filled in as landfill disposal sites. Most liquid hazardous wastes and a significant fraction of solids are placed in surface impoundments in some stage of treatment, storage, or disposal.

A surface impoundment may consist of an excavated "pit," a structure formed with dikes, or a combination thereof. The construction is similar to that discussed above for landfills in that the bottom and walls should be impermeable to liquids and provision must be made for leachate collection. The chemical and mechanical challenges to liner materials in surface impoundments are severe, so proper geological siting and construction with floors and walls composed of low-permeability soil and clay are important in preventing pollution from these installations.

Deep-well Disposal of Liquids

Deep-well disposal of liquids consists of their injection under pressure to underground strata isolated by impermeable rock strata from aquifers.[9] Early experience with this method was gained in the petroleum industry where disposal is required of large quantities of saline wastewater coproduced with crude oil. The method was later extended to the chemical industry for the disposal of brines, acids, heavy metal solutions, organic liquids, and other liquids.

A number of factors must be considered in deep-well disposal. Wastes are injected into a region of elevated temperature and pressure, which may cause chemical reactions to occur involving the waste constituents and the mineral strata. Oils, solids, and gases in the liquid wastes can cause problems such as clogging. Corrosion may be severe. Microorganisms may have some effects. Most problems from these causes can be mitigated by proper waste pretreatment.

The most serious consideration involving deep-well disposal is the potential contamination of groundwater. Although injection is made into permeable saltwater aquifers presumably isolated from aquifers that contain potable water, contamination may occur. Major routes of contamination include fractures, faults, and other wells. The disposal well, itself, can act as a route for contamination if it is not properly constructed and cased or if it is damaged.

15.5. *IN SITU* TREATMENT

***In situ* treatment** refers to waste treatment processes that can be applied to wastes disposed aboveground, in a landfill, or in a surface impoundment without removing the waste. In those limited cases where it can be practiced, *in situ* treatment is highly desirable as a waste site remediation option. For *in situ* treatment to work, any reagents must be efficiently transferred to wastes. Wastes held in porous sandy soil or coarse silt best meet this requirement. Several aspects of *in situ* treatment are discussed here.

Immobilization *in situ*

In situ immobilization is used to convert wastes to insoluble forms that will not leach from the disposal site. Heavy metal contaminants can be immobilized by chemical precipitation as the sulfides. Sulfides of metals such as lead, cadmium, zinc, and mercury are very insoluble. There are several disadvantages to this technique. Although highly mobile hydrogen sulfide gas effectively precipitates heavy metals, it can be a dangerous reagent to use because of its toxicity. A solution of sodium sulfide, Na_2S, maintained at a high pH by excess sodium hydroxide can be used to immobilize heavy metals as the sulfides. It is important to have sufficient alkali to maintain a basic solution to prevent the production of hydrogen sulfide from waste acid that might be present by the reaction

$$2H^+ + S^{2-} \rightarrow H_2S \qquad (15.5.1)$$

- $2H^+$: Hydrogen ion from waste acid present
- S^{2-}: Sulfide from precipitant
- H_2S: Gaseous hydrogen sulfide

Another disadvantage is that excess sulfide may become a groundwater pollutant. Although precipitated metal sulfides should remain as solids in the anaerobic conditions of a landfill, unintentional exposure to air can result in oxidation of the sulfide and remobilization of the metals as soluble sulfate salts.

Oxidation and reduction reactions can be used to immobilize heavy metals *in situ*. Oxidation of soluble Fe^{2+} and Mn^{2+} to their insoluble hydrous oxides, $Fe_2O_3 \cdot xH_2O$ and $MnO_2 \cdot xH_2O$, respectively, can precipitate these metal ions and coprecipitate other heavy metal ions. However, subsurface reducing conditions could result in reformation of soluble reduced species. Reduction can be used *in situ* to convert soluble, toxic chromate to insoluble chromium(III).

Chelation may convert metal ions to less mobile forms, although with most agents, chelation has the opposite effect. A chelating agent called Tetran is supposed to form metal chelates that are strongly bound to clay minerals.

Solidification *in situ*

In situ solidification can be used as a remedial measure at hazardous waste sites. One approach is to inject soluble silicates followed by reagents that cause them to solidify. For example, injection of soluble sodium silicate followed by calcium chloride or lime would form solid calcium silicate.

Detoxification *in situ*

When only one or a limited number of harmful constituents is present in a waste disposal site, it may be practical to consider detoxification *in situ*. This approach is most practical for organic contaminants including pesticides (organophosphate esters and carbamates), amides, and esters. Among the chemical and biochemical processes that can detoxify such materials are chemical and enzymatic oxidation, reduction, and hydrolysis. Chemical oxidants that have been proposed for this purpose include hydrogen peroxide (for some wastes, oxidation with peroxide requires an Fe^{2+} catalyst), ozone, and hypochlorite.

Enzyme extracts collected from microbial cultures and purified have been considered for *in situ* detoxification. One cell-free enzyme that has been used for detoxification of organophosphate insecticides is parathion hydrolase. The hostile environment of a chemical waste landfill, including the presence of enzyme-inhibiting heavy metal ions, is detrimental to biochemical approaches to *in situ* treatment. Furthermore, most sites contain a mixture of hazardous constituents, which might require several different enzymes for their detoxification.

Permeable Bed Treatment

Some groundwater plumes contaminated by dissolved wastes can be treated by a permeable bed of material placed in a trench through which the groundwater must flow. Acidic wastes and heavy metals are particularly amenable to such treatment. Limestone contained in a permeable bed neutralizes acid and precipitates some kinds of heavy metals as hydroxides or carbonates. Synthetic ion exchange resins can be used in a permeable bed to retain heavy metals and even some anionic species, although competition with ionic species naturally present in the groundwater can cause some problems with the use of ion exchangers. Activated carbon in a permeable bed will remove some organics, especially less soluble, higher molecular mass organic compounds.

Permeable bed treatment requires relatively large quantities of reagent, which argues against the use of activated carbon and ion exchange resins. In such an application, it is unlikely that either of these materials could be reclaimed and regenerated as is done when they are used in columns to treat wastewater. Furthermore, ions taken up by ion exchangers and organic species retained by activated carbon may be released at a later time, causing subsequent problems. Finally, a permeable bed that has been truly effective in collecting waste materials may, itself, be considered a hazardous waste requiring special treatment and disposal.

In situ Thermal Processes

Heating of wastes *in situ* can be used to remove or destroy some kinds of hazardous substances. Both steam injection and radio frequency heating have been proposed for this purpose. Volatile wastes brought to the surface by heating can be collected and held as condensed liquids or by activated carbon.

One approach to immobilizing wastes *in situ* is high temperature vitrification using electrical heating. This process involves placing conducting graphite between two electrodes located on the surface and passing an electrical current between the electrodes. In principle, the graphite becomes very hot and "melts" into the soil leaving a glassy slag in its path. Volatile species evolved are collected and, if the operation is successful, a nonleachable slag is left in place. It is easy to imagine problems that might occur, including difficulties in getting a uniform melt, problems from groundwater infiltration, and very high consumption of electricity.

15.6. LEACHATE AND GAS EMISSIONS

Leachate

The production of contaminated leachate is a possibility with most disposal sites.[10] Therefore, new hazardous waste landfills require leachate collection and treatment systems and many older sites are required to have such systems retrofitted to them. In the U.S. under the Minimum Technological Requirements specified in Section 202 of the 1984 RCRA amendments, leachate collections are required for new hazardous waste landfills and expansions of existing ones. A "pump-and-treat" system is one of the more common remedial actions required. Even when more permanent remedies are contemplated, leachate treatment may be required as an intermediate measure to prevent imminent danger to groundwater supplies.

The design and construction of leachate collection systems are primarily engineering concerns and will not be addressed in detail here. Modern hazardous waste landfills typically have dual leachate collection systems, one located between the two impermeable liners required for the bottom and sides of the landfill and another just above the top liner of the double-liner system. The upper leachate collection system is called the primary leachate collection system, and the bottom is called the secondary leachate collection system. Leachate is collected in perforated pipes that are imbedded in granular drain material. In a properly designed system, leachate does not collect to a depth greater than 1 foot over the drainage pipes, so hydraulic pressures on the landfill linings are low, as is the potential for leakage. The landfill is designed so that the leachate flows by gravity to sumps, where it can be pumped out for treatment. It should be noted though that a leachate collection system is relatively easy to design for a new installation, but may be very difficult for an existing one from which leachate may flow in numerous and unpredictable paths.

Chemical and biochemical processes have the potential to cause some problems for leachate collection systems. One such problem is clogging by insoluble manganese(IV) and iron(III) hydrated oxides upon exposure to air. This phenomenon has occurred in drainage systems installed to remove water from water-saturated soils. Exposed to air in drainage pipes, dissolved soluble Mn^{2+} and Fe^{2+} are oxidized to insoluble deposits of the hydrated oxides, $Fe_2O_3{\bullet}xH_2O$ and $MnO_2{\bullet}xH_2O$, respectively, and clogging of the drainage pipes results. A similar phenomenon occurs during the pumping of groundwater containing dissolved iron(II) from wells when deposits of $Fe_2O_3{\bullet}xH_2O$ form on the well walls, hindering water flow into the well.

Leachate consists of water that has become contaminated by wastes as it passed through a waste disposal site. It contains waste constituents that are soluble, not retained by soil, and not chemically or biochemically degraded. Some potentially harmful leachate constituents are products of chemical or biochemical transformations of wastes.

The best approach to leachate management is to prevent its production by limiting infiltration of water into the site. Rates of leachate production may be very low when sites are selected, designed, and constructed with minimal production of leachate as a major objective. A well-maintained, low-permeability cap over the landfill is very important for leachate minimization.

Hazardous Waste Leachate Treatment

The first step in treating leachate is to fully characterize it, particularly with a thorough chemical analysis of possible waste constituents and their chemical and metabolic products. The biodegradability of leachate constituents should also be determined.

The options available for the treatment of hazardous waste leachate are generally those that can be used for industrial wastewaters.[11] They are summarized briefly below.

One of two major ways of removing organic wastes is biological treatment by an activated sludge, or related process (see Section 14.16 and Figure 14.8). It may be necessary to acclimate microorganisms to the degradation of hazardous wastes constituents that are not normally biodegradable (see biodegradation of phenols, Section 14.15). Possible hazards of biotreatment sludges, such as those containing excessive levels of heavy metal ions, should be considered. The other major process for the removal of organics from hazardous waste leachate is sorption by activated carbon (see Section 14.2), usually in columns of granular activated carbon. Activated carbon and biological treatment can be combined with the use of powdered activated carbon in the activated sludge process. The powder sorbs some constituents that may be toxic to microorganisms and is collected with the sludge. A major consideration with the use of activated carbon on hazardous waste leachate is the hazard that spent activated carbon may present from the wastes it retains. These hazards may include those of toxicity or reactivity, such as those posed by explosives manufacture wastes sorbed to activated carbon. Regeneration of the carbon is expensive and can be hazardous in some cases.

Hazardous waste leachate can be treated by a variety of chemical processes, including acid/base neutralization, precipitation, and oxidation/reduction. In some cases these treatment steps must precede biological treatment; for example, leachate exhibiting extremes of pH must be neutralized for microorganisms to thrive in it. Cyanide in the leachate may be oxidized with chlorine and organics with ozone, hydrogen peroxide promoted with ultraviolet radiation, or dissolved oxygen at high temperatures and pressures. Heavy metals may be precipitated with base, carbonate, or sulfide.

Leachate can be treated by a variety of physical processes. In some cases, simple density separation and sedimentation can be used to remove water-immiscible liquids and solids. Filtration is frequently required and flotation (see Section 14.2) may be useful. Leachate solutes can be concentrated by evaporation, distillation, and membrane processes, including reverse osmosis, hyperfiltration, and ultrafiltration. Organic constituents can be removed by solvent extraction, air stripping, or steam stripping.

Synthetic resins are useful for removing hazardous solutes from hazardous waste leachate. Organophilic resins have proven useful for the removal of alcohols; aldehydes; ketones; hydrocarbons; chlorinated alkanes, alkenes, and aryl compounds; esters, including phthalate esters; and pesticides. Cation exchange resins are effective for the removal of heavy metals.

Gas Emissions

In the presence of biodegradable wastes, designated here as $\{CH_2O\}$, methane and carbon dioxide gases are produced in landfills by anaerobic degradation as represented by the following reaction:

$$2\{CH_2O\} \xrightarrow[\text{microbial processes}]{\text{Anaerobic}} CH_4 + CO_2 \qquad (15.6.1)$$

Gases may also be produced by chemical processes with improperly pretreated wastes, as would occur in the hydrolysis of calcium carbide to produce acetylene.

$$CaC_2 + 2H_2O \rightarrow C_2H_2 + Ca(OH)_2 \quad (15.6.2)$$

Odorous and toxic hydrogen sulfide, H_2S, may be generated by the chemical reaction of sulfides with acids or by the biochemical reduction of sulfate by anaerobic bacteria (*Desulfovibrio*) in the presence of biodegradable organic matter.

$$SO_4^{2-} + 2\{CH_2O\} + 2H^+ \xrightarrow[\text{bacteria}]{\text{Anaerobic}} H_2S + 2CO_2 + 2H_2O \quad (15.6.3)$$

Gases such as these pose hazards of toxicity and flammability, as well as having the potential to form explosive mixtures with air. Furthermore, gases permeating through landfilled hazardous waste may carry along waste vapors, such as those of volatile aryl compounds and low-molecular-mass chlorinated hydrocarbons. Of these, the ones of most concern are benzene, carbon tetrachloride, chloroform, 1,2-dibromoethane, 1,2-dichloroethane, dichloromethane, tetrachloroethane, 1,1,1-trichloroethane, trichloroethylene, and vinyl chloride.[11] Because of the hazards from these and other volatile species, it is important to minimize production of gases and when gas production is likely, to provide for venting, and even treatment, of the gas. Landfill gas can be collected by systems of perforated or slotted pipes placed horizontally below the landfill cap and connected to vertical riser pipes through which the gas is released. An extraction blower can be installed to pull the gas from the landfill. If the gas is too hazardous to vent to the atmosphere, it may be flared or run through a gas incinerator. Activated carbon can be used to remove hazardous organic vapor constituents from the gas.

LITERATURE CITED

1. Wiles, Carlton C., "Solidification and Stabilization Technology," Section 7.3 in *Standard Handbook of Hazardous Waste Treatment and Disposal*, 2nd ed., Harry M. Freeman, Ed., McGraw-Hill Book Company, New York, 1989, pp. 7.31–7.46.

2. Manahan, Stanley E., "Humic Substances and the Fates of Hazardous Waste Chemicals," Chapter 6 in *Influence of Aquatic Humic Substances on Fate and Treatment of Pollutants*, American Chemical Society, Washington D.C., 1988.

3. "Direct Waste Treatment," Section 10 in *Remedial Action at Waste Disposal Sites*, EPA/625/6–85/006, U.S. Environmental Protection Agency Hazardous Waste Engineering Research Laboratory, Cincinnati, OH, 1985, pp.10-1–10-151.

4. "Destruction and Immobilization of Metal-Contaminated PCB Sludges by Gasification on an Activated Char from Subbituminous Coal," Stanley E. Manahan, Shubhender Kapila, Chris Cady, and David Larsen, preprint extended abstract of papers presented before the Division of Environmental Chemistry, American Chemical Society, Dallas, Texas, April, 1989.

5. Clements, J. A. and C. M. Griffiths, "Solidification Processes," Chapter 5 in *Hazardous Waste Management Handbook*, Butterworths, London, 1985, pp. 146–166.

6. "Oxidation-Reduction," Chapter 4 in *Environmental Chemistry*, 6th ed., Stanley E. Manahan, CRC Press/Lewis Publishers, Boca Raton, FL, 1994, pp. 87–110.

7. Brown, K. W., David C. Anderson, and James C. Thomas, "Aboveground Disposal," Section 10.5 in *Standard Handbook of Hazardous Waste Treatment and Disposal*, 2nd ed., Harry M. Freeman, Ed., McGraw-Hill, New York, 1998, pp. 10.66–10.75.

8. Leung, Charles W. and David E. Ross, "Hazardous Waste Landfill Construction," Section 10.1 in *Standard Handbook of Hazardous Waste Treatment and Disposal*, Harry M. Freeman, Ed., McGraw-Hill, New York, 1998, pp. 10.3–10.30.

9. Warner, Don L., "Subsurface Injection of Liquid Hazardous Wastes," Section 10.3 in *Standard Handbook of Hazardous Waste Treatment and Disposal*, Harry M. Freeman, Ed., McGraw-Hill, New York, 1998, pp. 10.46–10.58.

10. "Hazardous Waste Management and Pollution Control," Chapter 1 in *Hazardous Materials and Waste Management*, Nicholas P. Cheremisinoff and Paul N. Cheremisinoff, Noyes Publications, Park Ridge, NJ, 1995, pp. 1–19.

11. "Wastewater Treatment Systems," Chapter 4 in *Hazardous Materials and Waste Management*, Nicholas P. Cheremisinoff and Paul N. Cheremisinoff, Noyes Publications, Park Ridge, NJ, 1995, pp. 53–70.

SUPPLEMENTARY REFERENCES

Anderson, W. C., *Innovative Site Remediation Technology*, Springer Verlag, New York, NY, 1995.

Armour, Margaret-Ann, *Hazardous Laboratory Chemicals Disposal Guide*, CRC Press/Lewis Publishers, Boca Raton, FL, 1996.

Barth, Edwin F., *Stabilization and Solidification of Hazardous Wastes* (Pollution Technology Review, No 186), Noyes Publications, Park Ridge, NJ, 1990.

Bentley, S. P., Ed., *Engineering Geology of Waste Disposal*, American Association of Petroleum Geologists, Houston, TX, 1995.

Bodocsi, A., Michael E. Ryan, and Ralph R. Rumer, Eds., *Barrier Containment Technologies for Environmental Remediation Applications*, John Wiley & Sons, New York, NY, 1995.

Epps, John A. and Chin-Fu Tsang, Eds., *Industrial Waste: Scientific and Engineering Aspects*, Academic Press, San Diego, CA, 1996.

Freeman, Harry and Eugene F. Harris, Eds., *Hazardous Waste Remediation: Innovative Treatment Technologies*, Technomic Publishing Co., Lancaster, PA, 1995.

Gilliam, Michael T. and Carlton G. Wiles, Eds., *Stabilization and Solidification of Hazardous, Radioactive, and Mixed Wastes* (ASTM Special Technical Publication, 1123), American Society for Testing and Materials, Philadelphia, PA, 1992.

Haas, Charles N. and Richard J. Vamos, *Hazardous and Industrial Waste Treatment*, Prentice Hall Press, New York, NY, 1995.

Hester, R. E. and R. M. Harrison, *Waste Treatment and Disposal* (Issues in Environmental Science and Technology, 3), Royal Society of Chemistry, London, U.K., 1995.

National Research Council, *Review and Evaluation of Alternative Chemical Disposal Technologies*, National Academy Press, Washington, D.C., 1997.

Reith, Charles C. and Bruce M. Thomson, Eds., *Deserts As Dumps?: The Disposal of Hazardous Materials in Arid Ecosystems*, University of New Mexico Press, Albuquerque, NM, 1993.

Watts, Richard J., *Hazardous Wastes: Sources, Pathways, Receptors*, John Wiley & Sons, New York, NY, 1997.

INDEX